ALGORITHMS AND ARCHITECTURES FOR REAL-TIME CONTROL 1998
(AARTC'98)

*A Proceedings volume from the 5th IFAC Workshop,
Cancun, México, 15 - 17 April 1998*

Edited by

D.F. GARCÍA NOCETTI, J. SOLANO GONZÁLEZ, P. ACEVEDO CONTLA
*Instituto de Investigaciones en Matemáticas Aplicadas y en Sistemas,
Universidad Nacional Autónoma de México, México*

and

P.J. FLEMING
*Department of Automatic Control and Systems Engineering,
University of Sheffield, UK*

Published for the

INTERNATIONAL FEDERATION OF AUTOMATIC CONTROL

by

PERGAMON
An Imprint of Elsevier Science

UK   Elsevier Science Ltd, The Boulevard, Langford Lane, Kidlington, Oxford, OX5 1GB, UK

USA   Elsevier Science Inc., 660 White Plains Road, Tarrytown, New York 10591-5153, USA

JAPAN   Elsevier Science Japan, Tsunashima Building Annex, 3-20-12 Yushima, Bunkyo-ku, Tokyo 113, Japan

Copyright © 1998 IFAC

First edition 1998

Library of Congress Cataloging in Publication Data

A catalogue record for this book is available from the Library of Congress

British Library Cataloguing in Publication Data

A catalogue record for this book is available from the British Library

ISBN 0-08-043235 2

Transferred to digital printing 2008

IFAC WORKSHOP ON ALGORITHMS AND ARCHITECTURES FOR REAL-TIME CONTROL 1998

Sponsored by
International Federation of Automatic Control (IFAC)
- Technical Committee on Algorithms and Architectures for Real-Time Control

Co-sponsored by
IFAC - Technical Committee on Fault Detection and Supervision of Technical Processes
(SAFEPROCESS)

Organized by
Universidad Nacional Autónoma de México (UNAM)
Asociación de México de Control Automática (AMCA)

Supported by
Instituto de Investigaciones en Matemáticas Aplicadas y en Sistemas (IIMAS-UNAM)
Dirección General de Asuntos del Personal Académico (DGAPA-UNAM)
Consejo Nacional de Ciencia y Tecnología (CONACYT)

International Programme Committee (IPC)

Fleming, P.J. (UK)　　(Chairman)	Man, K. (HK)
Aggarwal, S. (USA)	Motus, L. (ESTONIA)
Ajtonyi, I. (H)	Naranjo, F. (CO)
Atherton, D.P. (UK)	Ng, T.S. (HK)
Bass, J. (UK)	Parra, V. (MEX)
Boullart, L. (B)	Pashkevich, A.P. (BYELORUSSIA)
Bulsari, A. (SF)	Patton, R. (UK)
Carelli, R. (RA)	Pereira, C. (BR)
Cipriano, A. (RCH)	Puente, J. (E)
Evans, R.J. (AUS)	Ruano, A.E. (P)
Ferreira, E. (BR)	Ruano, M.G. (P)
García Nocetti, D.F. (MEX)	Sa da Costa, J. (P)
Goodall, R.M. (UK)	Sanchez, J. (MEX)
Halang, W.A. (D)	Sanz, R. (E)
Hernandez, V. (E)	Savage, J. (MEX)
Holding, D.J. (UK)	Shin, K.G. (USA)
Ibarra, J.M. (MEX)	Skubich, J. (F)
Ionescu, C. (R)	Solano, J. (MEX)
Irwin, G.W. (UK)	Spong, M.W. (USA)
Kopacek, P. (A)	Tokhi, M. (UK)
Kwon, W.H. (ROK)	Tyrrell, A. (UK)
Lauwereins, R. (B)	Tzafestas, S. (GR)
Lewis, F. (USA)	Verde, C. (MEX)
Levy, D.C. (AUS)	Zalzala, A.M.S. (UK)
Magnani, G. (I)	Zomaya, A.Y. (AUS)

National Organizing Committee (NOC)
García-Nocetti, D.F.　　(Chairman)
Solano, J.
Acevedo, P.
Ibarra, J.M.

FOREWORD

The 5th IFAC Workshop on Algorithms and Architectures for Real-Time Control (AARTC'98) was organized under the auspices of the IFAC Technical Committee. This Committee is concerned with the use of emerging software and hardware developments in real-time control. Previous AARTC Workshops have been held at Bangor-U.K. (1991), Seoul-Korea (1992), Ostend-Belgium (1995) and Vilamoura-Portugal (1997). In 1998 the Workshop was held at Cancun-Mexico, being the first time this event took place on the American continent. It was organized by the National Autonomous University of Mexico (UNAM) and the Mexican Society of Automatic Control (AMCA), the Mexican IFAC NMO.

The AARTC'98 Technical Programme consisted of seventeen sessions, covering major areas of software hardware and applications for real- time control, namely robotics, modeling and control, software design tools and methodologies, industrial process control and manufacturing systems, parallel and distributed systems, non-linear control systems, neural networks, parallel and distributed algorithms for real-time signal processing and control, transport applications, algorithms, fault tolerant systems and fuzzy control.

The contributions were selected from a large number of high-quality full draft papers and late breaking paper contributions (consisting of extended abstracts) presenting very recent research work. The programme also included two Plenary Sessions given by two eminent international researchers. We were honored to have Professor Rolf Isermann (Technical University of Darmstadt, Germany) who presented a Keynote Lecture on "Hardware-in-the-loop Simulation for the Design and Testing of Control Systems", and Professor Bernard Widrow (Stanford University, U.S.A.) whose Keynote Lecture addressed "Adaptive Inverse Control Based on Nonlinear Adaptive Filtering".

We are grateful to Professor Ron Patton for the co-sponsorship of his IFAC Technical Committee on Fault Detection and Supervision of Technical Processes (SAFEPROCESS). We would also like to record our special thanks to the "Instituto de Investigaciones en Matemáticas Aplicadas y en Sistemas -IIMAS", the "Dirección General de Asuntos del Personal Académico - DGAPA and the "Consejo Nacional de Ciencia y Tecnología - CONACYT" for their generous support and facilities provided.

We are most grateful to the members of the International Programme Committee - IPC who did a remarkable job in reviewing the papers submitted to the Workshop within the established schedule. The members of the National Organizing Committee, Dr. Julio Solano, Dr. Pedro Acevedo deserve a special mention for their enormous efforts and their dedication to ensuring that AARTC 98 was a success. Also special thanks to all the technical and administrative staff of the "Departamento de Ingeniería de Sistemas Computacionales y Automatización - DISCA, IIMAS" for their support and assistance.

D.F. García Nocetti
P.J. Fleming

CONTENTS

HARDWARE-IN-THE-LOOP SIMULATION FOR THE DESIGN AND TESTING OF ENGINE-CONTROL SYSTEMS

R. Isermann, J. Schaffnit, S. Sinsel

Institute of Automatic Control, Laboratory of Control Engineering and Process Automation
Technical University of Darmstadt, Landgraf-Georg-Str. 4, D-64283 Darmstadt, Germany
e-mail: risermann@irt.tu-darmstadt.de

Abstract: For the design, implementation and testing of control systems increasingly hardware-in-the-loop (HIL) simulation is required, where parts of the control loop components are real hardware and parts are simulated. Usually, the process is simulated because it is not available (simultaneous engineering), or experiments with the real process are too costly or require too much time. The sensors and actuators may be simulated or real and the controller is real hardware and software. The real-time requirements for the simulation depends on the time scale of the process and the simulated components. The contribution gives first an overview of the various kinds of real-time and HIL simulation. Then, two cases are considered. First the HIL simulation for relatively slow processes, like in basic industries or heating systems. Here, the simulation-speed may be limited either by the complexity of the processes or by the real controller hardware. Then, the HIL simulation of combustion engines both with transputers and digital signal processors is shown in detail. The required models for 6- and 8-cylinder diesel engines are described, including fuel injection and burning, pressure development, torque generation at the crankshaft, exhaust turbocharger dynamics and the vehicle dynamics. The HIL-simulator test bench consists of a real-time computer system, sensor-interface, actuator interface, real injection pumps and the real control unit. Comparisons of real-time simulation with measurements on real diesel engines and trucks are shown. The goal of the HIL system is to develop new control algorithms and to investigate the effect of faults in sensors and actuators and the engine itself. *Copyright © 1998 IFAC*

Keywords: real-time simulation, hardware-in-the-loop simulation, diesel engine, turbocharger, vehicle dynamics, transputers, digital signal processors, simulator testbench

1. INTRODUCTION

The development of many products and processes is characterized by the integration with digital control systems. Herewith the integration is performed by the components (integration of hardware) and by the functions (integration of software). Because of the increasing complexity and the mutual relationships between the design of the processes and the design of the control system, increasingly computer aided methods for modeling, simulation and also the design itself are required. This development is also influenced by the decreasing development times (time to market) and the increasing quality, reliability and safety requirements. Typical examples are the design of *mechatronic systems* (e.g. precision mechanic devices, machines and vehicles), (Isermann, 1997).

With regard to the required speed of the computation *simulation methods* can be subdivided in

- *simulation without* (hard) *time limitation*
- *real-time simulation*
- *simulation faster than real-time*

Some application examples are given in Fig. 1. Herewith, *real-time simulation* means that the simulation of a component is performed such that the input and output signals show the same time dependent values as the real dynamically operating component. This becomes a computational problem for processes which have fast dynamics compared to the required algorithms and calculation speed.

Different kinds of real-time simulation methods are shown in Fig. 2. The reason for the real-time requirement is mostly that one part of the investigated system is not simulated, but real. Three cases can be distinguished:

(1) The *real-process* can be operated together with the *simulated control* by using another hardware

than the final hardware. This is also called "control prototyping".

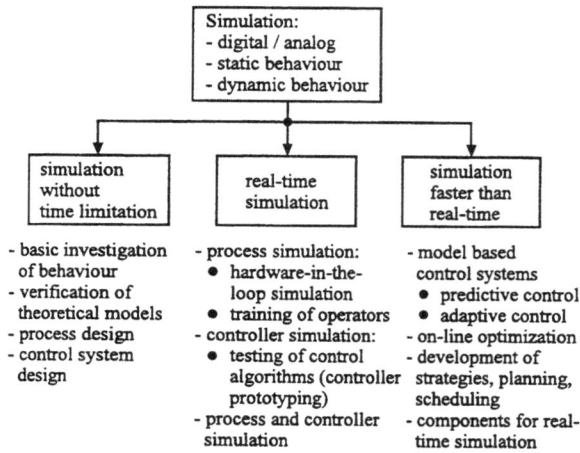

Fig. 1. Classification of simulation methods with regard to the speed and application examples

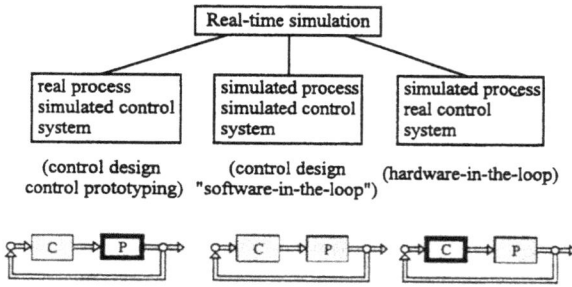

Fig. 2. Classification of real-time simulation

(2) The *simulated process* can be operated with the *real control hardware*, which is called "hardware-in-the-loop simulation"

(3) The *simulated process* is run with the *simulated control* in real-time. This may be required if the final hardware is not available or if a design step before the hardware-in-the-loop simulation is considered.

In the following the *hardware-in-the-loop simulation* is considered.

2 HARDWARE-IN-THE-LOOP SIMULATION

2.1 Architectures

The *hardware-in-the-loop* simulation (HIL) is characterized by operating real components in connection with real-time simulated components. Usually, the control system hardware and software is the real system, as used for series production. The controlled process, consisting of actuators, physical processes and sensors can then either be simulated or parts of them may be real components, compare Fig. 3. Table 1 shows some possible combinations. In general mixtures of the shown cases are realized. Frequently some actuators are real and the process and the sensors are simulated. The reason is that actuators and the control hardware very often form one integrated subsystem or that actua-

tors are difficult to model precisely and to simulate in real time. (The use of real sensors together with a simulated process may require considerable realization efforts, because the physical sensor input does not exist and must be generated artificially). In order to change or redesign some functions of the control hardware or software a bypass unit can be connected to the basic control hardware. Hence, hardware-in-the-loop simulators may also contain partially simulated (emulated) control functions.

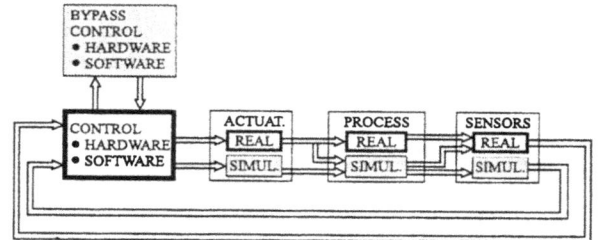

Fig. 3. Hardware-in-the-loop simulation: possible hybrid structures

Table 1   Hybrid real and simulated process components for hardware-in-the-loop simulation

cases	actuators		process		sensors	
	real	simul.	real	simul.	real	simul.
1	x	-	-	x	-	x
2	x	-	-	x	x	-
3	x	-	x	-	x	-
4	-	x	-	x	-	x
5	-	x	-	x	x	-

The advantages of the hardware-in-the-loop simulation are generally:

- Design and testing of the control hardware and software without operating a real process ("moving the process field into the laboratory")
- Testing of the control hardware and software under extreme environmental conditions in the laboratory (e.g. high/low temperature, high accelerations and mechanical shocks, aggressive media, electro-magnetic compatibility)
- Testing of the effects of faults and failures of actuators, sensors and computers on the overall system
- Operating and testing of extreme and dangerous operating conditions
- Reproducible experiments, frequently repeatable
- Easy operation with different man-machine interfaces (cockpit-design and training of operators)
- Saving of cost and development time

2.2 Historical development

First approaches of HIL simulation were probably realized for (real-time) *flight simulation*, where first goals were to simulate the instruments with a fixed cockpit, "linktrainer" (1936), (N.N., 1964; Marienfeld, 1965) and lateron, in addition to move a cockpit according to aircraft motions, e.g. for training of pilots. Here, the cockpit and the pilot was real and the motions were generated by electrical and hydraulic actuators. First generations used analog tube controllers and analog motion simulations which were lateron replaced by analog and then process computers (N.N., 1953; Anderson, 1962).

HIL motion simulators were also built for dynamic testing of *vehicle components* (e.g. suspension, body) with hydraulic or electrical actuators (testing machines). Here, for example the excitation of the wheels by the road surface is simulated. Another interesting HIL motion simulation is the *vehicle driving simulator* (Drosdol, and Panik, 1985). Also *dynamic motor teststands* where the engine is real and the vehicle and gear is simulated by another hardware, an electrical DC or AC motor, together with a digital process computer, is a special kind of HIL simulator (Thun, 1987; Pfeiffer, 1997).

With the development of digital electronic control systems for vehicles, as e.g. ABS (antiblock braking system) for brakes, ASR (TCS: traction control system), for the drive chain and automatic gears, the HIL simulators followed next steps of development (Kempf, *et al.*, 1987; Huber, *et al.*, 1988). First versions used high performance workstations and process computers. However, the amount of real-time simulation was very limited (Klinker, 1992; Wagner, and Furry, 1992). The availability of *parallel computers*, in the form of transputers (RISC processors with on-chip RAM and high speed communication links) and digital signal processors (DSP: microprocessors with efficient operation of additions and multiplications by parallel processing with different memories and buses for the program and the data) then opened ways for the real-time simulation of complete hydraulic systems, sensors, actuators, suspension systems, see e.g. (Fennel, *et al.* 1992; Hanselmann, 1993). Further research then has shown how more comprehensive mechanical systems can be simulated in real-time by parallel computers like *multibody-systems* (Schaefer, 1993; Rieger, and Schiehlen, 1995; Kortuem, *et al.*, 1992), *brake systems* (Sailer, and Essers, 1996; Sailer, 1996) and *combustion engines* (Savaglio, 1993; Kimura, and Maeda, 1996; Woermann, 1994; Sinsel, and Isermann, 1997)

Some of the HIL simulation were in-house developments of companies especially in the fields of aircraft and automobiles and were only published partially or not at all. Typical configurations consist of either transputer or DSP cards, a host PC and special interface cards for actuators and sensors. Off-the-shelf software exists for the basic operating system and some special processes, sensors, actuators. Commercially availabe HIL simulators are for example driscribed in (Hanselmann, 1993; ETAS, 1997).

2.3 HIL simulation for slow and fast processes

Until now HIL simulation with real controller hardware was mostly realized for relatively *fast processes* like aircraft, automobiles or combustion engines and

their components. The sampling time of the controller is about $T_{0c} \approx 1$ to 10 ms (sampling frequency $f_{0c} \approx 100$ Hz to 1 kHz) and the sampling time for the simulation of the mechanical processes $T_{0p} \approx 0.5$ to 10ms.
The limits in the real-time simulation of these relatively fast processes are given by

- possibility of paralleling algorithms
- kind of integration method
- stiffness of simulated differential equation systems (expressed in ratio: largest/smallest eigenvalue)

HIL simulators can also be used to develop and test real controller hardware and software for *slow processes*, i.e. processes in the power, chemical and basic industries. The advantages of "moving the process field into the laboratory" mentioned in section 2.1 are the same. However, in order to save observation and development time the simulated process should run faster than in real-time, i.e. to simulate one month operating time in one day, one hour or even some minutes. Then, the limits are usually not given by the speed of process simulation but by the possible speed-up of the real controller hardware. Examples for these limits are

- speed-up of timers, capture-compare units and watch dog timers
- speed-up of process interfaces (ADC, DAC)
- speed-up of real actuators
- time constants of analog components

However, even speed-up factors by 5 may safe considerable time for testing, debugging, compared to field tests. By this HIL simulation for slow processes not only development time can be saved but also commissioning time at the plant.

In the following the HIL simulation of a very fast process is shown, the dynamic simulation of Diesel engines with real-time simulation of each cylinder working cycle, drive chain and vehicle.

3. HIL SIMULATION OF DIESEL ENGINES WITH TURBOCHARGER

3.1 Process models for real-time simulation of combustion engines

For real-time simulation, dynamic models for the subsystems Diesel engine, turbocharger and vehicle longitudinal dynamics are required, Fig. 4. Several compromises have to be made with regard to the model complexity and calculation time. For an implementation on a multiprocessor system, a parallel structure of the overall model is aimed. To ensure the users acceptance of the simulator, the adaptation to the real system must be possible with reasonable effort. There are mainly two ways to develop mathematical process models:

Theoretical modeling is based on physical laws, expressed by equations. After simplifying assumptions, these equations are stated for single process elements.

They can be subdivided in balance equations for storage of mass, energy and momentum, constitutive equations for sources, transformers and converters, and phenomenological equations for irreversible processes like dissipative elements or sinks. The interconnections of the process elements are described by continuity equations (node law) and compatibility equations (closed circuit law). Based on these equations, an overall model can be calculated. For lumped parameter systems, the process model can be represented in state space form or in input/output form, i.e. differential equations or transfer functions. For distributed parameter systems, in general partial differential equations are obtained. The development of theoretical models is partially supported by domain specific software tools (e.g. ADAMS, SPICE). First object-oriented languages for multidisciplinary modeling are Omola (Anderson, 1990) Dymola (Elmqvist, 1978) and Modelica (N.N., 1997).

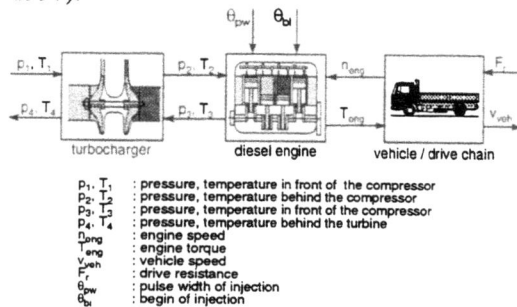

Fig. 4. Simulated process divided into the submodels Diesel engine, turbocharger and vehicle

The second way is *experimental modeling* or *identification*. Here, input and output signals are measured and for a selected class of models, the model is adapted to the process behaviour by minimization of an error measure between the process and the model. For *linear models*, identification methods have reached a very mature state, e.g. like methods of Fourier analysis, correlation, and parameter estimation. For *nonlinear models*, different approaches have been formulated. Classical methods are mostly based on polynomial approximators, e.g. Hammerstein, Wiener or nonlinear difference equation (NDE) models. Neural networks are especially attractive for the general task of nonlinear dynamic system identification due to their universal approximation capabilities.

The advantage of system identification is the less time-consuming development process. However, the quality of the resulting models highly depends on the appropriate model structure assumptions and the used measurement data. Theoretical models often have a high complexity because of their physical orientation. However, there is the possibility of model reduction, e.g. by simplifying to input/output models with less calculating time.

Sometimes the structure of the mathematical model can be determined in an analytically way, but several parameters are unknown. In this case, the combination with a parameter identification, e.g. parameter estima-

tion, is convenient.

For given equations several simulation software tools do exist like MATLAB/SIMULINK, ACSL or MATRIX$_x$.

For the development of the engine simulator, a combined theoretical-experimental way was taken. This will be illustrated by presenting the different submodels in the following.

3.2 Diesel engine model

In analogy to the real combustion engine, the torque at the crankshaft is determined in the model by superimposing the single dynamic cylinder torques. For the calculation of these cylinder torques, a special torque model was developed with regard to the following aims:

- restricted model complexity
- exact calculation of the average cylinder torque over one working cycle
- good reproduction of the oscillating cylinder torque during the working cycle
- easy adaptation of the model to the real engine
- inclusion of determined or measured multidimensional engine maps

A calculation of a complete thermodynamical cycle is not feasible due to the calculation time restrictions and the big effort for the adaptation. On the other hand, a simple static model based on a simple engine operating map (average torque) is not sufficient for dynamical investigations. The presented torque model considers a compromise between both variants, Fig. 5.

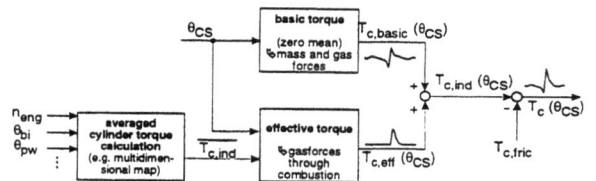

Fig. 5. Cylinder torque generation for real-time simulation

The approach is based on the idea to decouple the calculation of the average torque and the dynamic indicated torque over one working cycle:

First, the average cylinder torque $\overline{T_{c,ind}}$ over one working cycle is determined with a multidimensional map. The map can be generated either by processing measured data (e.g. on an engine test stand) or by using suitable software tools (in general a calculation of a thermodynamical cycle (Pischinger, 1989; Zapf, 1970)). Instead of look-up tables, neural networks can be used. They require less parameters and can be adapted if new data are available.

In the following, the dynamic indicated cylinder torque $T_{c,ind}(\theta_{CS})$ for one working cycle in dependence on the crank angle θ_{CS} is determined based on a given average engine torque. It consists of a zero mean *basic torque* $T_{c,basic}(\theta_{CS})$, describing the mass and gas forces and an *effective torque* $T_{c,eff}(\theta_{CS})$, describing the gas forces through combustion. The mean value of the effective part is equivalent to the average cylinder torque $T_{c,ind}$. This approach ensures that the mean value of the dynamically indicated cylinder torque $T_{c,ind}(\theta_{CS})$ is determined correctly, even if the modeling of the dynamics is only approximated. The dynamic cylinder torque $T_c(\theta_{CS})$ at the crankshaft is given by the difference of $T_{c,ind}(\theta_{CS})$ and the cylinder friction torque $T_{c,fric}$.

Basic torque. The basic torque $T_{c,basic}(\theta_{CS})$ is determined with a simplified combustion chamber model. First, the *basic cylinder pressure* $p_{c,basic}(\theta_{CS})$ is calculated which is a approximation of the cylinder pressure for the non-fired engine. During the compression and combustion cycle, $p_{c,basic}(\theta_{CS})$ follows a polytropic change of state. During the intake and exhaust cycle, $p_{c,basic}(\theta_{CS})$ is equivalent to the pressure at the beginning of compression $p_{c,0}$:

$$p_{c,basic}(\theta_{CS}) = p_{c,0}$$
$$\forall \ \theta_{CS} \in [0°,180°] \vee [540°,720°]$$
$$p_{c,basic}(\theta_{CS}) = p_{c,0}\left(\frac{V_{c,0}}{V_c(\theta_{CS})}\right)^n \qquad (1)$$
$$\forall \ \theta_{CS} \in [180°,540°]$$

$V_c(\theta_{CS})$ denotes the current cylinder volume, $V_{z,0}$ the cylinder volume at the begin of compression and n the polytropic coefficient. Fig. 6 represents the typical basic pressure for one working cycle. For comparison, the measured cylinder pressure $p_{c,ind}(\theta_{CS})$ for the fired operation is shown. The difference of both curves is the part, which is used for generating the effective torque. The pressure at the beginning of compression $p_{c,0}$ essentially depends on the charge-air pressure and can be determined by a simplified calculation of the charge cycle (Zapf, 1970).

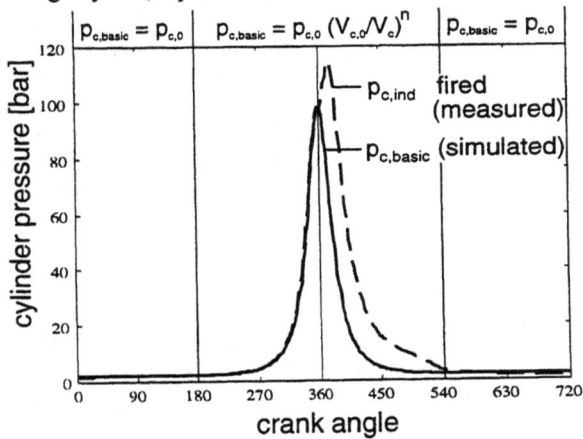

Fig. 6. Comparison between $p_{c,basic}(\theta_{CS})$ and the measured cylinder pressure $p_{c,ind}(\theta_{CS})$ for fired operation

The basic pressure is converted to the basic torque $T_{c,basic}(\theta_{CS})$ by using the mechanical laws of the crankgear (Sinsel, *et al.*, 1997; Maass, and Klier, 1981). Herewith, the resulting gas forces and the oscillating mass forces are taken into consideration.

Effective torque. The effective torque $T_{c,eff}(\theta_{CS})$ is given by the difference of the indicated cylinder torque $T_{c,ind}(\theta_{CS})$ and the zero mean basic torque $T_{c,basic}(\theta_{CS})$:

$$T_{c,eff}(\theta_{CS}) = T_{c,ind}(\theta_{CS}) - T_{c,basic}(\theta_{CS}) \qquad (2)$$

This part of the torque essentially results by the combustion during the expansion cycle. Fig. 7 shows the subdivision of the indicated torque in basic and effective torque. The represented indicated torque is based on measured cylinder pressure data acquired on a truck engine (Mercedes-Benz OM 402). $T_{c,eff}(\theta_{CS})$ can be approximated by the following equation:

$$\hat{T}_{c,eff}(\theta_{CS}) = a \cdot {\theta'_{CS}}^2 \cdot e^{(-b \cdot \theta'_{CS})},$$
$$\text{with } \theta'_{CS} = \theta_{CS} - 360° \qquad (3)$$

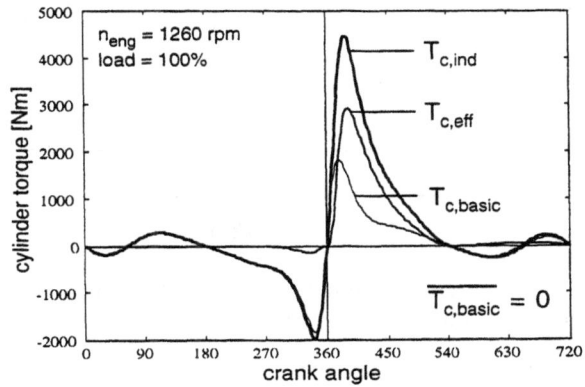

Fig. 7. Cylinder torque subdivided into basic and effective torque

In Eq. 3 it is assumed that the effective torque is equal to zero, except in the combustion cycle. The parameters a und b have to satisfy the following conditions:

1. Because $T_{c,basic}(\theta_{CS})$ is zero mean, the average of the approximated effective torque must be equivalent to the average of the indicated cylinder torque:

$$\frac{1}{720°} \cdot \int_0^{720°} \hat{T}_{c,eff}(\theta_{CS}) \ d\theta_{CS} = \overline{T_{c,ind}} \qquad (4)$$

2. The maximum of the approximating function is at crank angle $\theta'_{CS} = \theta'_{CS,max}$. While the crank angle for maximum pressure is influenced by the engine load, engine speed and begin of injection, the crank angle for maximum torque is nearly independent of these quantities. This fact is due to the mechanics of the crankgear. Thus $\theta'_{CS,max}$ can be assumed as a constant.

$$\frac{d}{d\theta'_{CS}} \hat{T}_{c,eff} \ |_{\theta'_{CS} = \theta'_{CS,max}} = 0 \qquad (5)$$

With Eq. (4) and (5), the parameters a und b can be obtained:

$$a = \frac{4 \cdot 720° \cdot \overline{T_{c,ind}}}{\left(\theta_{CS,max}^{'}\right)^3} \qquad b = \frac{2}{\theta_{CS,max}^{'}} \qquad (6)$$

Using Eq. (3) and (6), $T_{c,eff}(\theta_{CS})$ can be calculated dependent on $\overline{T_{c,ind}}$. Fig. 8 shows the effective torque simulated and derived from measurements of the cylinder pressure. Using this approach the dynamics of the overall cylinder torque $T_{c,ind}(\theta_{CS})$ can be reproduced with sufficient accuracy. In Fig. 9, the simulated and reconstructed cylinder torque for one working cycle is depicted. This good agreement can also be obtained at other speed/torque operation points. Only at low loads, little deviations occur due to the simplified model assumptions mentioned above.

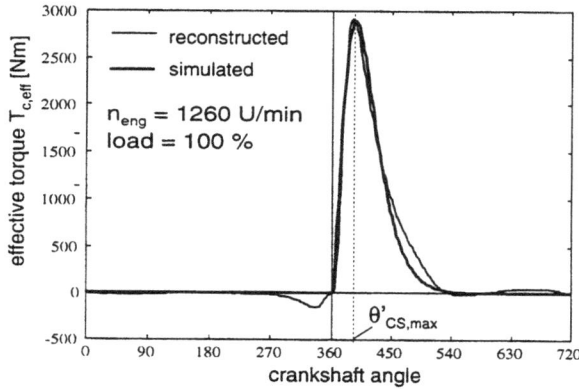

Fig. 8. Effective cylinder torque: Simulated and derived from measurements

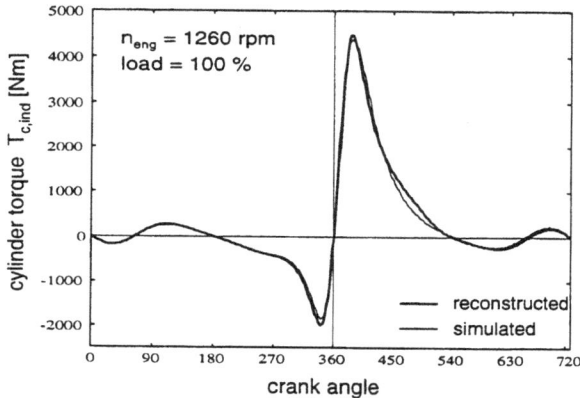

Fig. 9. Indicated cylinder torque: Simulated and derived from measurements

3.3 Turbocharger model

Fig. 10 schematically represents the exhaust turbocharger with wastegate. The charging process has a nonlinear static input/output behaviour as well as a strong dependency on the dynamics on the operating point.

In general, the static behaviour of the turbocharger may be sufficiently described by characteristic maps (look-up tables) of the compressor and turbine. However, if the dynamics of the turbocharger need to be considered, basic mechanical and thermodynamical

modelling is required, see (Zinner, 1985; Pucher, 1985).

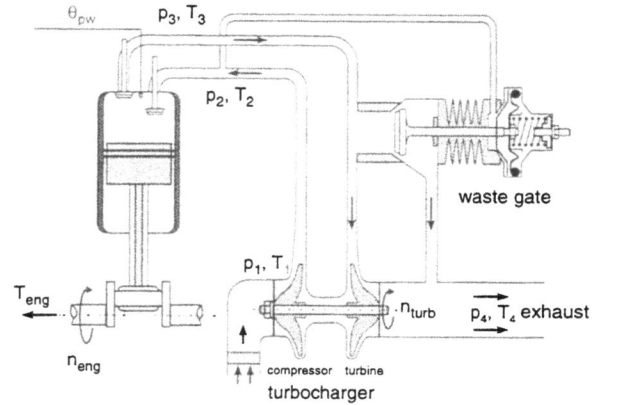

Fig. 10. Diesel engine with exhaust turbocharger and waste gate

Practical applications have shown that these methods are capable of reproducing the characteristic dynamic behaviour of the turbocharger. The model quality, however, essentially depends on the accurate knowledge of several process parameters, which have to be labouriously derived or estimated, in most cases by analogy considerations. A further disadvantage is the considerable computational effort due to the complexity of those methods.

For the present application a simple, easy to identify input/output model suitable for real-time simulation is required. Therefore, a novel approach for developing a dynamic model of an exhaust turbocharger was taken using an artificial neural network called LOLIMOT (Nelles, et al., 1996; Nelles, and Isermann, 1996). This is a special kind of basis functions network, complying with the demanded properties to a high degree. Only the recorded data of input/output measurements of the real process are required. Thus, no theoretical knowledge of the process is necessary. For a hardware-in-the-loop simulation, the pulse width θ_{pw} and the engine speed n_{eng} are chosen as inputs while the charge-air pressure p_2 is the output. The sampling time is 0.2s. It was found by trial and error that the turbocharger can be described sufficiently well by assuming a second order process. Therefore the charge-air pressure $p_2(k)$ at time instant k is modelled by the following equation:

$$k) = \\ {}_{pw}(k),\theta_{pw}(k-1),n_{eng}(k),n_{eng}(k-1),p_2(k-1),p_2(k- \qquad (7)$$

The training data was generated by a special driving cycle to excite the system with all amplitudes and frequencies of interest. Therefore, it covers the input space of the function $f(\circ)$ well. The measurements were recorded on a flat test track. Also, in order to operate the engine in high load ranges, the truck was driven with the largest load. For validation, special driving cycles were recorded which reproduce realistic conditions in urban and interstate traffic. LOLIMOT was trained until no substantial improvement could be reached by further increasing the model's complexity.

All local linear models have one slow and one fast stable real pole. The slow poles range from 0.12 to 0.37s⁻¹ (representing time constants between 0.6s and 1.2s) and the fast poles range from 0.71 to 0.84s⁻¹ in the z-domain. This is consistent with the physical knowledge about the turbocharger. The highest gain of all local linear models is five times larger than the lowest gain. This underlines the strongly nonlinear static and dynamic process characteristics.

Fig. 11 shows the validation performance. The data were measured by driving the truck in urban traffic. The maximum output deviation between simulated and measured charge-air pressure was less than 0.1 bar.

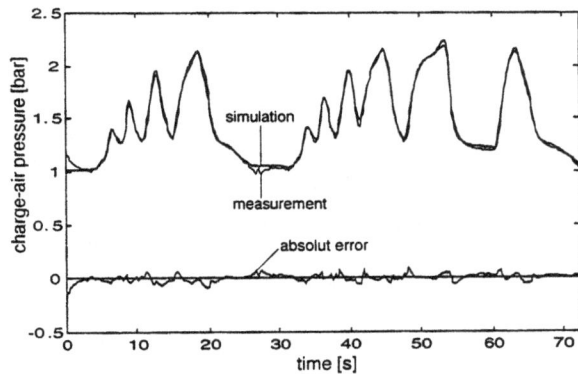

Fig. 11. Comparison between simulated (dynamic neural network) and measured charge-air pressure

The identification of the turbocharger with LOLIMOT as an universal approximator required less than a minute on a Pentium PC. The model development time was only one day including the trial and error approach for dynamic order determination (excluding the time for measurements).

3.4 Vehicle model

For operating the engine model with a realistic load, a suitable model for the vehicle longitudinal dynamics is required (Pfeiffer, 1997; Germann, 1996; Isermann, et al., 1996). The model includes the drive chain and the vehicle body, Fig. 12. Because it is not necessary to simulate the drive chain in all details for this application, a two- mass-oscillator model is implemented. Simple models for clutch, gearbox, brake and hydraulic retarder are integrated. The dynamic behaviour of the drive chain can be adapted to the real vehicle either by parametrization of the elasticity and the damping or by specifying the dominant natural frequency and the relaxation time, respectively. The driving resistance consists of rolling resistance, climbing resistance and aerodynamic drag.

3.5 HIL Simulator

In this section a simulator test bench for the investigation of the engine control system of the new Mercedes-Benz truck engine series 500 and 900 is presented. First a short survey of the engine management is given.

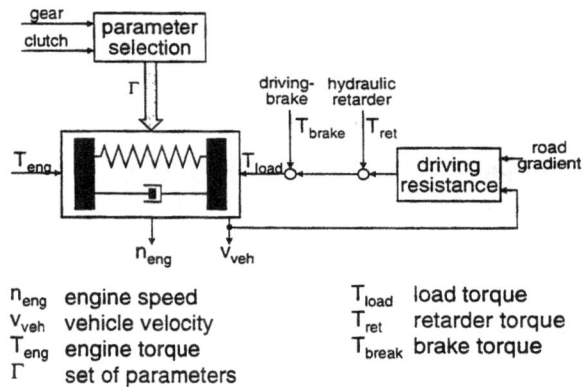

n_{eng}	engine speed	T_{load}	load torque
v_{veh}	vehicle velocity	T_{ret}	retarder torque
T_{eng}	engine torque	T_{break}	brake torque
Γ	set of parameters		

Fig. 12. Vehicle model

Electronic Engine Management. The electronic engine management is part of the Integrated Electronic System (IES) and consists of two control units connected by a CAN-bus interface, Fig. 13.

The *vehicle-engine-control unit* (FMR) contains all functions concerning the vehicle, e.g.:

- detection of the accelerator-pedal position
- engine brake control
- cruise control
- reduction of the engine torque in conjunction with anti-slip-control
- output of a reference value for the engine speed for speed control of accessories

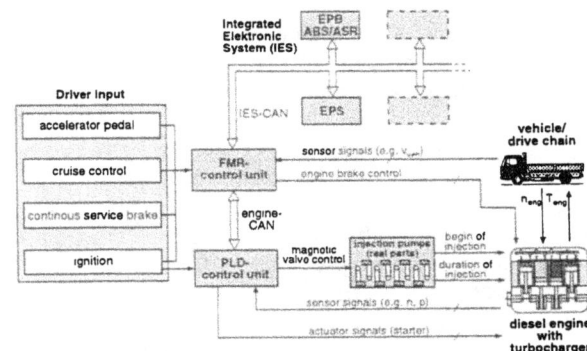

Fig. 13. Engine control system of the new Mercedes-Benz truck engine series 500 and 900

The FMR control unit calculates the reference values for the engine torque, brake torque and engine speed dependent on the drivers command, the sensor signals and the information from other units of the IES. The data is transmitted to the *pump-line-nozzle control unit* (PLD) via CAN-bus. The PLD control unit contains the engine control functions, especially the precise control of the magnetic injection valves. For that, the optimal beginning and duration (pulse width) of injection for each cylinder dependent on the reference values from the FMR control unit are calculated. By means of multidimensional maps and different sensor signals (sensors for engine speed, charge- air pressure, temperatures of fuel, coolant and charge- air) an optimal operating condition for fuel consumption, emissions and driving comfort is achieved.

7

Setup of the simulator test bench. Fig. 14 shows the setup of the simulator test bench for the investigation of the engine control system as mentioned above. It may be subdivided in the following parts:

- real-time computer system including I/O-modules
- periphery, consisting of the sensor and actuator interface
- PLD control unit including real actuator components, stand-alone or in combination with the real FMR
- PC with graphical user interface
- control panel

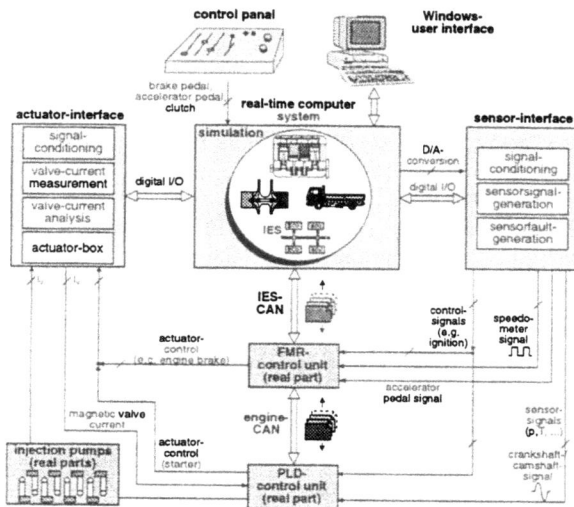

Fig. 14. Setup of the simulator test bench with simulated engine and vehicle and with real engine control unit and real injection actuators

Real-time computer system (Version 1 and 2) To fulfill the real-time constraints necessary for a HIL simulation, a powerful computer hardware is required. For the first version of the simulator test bench, a transputer system with 11 processors (T805) was developed. 4 transputers were used to calculate the simulation models and 7 for I/O communication. The models for the different subsystems are calculated in parallel to increase the computing power. The real-time computer system for the second simulator version is based on a dSPACE-system equipped with digital signal processors and a DEC Alpha processor. This system has the advantage of higher computing power which makes a parallel calculation unnecessary. It offers also the possibility to realize all the models in MATLAB/SIMULINK to use all the benefits of a graphical simulation environment. Special I/O-modules (Digital-I/O-module, D/A converter, A/D converter and CAN-interface) are used for the communication with the periphery.

Periphery The coupling of the simulator and the control unit is implemented with special periphery which can be subdivided into a sensor and an actuator interface. The *sensor interface* generates the necessary sensor signals like temperatures and pressures. The pulses of the camshaft and crankshaft inductive speed sensors are generated with a board specifically designed for high speed signal generation. For that purpose a lookup-table with the pulse-signals versus the crankshaft-angle is stored off-line in memory. During the simulation, the signals are periodically read out, synchronous to the simulated engine speed. This realization guarantees a high flexibility in forming the pulses and adapting different gear wheels. The sensor interface also contains a relay electronic to simulate sensor faults like interruptions and short circuits.

The *actuator interface* mainly consists of the injection pumps (pump-line-nozzle injection system) which are integrated in the simulation test bench as real components, because the combination of the PLD control unit and the injection pumps represent a mechatronic unit which is difficult to model. A special electronic device measures the magnetic valve currents to reconstruct the real valve opening time and to determine the pulse width and the beginning of injection. These quantities are transferred to the real-time computer system for engine simulation. By this way the real behaviour of the injection pumps is included.

PLD Control Unit and FMR Control Unit as Test Piece
The simulator test bench was set up with the objective of testing the PLD control unit stand alone or in combination with the FMR control unit. In the first case, the necessary FMR functions are simulated by the computer system. The data transfer is done via the engine CAN-bus. In the second case the PLD and the FMR control units are connected directly via the engine CAN-bus. The computer system emulates other IES units in this operation mode by transmitting the data via the IES-CAN-bus to the FMR.

PC with Graphical User Interface For an efficient use of the simulator, a comfortable experimental environment is needed. The simulator operation is performed with a windows user interface on the host-PC which copies the functionality of a real truck-cockpit. All relevant simulation quantities can be visualized on-line or be recorded for off-line analysis. To ensure reproducible results a driver simulation is implemented which can automatically follow a given speed cycle by operating the gas, brake, clutch and gear.

Control Panel As alternative, an interactive "driving" of the simulator can be performed manuelly with a control panel where the most important cockpit functionalities are realized.

3.6 Simulation results

In the following three HIL simulation examples for an 8-cylinder truck engine (420 kW) are represented in order to document the applicability and the performance of the simulator.

Fig. 15 demonstrates the effect of switching off a single injection pump valve. The gearbox is in neutral position and at the beginning the engine runs with idle speed. The cyclic decrease of the engine torque and the engine speed after to the fuel shut off can directly be seen. The control unit gradually compensates for the missing torque of one cylinder by increasing the pulse width in order to keep the desired idle speed.

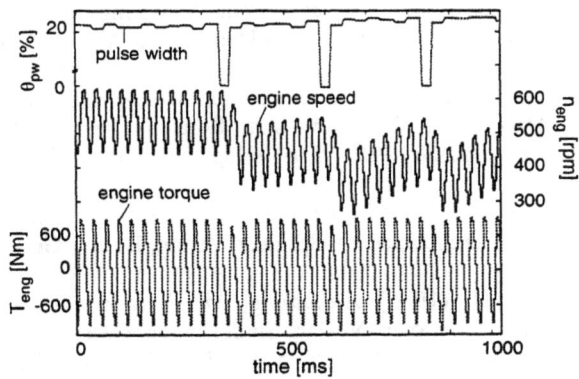

Fig. 15. HIL simulation of switching off a single injection pump valve.

A full power acceleration of a 40 tons truck including two gear shifts (1) is depicted in Fig. 16. The following effects can be observed:

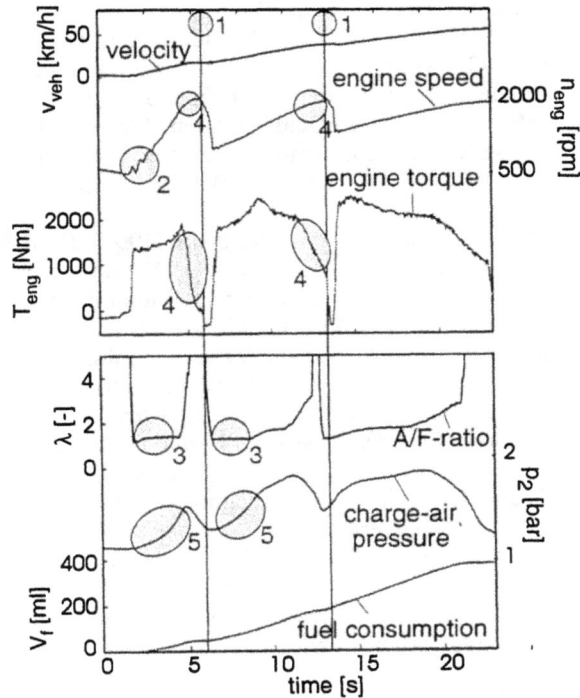

Fig. 16. HIL simulation of a full power acceleration of a 40 tons truck

- oscillations in the drive chain (2)
- limitation of soot (3)
- maximum speed limit regulation (4)
- lagged reaction of the turbo charging pressure (5)

In Fig. 17 the simulation results of a cruise control experiment are represented. At the beginning, the 40-tons truck runs at a constant speed of 60 km/h on a flat road. One can observe how the cruise control decreases and increases the engine torque as a reaction of changing the road gradient (first downhill, then uphill) in order to keep the reference speed.

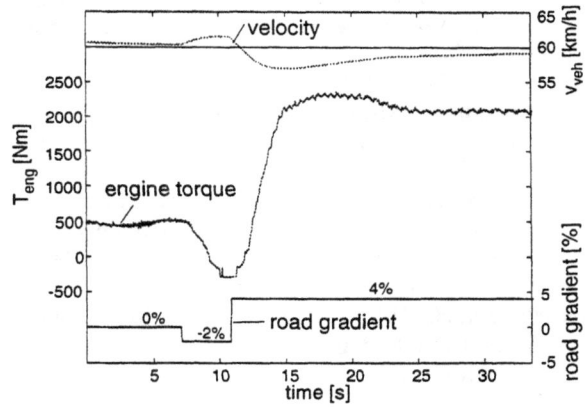

Fig. 17. HIL simulation for the investigation of cruise control

4. CONCLUSION

Fulfilling the future constraints of reducing development time and cost will require an increased use of real-time simulations. After some general considerations of real-time simulators, architectures and historical development of hardware-in-the-loop simulation was considered. Several examples during recent years show that hardware-in-the-loop simulation is an efficient tool for the development and testing of engine control systems.

As an application example, a simulator test bench for the investigation of the engine management system of the new Mercedes-Benz truck engine series is described. Several experimental results demonstrated the possibility of testing engine control functions without the use of a real engine. The HIL-simulator proved to be an efficient tool to develop new control functions, to test the software and hardware of the engine electronics and to investigate the behaviour with faults in the sensors, actuators and the engine.

REFERENCES

Anderson, O.K. (1962). Design tool (Research and Development) Simulator Survey Report. *SAE Nat. Aerospace Engng. Manufacturing Meeting,* Los Angeles, Oct. 1962.

Andersson M. (1990). *Omola - An Object-Oriented Language for Model Representation,* Lic Tech thesis TFRT-3208, Department of Automatic Control, Lund Institute of Technology, Lund, Sweden.

N.N. (1953). The Redifon Comet Flight Simulator. *De Haveland Gazette,* **Vol. August 1953,** pp. 139-141.

Drosdol, J. and F. Panik (1985). The Daimler-Benz driving simulator. SAE 850334.

Elmqvist, H. (1978). *A Structured Model Language for Large Continuous Systems.* Ph.D. Dissertation, Report CODEN: LUTFD2/(TFRT-1015), Department of Automatic Control, Lund Institute of Technology, Lund, Sweden.

ETAS (1997). *LabCar. Hardware-in-the-Loop Testsysteme für Steuergeräte*. ETAS, Schwieberdingen, Germany.

Fennel, H., S. Mahr and R. Schleysing (1992). Transputer-based real-time simulator - a high performance tool for ABS and TCS development. SAE 920643.

Germann S. (1996). *Modellbildung und modellgestützte Regelung der Fahrzeuglängsdynamik*. Dissertation TU Darmstadt. Fortschr.-Ber. VDI Reihe 12 Nr. 309. VDI-Verlag, Düsseldorf, Germany.

Hanselmann, H. (1993). Hardware-in-the-loop simulation as a standard approach for development, customization, and production test of ECU´s. SAE 931953.

Huber, W. et al. (1988). Simulation, performance and quality evaluation of ABS and ASR. SAE 880323.

N.N. (1964). Schulung und Praxis mit Flugsimulator. *INTERAVIA*, **Vol. 19**, No. 8, pp. 1104-1107

Isermann R., S. Germann, M. Würtenberger, C. Halfmann and H. Holzmann (1996). Model-Based Control and Supervision of Vehicle Dynamics. *Fisita Congress*, Prag, Czech Republic, 17-21 June 1996.

Isermann, R. (1997). Mechatronic systems - A challenge for control engineering - . *Proceedings of the 1997 American Control Conference*, Albuquerque, USA, 4-6 June 1997, Vol. 5, pp. 2617-2632.

Isermann, R., S. Sinsel and J. Schaffnit (1997). Hardware-in-the-loop simulation of Diesel-engines for the development of engine control of Diesel-engines for the development of engine control systems. *Proceedings of the 4th IFAC-Workshop on Algorithms and Architecture for Real-Time Control, AARTC´97*, Vilamoura, Portugal, 9-11 April 1997, pp. 90-92.

Kempf, D., L.S. Bonderson and L.I. Slafer (1987). Real time simulation for application to ABS development. SAE 870336.

Kimura, A. and I. Meda (1996). Development of engine control system using real time simulator. *Proceedings of IEEE Int. Symposium Comp. Aid. Contr. Systems Design*, Dearborn, 15-18. Sept. 1996, pp. 157-163.

Klinker, W. (1992). Hardware-in-the-loop. *SUN News*, **October 92**, pp. 21-23.

Kortuem, W., R.S. Sharp and A.D. Departer (1992). *Review of multibody computer codes for vehicle system dynamics: problem requirement codes and experiences with benchmark problems*. Carl Cranz Gesellschaft, Oberpfaffenhofen, Germany.

Maass H. and H. Klier (1981). *Kräfte, Momente und deren Ausgleich in der Verbrennungskraftmaschine*. Springer-Verlag, Berlin, Germany.

Marienfeld, H. (1965). Flugsimulation. Dissertation TU Darmstadt.

Nelles O. and R. Isermann (1996). Basis Function Networks for Interpolation of Local Linear Models. *Proceedings of the 35th IEEE Conf. on Decision and Control*, Kobe, Japan, Dec. 1996, **Vol. 1**. pp. 470-475.

Nelles, O., S. Sinsel and R. Isermann (1996). Local Basis Function Networks for Identification of a Turbocharger. *Proceedings of the UKACC*

International Conference on Control '96, Exeter, UK, 2-5 Sept. 1996, **Vol. 1**, pp. 7-12.

N.N. (1997). Modelica, A Unified Object-Oriented Language for Physical Systems Modeling. *Modelica homepage*: http://www.Dynasim.se/ Modelica/

Pfeiffer, K. (1997). *Fahrsimulation eines Kraftfahrzeugs mit einem dynamischen Motorprüfstand*. Dissertation TU-Darmstadt. Fortschr.-Ber. VDI Reihe 12 Nr. 336. VDI-Verlag, Düsseldorf, Germany.

Pischinger, R. (1989) *Thermodynamik der Verbrennungskraftmaschine*. Springer-Verlag, Berlin, Germany.

Pucher, H. (1985) *Aufladung von Verbrennungsmotoren*. Expert-Verlag, Sindelfingen, Germany.

Rieger, K. and W. Schiehlen (1995). Echtzeitsimulation eines Fahrzeugmodells mit aktiver Federung. *VDI-Bericht Nr. 1189*, pp. 17-34. VDI-Verlag, Düsseldorf, Germany.

Sailer, U. (1996). *Nutzfahrzeug-Echtzeitsimulation auf Parallelrechnern mit Hardware-in-the-Loop*. Expert-Verlag, Renningen-Malmsheim, Germany.

Sailer, U. and U. Essers (1996). Parallelverarbeitung als Weg zur Echtzeitsimulation in der Kfz-Steuergeräteentwicklung. *VDI-Bericht Nr. 1283*, pp. 99-116. VDI-Verlag, Düsseldorf, Germany

Savaglio, C. (1993). Hardware-in-the-loop simulation - an engine controller implementation. SAE 930204.

Schaefer, P. (1993). Echtzeitsimulation aktiver Mehrkörpersysteme auf Transputernetzen. Fortschr.-Ber. VDI Reihe 11 Nr. 202. VDI-Verlag, Düsseldorf, Germany.

Sinsel, S., J. Schaffnit and R. Isermann (1997). Hardware-in-the-Loop Simulation von Dieselmotoren für die Entwicklung moderner Motormanagementsysteme. *VDI-Tagung*, Moers, Germany.

Spindler, W. and M. Doll (1990). Ermittlung des stationären und des instationären Betriebsverhaltens von kleinen schnellaufenden Dieselmotoren. *Forschungsberichte Verbrennungskraftmaschinen*: Heft 454, FVV, Gesamtprozeßanalyse-Thermodynamik PKW-Dieselmotoren, Vorhaben 384, Abschlußbericht.

Suetomi, T. et al. (1991). The driving simulator with large amplitude motion system. SAE 910113.

Thun, H.-J. v. (1987). Dynamischer Verbrennungsmotorenprüfstand mit Echtzeitsimulation des Kraftfahrzeug-Antriebsstranges. *Automobiltechnische Zeitschrift - ATZ*, **Vol. 89**, No. 1.

Wagner, J.R. and J.S. Furry (1992). A real time simulation environment for the verification of automotive electronic controller software. *International Journal of Vehicle Design*, **Vol. 13**, No. 4, pp. 365-377

Woermann, R. (1994). Ein Beitrag zur Echtzeitsimulation technischer Systeme hoher Dynamik mit diskreten Modellen. Dissertation Universität Gesamthochschule Kassel.

Zapf, H. (1970). Untersuchungen zur Vorausberechnung der Ladungsendtemperatur in Viertakt-Dieselmotoren. *Motortechnische Zeitschrift - MTZ*, **Vol. 12**.

Zinner, K. A. (1985). *Aufladung von Verbrennungsmotoren*. Springer-Verlag, Berln, Germany.

DECENTRALIZED IMPLEMENTATION OF REAL-TIME SYSTEMS USING TIME PETRI NETS. APPLICATION TO MOBILE ROBOT CONTROL.

F.J. García* J.L. Villarroel**

*Universidad de La Rioja, Dpto. Matemáticas y Computación,
C/ Luis de Ulloa s.n., 26004 Logroño, Spain*
**CPS, Universidad de Zaragoza, Dpto. de Informática e Ing. de
Sistemas, C/ Maria de Luna 3, 50015 Zaragoza, Spain*

Abstract: Time Petri nets are used as the formalism for the whole life cycle of real-time systems. We present how to model real-time systems using this formalism and we focus our work on the code generation for these systems. The first step of this implementation technique consists of the extraction of the processes (states machines) embedded in the net, each of which is implemented in an Ada task. The result of the application of this technique is a set of concurrent processes coupled by means of synchronous or asynchronous communications, with the same behaviour as the model. *Copyright © 1998 IFAC*

Keywords: Formal methods, Petri-nets, modeling, real time, Ada tasking

1. INTRODUCTION

In this paper we continue the work started with García and Villarroel (1996) in which a way of modeling real-time systems (using time Petri nets) and an automatic technique (through the use of an interpreter) for code generation was shown. Our objective was and is to improve the reliability in all steps of the life cycle of a real-time system, by means of a unique formal method during it: time Petri nets.

Petri nets have been widely used for modeling and analyzing discrete event systems because of the possibility of modeling concurrency, resource sharing, synchronizations, ... However, ordinary Petri nets are not suitable either for the modeling or the analysis of real-time systems, due to the impossibility of including time features in the model. This is the reason why in this paper a time extension of Petri nets, Time Petri Nets (TPN), is used. TPNs are useful in order to develop reliable real-time software due to the possibility of modeling time-outs, synchronizations, concurrence, periodic or aperiodic communicating processes.

Other references can be found in the literature where time Petri nets are used in relation to real-time systems. Some times they are used for the modeling and analyzing communication or control systems (Berthomieu and Diaz (1991); Buy and Sloan (1994); Aalst and Odijk (1995)). On other occasions (Shatz et al. (1996)) Petri nets are used to generate nets models of concurrent programs (tasking programs) with the purpose of an analysis of the concurrent behaviour, especially deadlock properties. Finally another line of research (Gedela and Shatz (1997)) has modeled Ada-tasking structures with time Petri nets as a way of defining precise behaviour for tasking semantics and providing support for automated analysis.

In this work we consider just the opposite approach: TPN are considered as the initial tool for the modeling of the real-time system. After analysis and validation, the implementation is generated from the TPN model, being this last point the main contribution of this paper. The use of the same formalism during the life cycle will allow the detection of bad properties in the early stages of the cycle and, indeed, will allow us not to restrict the structure of systems in order to analyze their temporal constraints. In this sense, the design flexibility is increased with respect to the use of classical analytic techniques such as RMA where, for example, in order to allow the analysis, the communications between the periodical tasks must take place through an intermediate server with no guarded entry. Moreover, the use of this formal method can allow us the automatic code generation (García and Villarroel (1996)), and so it avoids making mistakes during the codification.

The scope of this paper is limited only to the modeling and the implementation of real-time systems. A Petri net implementation is a program which simulates the firing of the net transitions. Adaptations of Petri net classical implementation techniques are used (Colom et al. (1986)) which can be split into centralized and decentralized ones. The former use a single coordinator process responsible for the control and firnig of the transitions of the net, which represent the operational part of the system. This technique (object of study in the related paper García and Villarroel (1996)) is subject to several problems related to the presence of the coordinator, which acts in every transition firing introducing an overload into the implemented system. In addition, the coordinator alone is responsible for the control of the implementation, sequentializing the control of the implemented system, which is in fact concurrent, and making it sensitive to faults, since if the coordinator fails the whole system fails too. The decentralized implementations try to solve these problems by splitting the net into several concurrent subnets. Ada 95 is used as the language for the implementation code and only monoprocessor platforms are considered.

2. MODELING REAL-TIME SYSTEMS USING TIME PETRI NETS

A Time Petri Net (Berthomieu and Diaz (1991)) is a tuple $(P, T; F, B, M_o, SIM)$, where $(P, T; F, B, M_o)$ defines a marked Petri net, the *underlying Petri net*; and SIM is the mapping called static interval $SIM : T \rightarrow \mathbb{Q}^* \times (\mathbb{Q}^* \cup \infty)$, where $\mathbb{Q}^*$ is the set of positive rational numbers. Thus, TPNs can be seen as Petri nets with labels: two time values (α_i, β_i) associated to transitions. Assuming that transition t_i was enabled at time θ_0, and is being continuously enabled, the first time value represents the minimum time, starting from θ_0 that t_i has to wait until it can be fired, and the second is the maximum time that t_i can be enabled without firing. So these two time values allow the calculation of the firing interval for each transition t_i in the net: $(\theta_0 + \alpha_i, \theta_0 + \beta_i)$. Once the transition is to be fired, the firing is instantaneous.

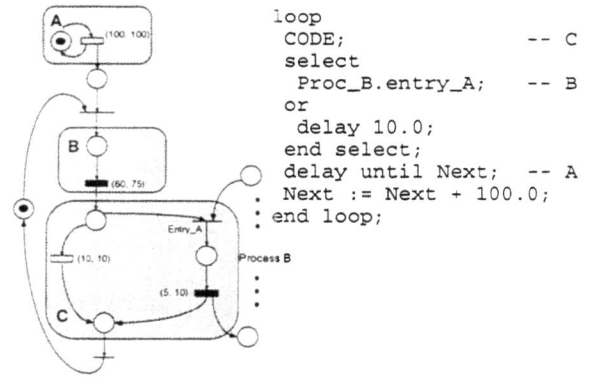

```
loop
  CODE;              -- C
  select
    Proc_B.entry_A;  -- B
  or
    delay 10.0;
  end select;
  delay until Next;  -- A
  Next := Next + 100.0;
end loop;
```

Fig. 1. Example of TPN model

As an example in fig.1, a TPN modeling a periodic process that executes a piece of code and communicates with another process is shown. This communication has an associated time-out. Three elements have been highlighted (a piece of Ada code with the same behaviour is provided for a better understanding of the model):

- Box B shows an action, i.e. code, to be executed by the process. The execution starts when the input place gets marked. The computation time of this activity is between (60,75) time units.
- Box A models the periodic activation of the process. Every 100 time units the transition fires and promotes the execution of the process.
- Box C shows a communication with another process which has an associated time-out. Let us suppose that the place is marked at time τ. If the transition labeled with entry_A does not fire (starts the communication) before $\tau + 10$ (expiration time of the time-out), then transition (10,10) will fire, aborting the starting of the communication.

Transitions in TPN have the same functionality. But the different situations that appear in a real-time system must be highlighted in our models. Therefore, and with the aim of implementing the model, we distinguish three kinds of transitions:

- CODE-Transitions (thick segments) together with its input place, represents the code associated to an activity, that starts its execution when the transition gets enabled, i.e. the input place gets marked. The two time values (α, β) represent the execution time of the activity. At best, the code execution finishes at time α, and at worst the execution will last β. The firing represents the end of the code.
- TIME-Ts. (empty thick segment) are transitions with an associated time event, e.g. a time-out. These transitions also have associated time information, described with an interval (α, α), where α represents the event time. The firing of this kind of transitions represents the occurrence of the event.

- SYCO-Ts. (thin segment) are transitions with no temporal meaning used to perform synchronizations (SY) and control (CO) tasks. The firing of a SYCO-T leads to plain state changes.

3. NET DECOMPOSITION

The idea of decentralized implementations is quite simple. In order to avoid the problems of centralized implementations (mentioned in the introduction) the use of the centralized coordinator must be avoided. Therefore the control of the net is split into several sequential subnets, each of which is implemented in a separate process, concurrent with the others. The identification of the concurrent processes embedded in the net and their inter-connection through a communication mechanism like a buffer or a rendezvous is the first step in the implementation. The basis is to merge into a single process a set of transitions in mutual exclusion (ME) with the others in the subnet (two transitions t_i, t_j are in mutual exclusion, $t_i M E t_j$, if they can not be fired simultaneously). A set of transitions which are in ME relationship are not concurrent, so they can be in the same process (π_i) without reducing the actual concurrence. For the computation of the sets of transitions in mutual exclusion the net without time information will be used. The reason is that two or more transitions which are in ME in the Petri net, remain in ME in the TPN, but the opposite it is not true. The application of temporal mutual exclusion is still under research. The decomposition techniques used in this work are based on Colom et al. (1986); Villarroel (1990). Through the computation of compatibility classes of transitions in ME a partition of the net is obtained solving a coverability problem. Each class is a set of transitions in ME that can be implemented as a sequential process (π). The ME of transitions can be achieved by a computation of *monomarked p-invariants*. Monomarked p-invariants are particularly interesting because they describe a set of places in ME. In fig. 2 either p_1, p_2, p_3 or p_4 is marked, but never two or more of them at the same time. Obviously, a set of places in ME implies a set of transitions in ME, the input and output transitions of the places. Unfortunately it is not always possible to cover a Petri net with a set of monomarked p-invariants. To solve this problem Villarroel (1990) proposes a technique (based in the concept of *pipeline*) to make a set of monomarked p-invariants which cover all transitions of the net.

A place p, with respect to a process π_i, can be either *private* (every input and output arc of p

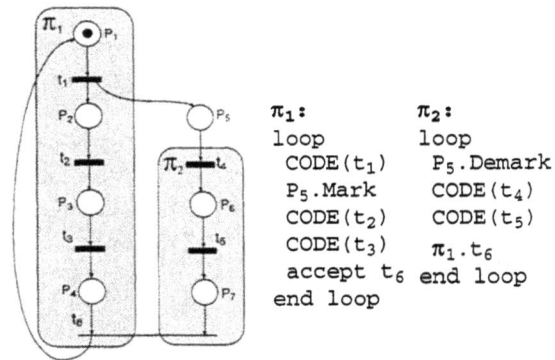

Fig. 2. Decomposition into two sequential processes. An asynchronous comm. in p_5, and a synchronous communication in t_6.

are connected to transitions belonging to π_i), or external (p is only connected to transitions not belonging to π_i), or shared (p is connected both to transitions belonging and not belonging to π_i). In fact, a shared place is modeling an asynchronous communication between two processes. A shared transition represents a synchronous communication (rendezvous). Moreover, it is possible to share sets of transitions and places grouped in a subnet, which are representing the execution of a piece of code in a rendezvous.

4. PROCESS IMPLEMENTATION

However it is achieved, a partition of the Petri net which covers every transition of the net can be found. The partition is made up of a set of sequential processes, each one having an associated p-invariant that can be used to describe the control flow of the process. In this way, only those transitions whose input place belonging to the p-invariant is marked, are able to fire. Each process can be implemented in an Ada task, using a **case** structure. Each place of the p-invariant describes a state which will be implemented at each branch of the **case**. The code associated with each branch depends basically on the output transition of the place (or transitions if there is a conflict). It is possible to improve the implementation using the single token of the p-invariant as if it were the program counter of the process. The flow of the token through the p-invariant defines the execution order of the transitions, avoiding the use of the **case** structure (e.g. see the code shown for the net in fig.2). Implementing each process in an Ada task, both control and operational parts of the set of transitions are integrated in the same process avoiding the use of a centralized coordinator.

The three kinds of transitions in our modeling approach will be implemented in a different manner. A SYCO-transition will be taken into

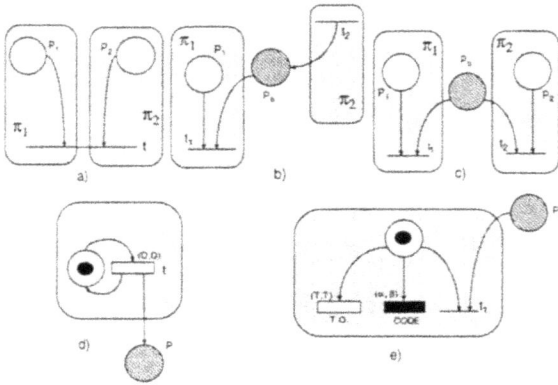

Fig. 3. a) Synchronous communication, transition t; b) asynchronous communication, place p_b; c) simple conflict; d) Periodical activator; e) Inner conflict inside a process;

account to make decisions inside a process or to perform synchronous communication between processes. A CODE-T involves the execution of their associated code. A TIME-T represents a delay in the execution of the process. As a first approximation the delay starts when the input place of the transition gets marked. When the delay expires the transition is fired. But this situation can provoke accumulative drift in the processes, e.g. due to preemption. To avoid it a time variable is associated with each process with TIME-Ts inside. This variable (Last_update) records the time at which the last marking update occurred in the process. This time is used in the computation of the expiration time of the delays. E.g., consider these implementations for the fig.3.d. (a periodical activation). The implementation on the left presents accumulative drift, solved on the right.

```
loop              Last_update := CLOCK;
  delay D;        loop
  P.Mark;           next := Last_update + D;
end loop;           delay until next;
                    Last_update := next;
                    P.Mark;
                  end loop;
```

Sometimes a transition or subnet can be shared between two processes. This situation represents a synchronous communication between both processes, and will be implemented with an Ada rendezvous (fig.3.a). The remaining non private places that are not included in a p-invariant act as asynchronous communications between processes that will be implemented with a buffer or a relay process (fig.3.b). There are likely to be several transitions at the output of a place (conflict situation). If the transitions belong to different processes, the place will act as a shared data between them. The descending processes will compete for the token of the place (fig.3.c). This situation is very common, since it is the natural way of modeling shared data or resources. Since

there are two or more processes involved in the synchronous or asynchronous communication, in a general case each one can be marked at a different time, thus, it is necessary to communicate the time Last_update of the processes.

If the place originating the conflict belongs to a p-invariant all its output transitions in conflict will be implemented in the same process. The simplest case is when several SYCO-Ts depend on the same place. This situation represents a choice, implementable with an if structure if there is no communication involved or with a select structure with several accept branches if there are communications. Other conflicts can appear between different kinds of transitions, as in fig.3.e, which models the execution of a code abortable by the expiration of a time-out, or a control action external to the process. For the implementation of this kind of conflicts we will use the Ada A.T.C. structure.

5. AN EXAMPLE

The technique presented in this paper, has been used for the code generation for the controller of a real-time mobile robot navigation application. Off-line planned trajectories and motions are modified in real-time to avoid obstacles, using a reactive behaviour. The information about the environment is provided to the control system of the robot by a rotating 3D laser sensor with two degrees of freedom.

The controller must perform three main activities: motion control, supervision and data sensor processing. The control loop has a sample period T=0.25 s obtained in the analysis phase so that the system meets all the temporal constraints. The communications between the robot and the controller have defined time-outs: 0.1 s in position reading and 0.1 s in setpoint sending. The objectives of the supervisor activity are: trace the real trajectory of the robot, test if the actual goal has been reached, update the actual goal point and manage the system alarms. At this point of application development, only alarms related to the communication time-outs have been taken into account. The firing of a communication time-out must stop the system within 0.1 s. The 3D laser sends a new scan each 0.1s. The controller must be capable of accepting and processing the sensor data at this rate. The communication with the laser has a 0.2 s time-out. Fig.4 shows the TPN that specifies the control and time restrictions of the real-time system (time information has been removed for clarity). This Petri net has been used in the analysis for the validation of system specifications and for scheduling the controller tasks in the processor by priority assignment (not

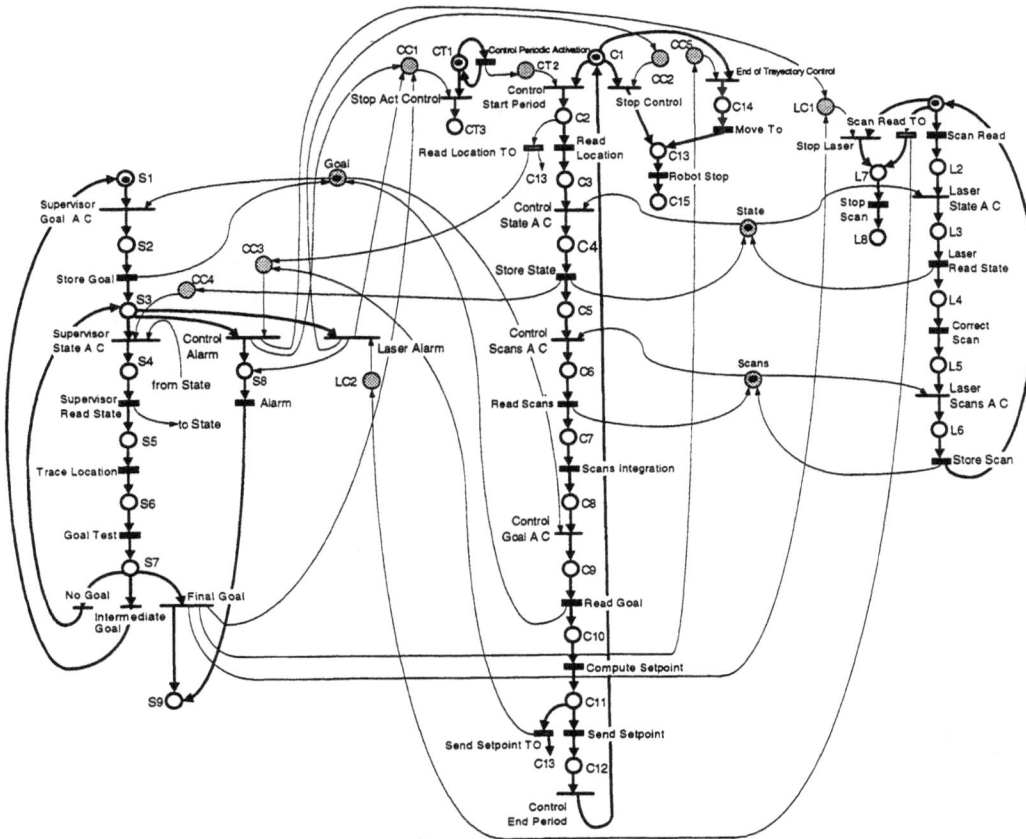

Fig. 4. Sequential processes are highlighted and communication places are shaded in grey

considered in this paper). The priority assignment has been performed using an heuristic method based on RMA techniques.

For the sequential process recognition a p-invariant computation has been developed. There are seven p-invariants: (I_1) Supervisor S_1. S_9, (I_2) Control $C_1..C_{15}$, (I_3) Activation CT_1, CT_3, (I_4) Laser $L_1..L_8$, (I_5) Protected Goal Goal, S_2, C_9, (I_6) Protected State State, L_3, C_4, S_4, (I_7) Protected Scans Scans, L_6, C_6. With these p-invariants the transition coverability problem can be solved, see fig 5. In this table it can be seen that there are four essential p-invariants (I_1, I_2, I_3, I_4), which cover all the net transitions (shaded in grey). The first set of transitions in ME (covered by I_1) corresponds to *Supervisor* process. The second (covered by I_2) corresponds to the *Control* process. The third (I_3), is the periodical activator of the Control process, the *Control Activation* process, and the fourth (I_4) is the process which deals with the laser, the *Laser* process. There are several places remaining which do not belong to any process, which are modeling asynchronous communications between the processes. The places are CC_4, CC_3, LC_2, CC_1, CT_2, LC_1, CC_2, CC_5, all of private destination, and *Goal*, *State*, *Scans*, of non

	I_1	I_2	I_3	I_4	I_5	I_6	I_7
Store_Goal	x				x		
Supervisor_Read_State	x					x	
Trace_Location	x						
Goal_Test	x						
Alarm	x						
Supervisor_Goal_Access_Control	x				x		
Supervisor_State_Access_Control	x					x	
No_Goal	x						
Intermediate_Goal	x						
Final_Goal	x						
Control_Alarm	x						
Laser_Alarm	x						
Move_To		x					
Robot_Stop		x					
Read_Location		x					
Store_State		x				x	
Read_Scans		x					x
Scans_Integration		x					
Read_Goal		x			x		
Compute_Setpoint		x					
Send_Setpoint		x					
Control_Scans_Access_Control		x					x
Control_Goal_Access_Control		x			x		
Control_State_Access_Control		x				x	
Control_Start_Period		x					
Stop_Control		x					
Control_End_Period		x					
End_of_trajectory_Control		x					
Read_Location_TO		x					
Send_Setpoint_TO		x					
Stop_Act_Control			x				
Control_Periodic_Activation			x				
Scan_Read				x			
Laser_Read_State				x		x	
Correct_Scan				x			
Store_Scan				x			x
Stop_Scan				x			
Laser_State_Access_Control				x		x	
Laser_Scans_Access_Control				x			x
Stop_Laser				x			
Scan_Read_TO				x			

Fig. 5. Covering table of transitions

private destination. E.g. the code corresponding to Control process is shown:

```
task body Control is
 Read_Location_TO: constant duration:=0.1;
 Send_Set_Point_TO:constant duration:=0.1;
begin
  loop
    select
      accept Control_Star_Period;
      select
        delay Read_Location_TO;
        CC3.Mark;    exit;
      then abort
        Read_Location;
      end select;
      State.Demark;  Store_State;
      State.Mark;    CC4.Mark;
      Scans.Demark;  Read_Scans;
      Scans.Mark;    Scans_Integration;
      Goal.Demark;   Read_Goal;
      Goal.Mark;     Compute_Setpoint;
      select
        delay Send_Set_Point_TO;
        CC3.Mark;    exit;
      then abort
        Send_Setpoint;
      end select;
    or
      accept Stop_Control;
      exit;
    or
      accept End_Of_Trajectory;
      Move_To;       exit;
    end select;
  end loop;
  Stop_Robot;
end;
```

6. CONCLUSIONS AND FUTURE WORK

Time Petri nets have been proposed as the formalism for the whole life cycle of real-time systems providing the following advantages: it allows an unambiguous and easy to understand system specification due to its graphical nature; it allows the verification and validation of the correction of the system in the early stages of the cycle; it allows a high modeling flexibility, since it will no longer be necessary to impose restrictions on the system in order to analyze the temporal behaviour and verify the timing constraints. Structural techniques have been applied in order to detect the concurrent sequential processes embedded in the net, avoiding the problems of centralized techniques. Moreover, the implementation is automatizable allowing us the automatic code generation, preventing us from making mistakes during the codification and simplifying the development of the system.

For the computation of sequential processes we have used the net without time information. But sometimes, due to time interpretation, a set of transitions which are not in ME in the underlying Petri net, are in the time Petri net. In future works we will try to detect and formalize this temporal mutual exclusion A further line of research will be the study of the shedulability of systems modeled with TPN since, up to now, it has been performed using heuristic rules. The appropriate priority assignment policy for a process consisting of a set of transitions of different priority must be studied: a static priority equal to the highest of the transition priority in the set, a dynamic priority depending on the transition which is currently fired, or an alternative priority assignment.

ACKNOWLEDGEMENTS

This work has been supported in part by project TAP97-0992-C02-01 from the CICYT of Spain.

References

Aalst, W. v. d. and Odijk, M. A. (1995). Analysis of railway stations by means of interval timed coloured petri nets. *Real-Time Systems*, 9(3):241–263.

Berthomieu, B. and Diaz, M. (1991). Modeling and verification of time dependent systems using time petri nets. *IEEE Trans. on Soft. Eng.*, 17(3):259–273.

Buy, U. and Sloan, R. (1994). Analysis of real-time programs with simple time petri nets. In *Proc. Int. Symp. on Software Testing and Analysis*, pages 228–239.

Colom, J., Silva, M., and Villarroel, J. (1986). On software implementation of petri nets and colored petri nets using high level concurrent languages. In *Proc. of 7th European Workshop on Application and Theory of Petri nets*, pages 207–241, Oxford, England.

García, F. J. and Villarroel, J. L. (1996). Modelling and ada implementation of real-time systems using time petri nets. In *Proc. 21st IFAC/IFIP Workshop on Real-Time Programming*, Gramado - RS, Brazil.

Gedela, R. and Shatz, S. (1997). Modelling of advanced tasking in ada-95: A Petri net perspective. In *Proc. 2nd Int. Workshop on Soft. Eng. for Parallel and Distributed Systems*, Boston, USA.

Shatz, S. M., Tu, S., Murata, T., and Duri, S. (1996). An application of petri net reduction for ada tasking deadlock analysis. *IEEE Transactions on Parallel and Distributed Systems*, 7(12):1307–1322.

Villarroel, J. L. (1990). *Integración Informática del Control en Sistemas Flexibles de Fabricación*. PhD thesis, Universidad de Zaragoza, María de Luna 3 E-50015 Zaragoza, España.

LOCAL BAYESIAN REGULARISATION OF PARSIMONIOUS NEUROFUZZY MODELS FOR REAL WORLD DYNAMIC PROCESSES

K.M. Bossley* M. Brown** C.J. Harris**

*Parallel Applications Centre, 2 Venture Road, Chilworth,
Southampton, SO16 7NP, UK.*
*** Image, Speech and Intelligent Systems Research Group,
Department of Electronics and Computer Science, University of
Southampton, SO17 1BJ, UK.***

Abstract: By combining properties of fuzzy systems and neural networks, neurofuzzy modelling is ideally suited to many system identification and data modelling applications. Recently, data-driven model construction algorithms have been developed to identify these models. These algorithms have proved essential for producing accurate parsimonious models. However, due to problems with sparse data and restricted model structures, models with high model variance are often produced. Thus resulting in models which generalise poorly.
In this paper local Bayesian inference techniques are applied to neurofuzzy models; multiple prior probability density functions are placed on the weights and superfluous model variance is controlled. This gives a form of regularisation where Bayesian estimation produces simple re-estimation formulae which identify a suitable bias/variance balance from the data. This approach is considered a post-processing step to model construction, the merits of which are demonstrated by the application to a real world data set. *Copyright © 1998 IFAC*

Keywords: System identification, neural networks, fuzzy logic and regularisation.

1. INTRODUCTION

Neurofuzzy systems have been developed to provide a modelling technique that utilises both linguistic and numerical knowledge. This is achieved by combining fuzzy systems and neural networks into one unified framework. The power behind these models is the existence of the direct equivalence between the union of a set of fuzzy rules of the form:

$$\text{IF } (\mathbf{x} \text{ is } \mathbf{A}^i) \text{ THEN } (y \text{ is } B^j) \ c_{ij}$$

and the weighted sum of the multi-dimensional fuzzy input membership functions, given by:

$$y(\mathbf{x}, \mathbf{w}) = \sum_{i=1}^{p} a_i(\mathbf{x}) w_i, \qquad (1)$$

where $(\mathbf{x} \text{ is } \mathbf{A}^i)$ and $(y \text{ is } B^j)$ represent linguistic expressions for the input and output respectively, c_{ij} is the rule confidence, $a_i(\mathbf{x})$ represents the multi-dimensional fuzzy membership function and w_i is its associated weight.

This relationship is fully explained in (Brown M. and Harris C.J., 1994), and allows model initialisation and validation via the linguistic interpretation, and simple model adaption and learning by using the generalised linear model representation, equation (1). The model's output can be calculated by a small number of simple mathematical operations and is hence these models are ideal for resource limited real-time applications. Recently similarities between these models and local modelling techniques have been exploited (Bossley K.M., 1997; Hunt J.K., Haas, R. and Brown M., 1995). Also, they have been applied to non-linear Kalman filtering (Wu Z.Q. and Harris C.J., 1996).

Conventionally neurofuzzy models are identified by designing the fuzzy rules from expert knowledge, and then exploiting neural network training techniques to further adapt the rule confidences. In real world modelling scenarios this approach is troubled with difficulties, as both numerical and expert knowledge may be limited. Coupled with the curse of dimensionality, data-driven neurofuzzy model identification algorithms have been developed (Bossley K.M., 1997). These techniques iteratively construct parsimonious neurofuzzy models based on a combination of the available *a priori* knowledge and empirical data. Despite the popularity and success of these algorithms, models with unnecessarily high model variance are often produced. The resulting redundant degrees of freedom tend to fit noise in the data (termed overfitting), thus degrading the interpretability and generalisation abilities of the models.

In this paper this problem is tackled by the application of Bayesian inferencing techniques which results in a powerful method for controlling these superfluous model parameters. First, model construction is summarised and then the local Bayesian estimation approach is described. Finally, the usefulness of this approach is demonstrated on a real data example. This technique successfully produces an accurate transparent model and highlights inadequacies in the data.

2. NEUROFUZZY MODEL CONSTRUCTION

Neurofuzzy construction algorithms strive to match the model's structure to the underlying function represented by the data. Due to the curse of dimensionality and the lack of expert knowledge, these algorithms have proved essential for neurofuzzy modelling. To successfully identify parsimonious model structures a simple additive decomposition of conventional models[1] is employed. A model consisting of U submodels is given by:

$$\hat{y}(\mathbf{x}, \mathbf{w}) = \sum_{u=1}^{U} \hat{y}_u(\mathbf{x}_u, \mathbf{w}_u) \qquad (2)$$

---

[1] This additive decomposition is the simplest variant on the classical neurofuzzy model, and construction algorithms have been developed which exploit other representations as described in (Bossley K.M., 1997).

where $\mathbf{x}_u \subset \mathbf{x}$ and $\mathbf{w}_u \subset \mathbf{w}$ are the submodels' inputs and weights, respectively. These neurofuzzy models remain linear with respect to their weights and hence given the model structure conventional linear optimisation techniques can be used to identify the non-linear input-output relationship.

The, well known, and fundamental principle governing data-driven system identification is the bias-variance trade-off. Under normal modelling assumptions, the expected error over all possible data sets can be expressed as (Geman S., Bienenstock E. and Doursat R., 1992):

$$\text{generalisation error} = \text{variance} + \text{bias}^2, \quad (3)$$

where the variance represents the sensitivity of the model to different data sets and the bias is the expected error between the current model and the best possible model within the given structure. The basic principle of model/system identification is to find a good balance between these sources of modelling error. It can be shown that (Ljung L., 1987; Ljung L., 1995) each parameter contributes equally to the variance, and hence by reducing the number of redundant parameters the bias and variance can be drawn towards their minimum. This is a manifestation of the *principle of parsimony*; the simplest possible model gives the best results.

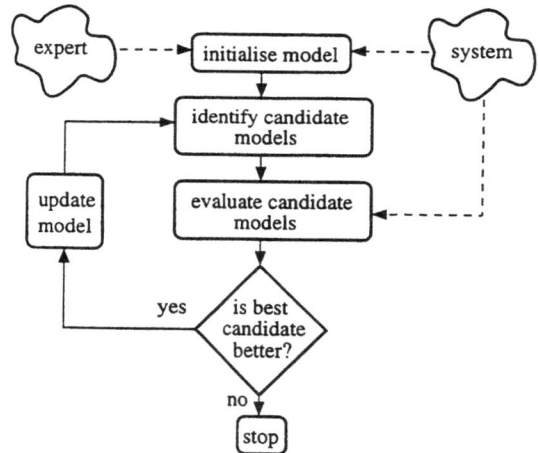

Fig. 1. The one-step ahead iterative search algorithm.

Based on the ASMOD algorithm developed by Kavli (Kavli T., 1992) an iterative one-step ahead construction algorithm has been developed, see figure (1). This search through the space of possible additive neurofuzzy models is guided by a model selection criteria which aims to balance the model's bias and variance. There are a number such criteria including Akiake's Information Criteria and the Minimum Descriptor Length (Gustafsson F. and Hjalmarsson H., 1995). A recent review given in (Gunn S.R., Brown M. and Bossley K.M., 1997) suggests that the Structural

Minimum Risk (SRM) gives superior (conservative) results. The resulting algorithm produces parsimonious additive neurofuzzy models with the following properties:

- only relevant regressors and interactions are used;
- improved generalisation abilities;
- enhanced transparency, with the identification of models which can be described by a small set of fuzzy rules, and low dimensional submodels allow visualisation of the model's components, see section (4).

This approach has proved successful, but when data is insufficiently distributed and noisy, and due to the structural symmetry enforced by the lattice structure of the submodels, models with high model variance can be produced.

Model variance can be controlled by placing prior p.d.f.s on the weights, such that those weights which are poorly identified by the data no longer have a detrimental effect on the model's generalisation error. This approach is maximum posterior (MAP) estimation, a framework to which Bayesian inferencing can be applied.

3. BAYESIAN WEIGHT IDENTIFICATION

This work extends the work of Mackay (Mackay D.J.C., 1991) to the application of Bayesian methods to additive neurofuzzy models. Given a training set, $\mathcal{D} = \{\mathbf{x}_t, y_t\}_{t=1}^{L}$, the weight vector $\mathbf{w}$ can be found by maximising the p.d.f. for the weights i.e.

$$P(\mathbf{w}|\mathcal{D}) = \frac{P(\{y_t\} \mid \{\mathbf{x}_t\}, \mathbf{w}) P(\mathbf{w})}{P(\{y_t\} \mid \{\mathbf{x}_t\})}. \quad (4)$$

This is maximum posterior (MAP) estimation. Traditionally inferences are made by maximising the *likelihood* function, $P(\{y_t\} \mid \{\mathbf{x}_t\}, \mathbf{w})$, which is maximum likelihood (ML) estimation. Studies have shown this can produce models which inadequately generalise across the input space as superfluous model parameters are insufficiently identified by the data. A solution to this high model variance is to assign a *prior* p.d.f., $P(\mathbf{w})$, on the weights and perform MAP estimation. Common priors include p.d.f.s which assign high likelihood to small weights and/or smooth model outputs, resulting in zero and second-order regularisation, respectively.

3.1 *Local Bayesian Inferencing*

For the additive model structure, equation (2), it proves advantageous to define independent *prior*

p.d.f.s on the weight vectors of the different submodels so the model variance of the submodels can be controlled independently. As overfitting is characterised by erratic model output surfaces a good model prior is the expected model's smoothness/curvature. Thus, a Gaussian *prior* is defined on each of the submodels u, given by:

$$P_u(\mathbf{w}_u) = \frac{\exp\left[-\alpha_u E_u^p(\mathbf{w}_u)\right]}{Z^p} \quad (5)$$

where Z^p is the normalisation coefficient, $E_u^p(\mathbf{w}_u)$, is the regulariser which represents the expected curvature of the model's output, and α_u controls the variance of this Gaussian. The MAP estimate of the weights becomes the weight vector that minimises the cost function, given by:

$$J(\mathbf{w}, \alpha, \beta) = \beta MSE + \sum_u^U \alpha_u E_u^p(\mathbf{w}_u) \quad (6)$$

where MSE is the conventional Mean Square Error (MSE) cost function and $\beta = 1/\sigma^2$ is the reciprocal of the variance of the noise. By penalising the MSE cost function the functions $E_u^p(\mathbf{w}_u)$ control redundant parameters. The regularisation coefficients control the effective complexity of the resulting model and hence the bias/variance trade-off.

An analytically attractive form of regulariser is one which is a quadratic function of the weights so linear optimisation techniques can be used to identify the weights. Fortunately, the expected curvature of the neurofuzzy submodels can be approximated by (Bossley K.M., 1997):

$$E_u^p(\mathbf{w}_u) = \frac{1}{2} \mathbf{w}_u^T \mathbf{K}_u \mathbf{w}_u \quad (7)$$

where the matrix $\mathbf{K}$ represents the square of the sum of the curvature, evaluated at the centres of the basis functions (the multi-dimensional fuzzy membership functions). Assuming the hyperparameters α and β are fixed, the cost function, equation (6), is a quadratic function of the weights and hence the optimal weight vector can be found by matrix inversion.

The major difficulty with regularisation is the identification of the regularisation coefficients i.e. determining the bias/variance trade-off. This balance is controlled by the *hyperparameters*, i.e. β and α. There are many different approaches to this task including cross-validation, non-linear model selection criteria, and a continuation of Bayesian estimation. The later approach known as evidence maximisation has proven very successful and is considered in this paper.

The identification of the multiple hyperparameters can be found by maximising their posterior distribution, given by:

$$P(\alpha, \beta | \mathcal{D}) = \frac{P(\{y(t)\} \mid \{\mathbf{x}(t)\}, \alpha, \beta) P(\alpha, \beta)}{P(\{y(t)\} \mid \{\mathbf{x}(t)\})}. \quad (8)$$

Assuming the prior for the hyperparameters to be uniform finding the MAP for the hyperparameters is equivalent to maximising their evidence, $P(\{y(t)\} \mid \{\mathbf{x}(t)\}, \alpha, \beta)$. This evidence is the normalising constant for the posterior for the weights (equation (4)) and results in the cost function:

$$E(\alpha, \beta) = -J(\mathbf{w}_{mp}, \alpha, \beta) - \frac{1}{2} \log \det(\mathbf{H}) +$$
$$\frac{L}{2} \log \beta + \sum_{u=1}^{U} \left[\frac{p_u}{2} \log \alpha_u + \frac{1}{2} \log \det(\mathbf{K}_u) \right] \quad (9)$$

where $\mathbf{w}_{mp}$ is the maximum posterior weight estimate, p_u is the number of weights in the uth submodel, $\mathbf{H}$ is the Hessian of the cost function, given by $\mathbf{H} = \beta \mathbf{A}^T \mathbf{A} + \mathbf{K}$, and $\mathbf{A}$ is the autocorrelation matrix. The hyperparameters are identified by maximising this non-linear cost function, which can be solved by the following simple re-estimation formulae (Bossley K.M., 1997):

$$\beta^{k+1} = \frac{N - \beta^k \, trace(\mathbf{H}^{-1} \mathbf{A}^T \mathbf{A})}{2 E_D(\mathbf{w}_{mp})}, \quad (10)$$
$$\alpha_u^{k+1} = \frac{p_u - \alpha_u^k \, trace(\mathbf{H}^{-1} \mathbf{K}'_u)}{2 E_W^u(\mathbf{w}_{mp})}.$$

where

$$\mathbf{K}'_u = \frac{\partial \mathbf{H}}{\partial \alpha_u}. \quad (11)$$

First the initial values for the hyperparameters are selected: an estimate of the reciprocal of the noise variance on the data can be used to set β, and the α_us are the confidences in the priors, typical values are 0.1 and 0.001. Then the current MAP weight estimate is calculated. Assuming the Hessian and MAP estimate are stationary with respect to the hyperparameters, new hyperparameters can be found using equations (10). This process is iterated until the hyperparameters have converged. Experience has shown the hyperparameters converge to a consistent solution.

3.2 *Error Bars*

Bayesian estimation, described above, allows the generation of error bars. These are metrics which statistically define confidences in the model's output. In regions of the input space where data is noisy and/or sparse the error bars on the output are high, and conversely in regions where the data is very rich a good approximation from the data can be inferred and hence a relatively high confidence in the output can be given (the associated error bars are low).

Error bars are defined as the variance of the distribution describing the model's output (Mackay D.J.C., 1991; Bishop C.M., 1995), which is given by Bossley K.M., (1997):

$$\sigma_y^2 = \frac{1}{\beta} + \mathbf{a}^T(\mathbf{x}) \mathbf{H}^{-1} \mathbf{a}(\mathbf{x}). \quad (12)$$

This variance is the unavoidable noise on the data plus the variance due to the uncertainty in the weights [2] which can be used to determine error bars. The resulting error bars represent directions in weight space which are poorly identified by the data, and are useful during model validation and for guiding active learning.

4. MODELLING EXAMPLE

In this section results of applying the neurofuzzy modelling approaches outlined in the previous sections, to a problem of engine torque prediction are presented. This study was carried out by Dr. M. Brown while on sabbatical at the Institute of Control Engineering, Darmstadt University of Technology, and the data was obtained from an engine, supplied by this Institute. This example serves as a particularly good benchmark for the identification algorithms as the true dynamics of the system can be approximated by a simple non-linear additive structure. Time histories and histograms of the data are shown in figure (2), and figure (3), respectively.

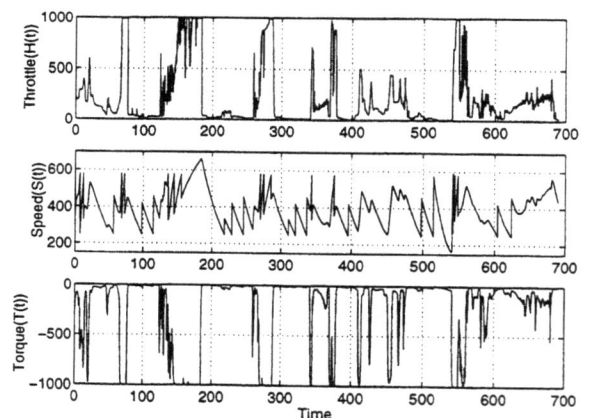

Fig. 2. The data used to identify the neurofuzzy model.

Using a regressor based on previous torques (T), engine speed (S) and throttle (H),

$$\mathbf{x}(t) = [T(t-1), \ldots, T(t-4), H(t-1), \quad (13)$$
$$\ldots, H(t-4), S(t-1), \ldots, S(t-4)],$$

---

[2] It is assumed here that the model and noise assumption can accurately represent the data, if this is not the case there will be a third component due to model mismatch.

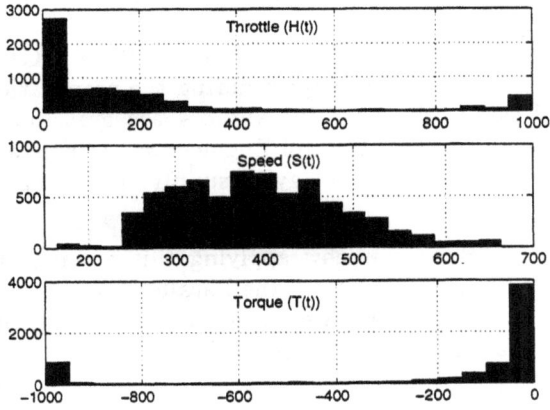

Fig. 3. Histograms representing the distribution of the data.

and no *a priori* knowledge, an additive neuro-fuzzy model estimating current torques $(T(t))$ was generated using the data-driven automatic model construction algorithms outlined in section (2). The following model structure was discovered from the data

$$\hat{y}(\mathbf{x}, \mathbf{w}) = \hat{y}_1(T(t-1)) + \qquad (14)$$
$$\hat{y}_2(H(t-1), S(t-1)),$$

which is simply the addition of a univariate and a bivariate submodel. This model structure agrees well with the physics of the process, and produces a good fit to the data. However, if the output surfaces of these models are observed the limitation of the model and maximum likelihood estimation are emphasised. Figure (4) shows these output surfaces; while $\hat{y}_1(.)$ is a linear relationship $\hat{y}_2(.)$ is non-linear and has clearly overfitted the data. This is due to both the distribution of the data across the input space of this submodel, as shown in figure (5a) and the number of degrees of freedom in the model. Note, visualisation of the output surfaces of these submodels proves a very powerful method for model validation and investigation and is made possible by the identification of parsimonious additive model structures.

As described in section (3) regularisation (Bayesian estimation) can be used to control redundant degrees of freedom. Second-order local regularisation, treated as a model post-processing step, is applied to the identified additive model. The first submodel is linear and is unaffected by regularisation but as local Bayesian estimation is employed the second submodel is regularised independently. The redundant degrees of freedom within this submodel are successfully controlled, see figure (5b). This model is a better match to the true system and possess better generalisation, in particular the improved extrapolation of the data can be observed.

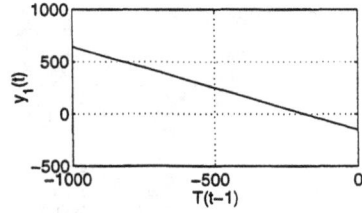

(a) submodel of $\hat{y}_1(T(t-1))$

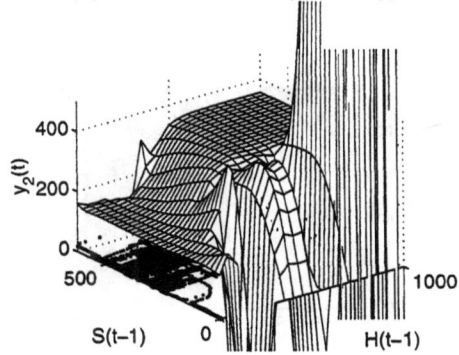

(b) submodel of $\hat{y}_2(H(t-1), S(t-1))$

Fig. 4. The output surfaces of the submodels of the neurofuzzy model identified by the construction algorithm.

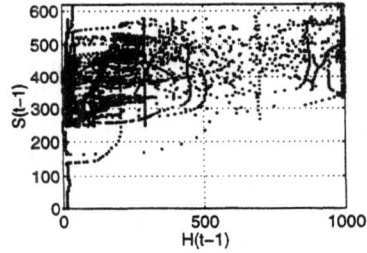

(a)

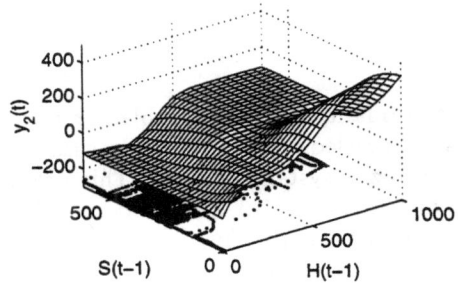

(b)

Fig. 5. (a) The distribution of the data across the input space of submodel $\hat{y}_2(H(t-1), S(t-1))$. (b) The output surface of the submodel $\hat{y}_2(.)$ after regularisation.

The model can be further validated by inspecting the variance of the predictive distribution. With the input to the first submodel $(T(t-1))$ held at -500, the variance across the submodel can be observed, see figure (6). The intensity plots of this variance, figure (a) and (b) show the

21

(a)

(b)

Fig. 6. Illustration of the variance of the predictive distribution across the input space of the second submodel (a) and (b) are the variances cut-off at 10 and 1, respectively.

variance cut-off at 10 and 1, respectively, which correspond to 1 and 0.1 percent of the range of the output. Note, the estimated variance of the noise on the data $(1/\beta)$ is 0.589. These error bars accurately match the distribution of the data and are greatest for high throttle and low torque. The true physical relationship modelled by this second submodel is a sigmodal type function, and the high variance (large error bars) in this region indicate the invalidity of the model.

5. CONCLUSION

Often the quality of the available numerical and linguistic knowledge conventionally used to identify neurofuzzy systems is poor. This problem is overcome by the use of advanced model identification algorithms presented in this paper. Parsimonious models are identified via data-driven construction algorithms which match the structure to the data and allow the application of neurofuzzy modelling to high-dimensional real world problems. Unfortunately, the inherent structure of neurofuzzy models can produce redundant degrees of freedom which are poorly identified by the data. As a solution to this problem Bayesian regularisation is applied, which smooths out any irregularities in the structure by controlling unidentified rules/weights. This is important in control and system identification scenarios where data may only be gathered around a collection of operating points.

ACKNOWLEDGEMENTS

The authors would like to thank EPSRC and GEC for their financial support during the preparation of this work. Also, the authors are grateful to the Institute of Control Engineering, Darmstadt University of Technology for supplying the engine data.

Kevin Bossley is now applying this research to condition monitoring at the Parallel Applications Centre and appreciates their support during the preparation of this paper.

6. REFERENCES

Bishop C.M. (1995). *Neural Networks for Pattern Recognition*. Clarendon Press. Oxford.

Bossley K.M. (1997). Neurofuzzy Modelling Approaches in System Identification. PhD thesis. Department of Electronics and Computer Science, University of Southampton.

Brown M. and Harris C.J. (1994). *Neurofuzzy Adaptive Modelling and Control*. Prentice Hall. Hemel Hempstead.

Geman S., Bienenstock E. and Doursat R. (1992). Neural networks and the bias/variance dilemma. *Neural Computation* **4**, 1–58.

Gunn S.R., Brown M. and Bossley K.M. (1997). Network performance assessment for neurofuzzy data modelling. Second International Symposium on Intelligent Data Analysis Birkbeck College, University of London, 4th-6th August.

Gustafsson F. and Hjalmarsson H. (1995). Twenty-one ML estimators for model selection. *Automatica* **10**(31), 1377–1392.

Hunt J.K., Haas, R. and Brown M. (1995). On the functional equivalence of fuzzy inference systems spline-based networks. *Int. J. Neural Systems* **6**(2), 171–184.

Kavli T. (1992). Learning Principles in Dynamic Control. PhD thesis. University of Oslo. Norway.

Ljung L. (1987). *System Identification: theory for the user*. Prentice Hall. Englewood Cliffs, New Jersey 07632.

Ljung L. (1995). System Identification. Technical Report LiTH-ISY-R-1763. Dept. Electrical Engineering, Linköping Univeristy.

Mackay D.J.C. (1991). Bayesian Methods for Adaptive Models. PhD thesis. California Institute of Technology, Pasadena, California.

Wu Z.Q. and Harris C.J. (1996). Adaptive neurofuzzy Kalman filter. In: *FUZZ-IEEE '96 - Proceedings of the fifth IEEE International Conference on Fuzzy Systems*. Vol. 2. New Orlean,USA. pp. 1344–1350.

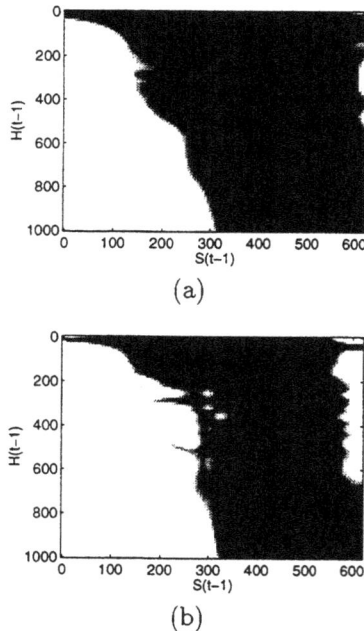

A MULTIPLE MODEL EXTENSION TO THE BOOTSTRAP FILTER FOR MANOEUVRING TARGETS

Shaun McGinnity and George Irwin

Advanced Control Engineering Research Group,
Dept. of Electrical and Electronic Engineering,
The Queen's University of Belfast,
Belfast, UK
BT9 5AH
s.mcginnity@qub.ac.uk, g.irwin@qub.ac.uk

Abstract: The bootstrap filter is an algorithm for implementing recursive Bayesian estimation in which the state probability density functions are approximated and evolved by updating and predicting a set of samples. An extension to this algorithm for application to the multiple model target tracking problem is presented as an alternative to the Interacting Multiple Model(IMM) algorithm. The Markovian transition probability matrix is used to approximate the model branching of the optimal solution to this problem. Simulation results from a standard manoeuvring target application suggest improved performance during non-manoeuvre periods and comparable performance during manoeuvres. Although these results are for a linear, Gaussian system the proposed method is more generally applicable. *Copyright © 1998 IFAC*

Keywords: manoeuvring target, estimation algorithms

1. INTRODUCTION

The Kalman filter is a solution to the general problem of optimal Bayesian stochastic state space estimation for the case of linear systems with additive Gaussian noises of known statistics. The solution is possible since closed form expressions are available for the Gaussian densities in the system. Further, the form of a Gaussian density is unchanged after the application of linear operators.

Recent research has been applied to the solution of Bayesian estimation for the more general case of nonlinear and non-Gaussian systems. Here there are usually no analytical solutions available, therefore numerical approximations to the system densities are required. The methods suggested have been classified as either global or local(Sorensen, 1974). The density approximations employed by local approaches, of which the most commonly used method is the extended Kalman filter, are accurate at only a single point in the state space. Global approaches, on the other hand, attempt to approximate the density functions over their complete region of significance and are therefore more complex. With increasing computing power global methods are receiving more attention.

This paper investigates the optimal solution of the multiple model approach to manoeuvring target tracking which constitutes a non-Gaussian estimation problem. The motion of the manoeuvring target can be considered as mainly constant velocity interspersed with accelerations, whose form, magnitude and onset time are usually unknown. When tracking with a single model the accuracy of the filter is good during the periods of motion matched to that assumed by the model, but deteriorates when the motion changes. Adaptive approaches, based on detecting changes from the quiescent straight line, constant velocity motion are common, but the delay inherent in identifying the manoeuvre can cause filter divergence. Also, once a manoeuvre is detected, it is difficult to determine what form of model the filter should switch to.

Multiple model approaches to manoeuvring target tracking operate a bank of filters in parallel, with each filter matched to a different form of target motion. The final state estimate is a weighted sum of the estimates from each filter. The approach is therefore "decision free" since the weightings are generally determined as the probability, based on the measurement, that each model is correct. Unfortunately, for linear, Gaussian systems the optimal solution to the multiple model approach requires an exponentially increasing number of Kalman filters and is therefore unpractical.

Several suboptimal algorithms have been devised which use a constant number of Kalman filters by merging the branched densities to a single Gaussian one with matched first and second moments. The most commonly used is the Interacting Multiple Model(IMM) algorithm(Bar-Shalom et al., 1989), (Munir and Atherton, 1995), (McGinnity and Irwin, 1997). Adopting a Bayesian approach, however, allows the number of filters to be kept constant whilst still maintaining the true system densities, although with increasing complexity. The Bayesian approach is also applicable to the nonlinear/non-Gaussian multiple model problem.

This paper begins by outlining the optimal solution to the linear, Gaussian multiple model problem with the sub-optimal IMM approach then illustrated. The bootstrap filter, a recently proposed global approach to Bayesian estimation is outlined and the extension of this to multiple models is then presented. Target tracking simulation results, comparing the Bayesian and suboptimal approaches, are given to illustrate the performance of the new Bayesian approach.

2. OPTIMAL MULTIPLE MODEL SOLUTION

The parameterised linear, Gaussian state space system is given by:

$$
\begin{aligned}
\underline{x}(k+1) &= A_\alpha \underline{x}(k) + B_\alpha \underline{u}_\alpha(k) + \underline{w}_\alpha(k) \\
\underline{y}(k+1) &= C_\alpha \underline{x}(k+1) + \underline{v}_\alpha(k+1)
\end{aligned}
\tag{1}
$$

where $\{\alpha : \alpha_i, i = 1, ..., N\}$ denotes the conditioning parameter and A_α, B_α and C_α are parameterised state transition, input and measurement matrices. The optimal, Bayesian, minimum variance estimate is given by the expected value of the posterior density, $p(\underline{x}(k)|Y^k)$:

$$
\hat{\underline{x}}(k|k) = E\langle \underline{x}(k)|Y^k\rangle = \int_{\underline{x}(k)} \underline{x}(k)p(\underline{x}(k)|Y^k)
\tag{2}
$$

where the set of measurements Y^k is defined as $\{\underline{y}(0), ..., \underline{y}(k)\}$. In the multiple model case this expands to the sum:

$$
\hat{\underline{x}}(k|k) = \sum_{j=1}^{N} \left[\int_{\underline{x}(k)} \underline{x}(k)p(\underline{x}(k)|\alpha(k) = \alpha_j, Y^k) \right]
\tag{3}
$$

where $p(\underline{x}(k)|\alpha(k) = \alpha_j, Y^k)$ are the model conditioned posterior densities. The multiple model branching will cause N^2 model conditioned prior densities to be formed at the next sample time:

$$
p(\underline{x}(k)|\alpha(k) = \alpha_j, \alpha(k+1) = \alpha_i, Y^k),
\tag{4}
$$
$$
i = 1, ..., N, j = 1, ..., N
$$

where the branching is determined by a Markovian probability transition matrix, H:

$$
h_{ij} = p(\alpha(k+1) = \alpha_i|\alpha(k) = \alpha_j)
\tag{5}
$$

In the linear, Gaussian case, the N^2 posterior densities at $k+1$:

$$
p(\underline{x}(k+1)|\alpha(k) = \alpha_j, \alpha(k+1) = \alpha_i, Y^{k+1}),
\tag{6}
$$
$$
i = 1, ..., N, j = 1, ..., N
$$

are found using N^2 Kalman filters matched to parameter, α_i. At the next sample time these posterior densities will again branch N ways, producing N^3 prior densities and requiring N^3 Kalman filters. Thus, in general, the optimal solution causes an exponential growth in the number of Kalman filters from one sample time to the next.

Several sub-optimal algorithms have been proposed which use a constant number of Kalman filters, the most common of which is the IMM algorithm(Bar-Shalom et al., 1989). This operates by approximating the merged model conditioned prior densities:

$$
p(\underline{x}(k)|\alpha(k+1) = \alpha_i, Y^k) =
\tag{7}
$$
$$
\sum_{j=1}^{N} p(\underline{x}(k)|\alpha(k) = \alpha_j, \alpha(k+1) = \alpha_i, Y^k)
$$
$$
p(\alpha(k) = \alpha_j|\alpha(k+1) = \alpha_i, Y^k)
$$

by a single Gaussian density with matched first and second moments. The number of Kalman filters is therefore kept equal to the number of models, N. By maintaining a Gaussian approximation, the IMM approach can be considered to be a local approximation to the true merged density.

The approach taken in this paper also forms the merged model conditioned prior density in (7) but propagates the true density using Bayesian estimation techniques. The Bayesian estimation algorithm used is the bootstrap filter.

3. BOOTSTRAP FILTER

The general nonlinear state space system is given by:

$$\underline{x}(k+1) = f(\underline{x}(k), \underline{w}(k))$$
$$\underline{y}(k+1) = h(\underline{x}(k+1), \underline{v}(k+1)) \qquad (8)$$

where $\underline{x}(k)$ and $\underline{y}(k)$ are state and measurement vectors, $\underline{w}(k)$ and $\underline{v}(k)$ are process and measurement noise sequences of known statistics and f and h are general nonlinear state transition and measurement functions respectively. Applying Bayes' rule to the posterior density illustrates the recursive evolution of the state vector densities from k to $k+1$:

$$p(\underline{x}(k+1)|Y^{k+1}) = \qquad (9)$$

$$\frac{p(\underline{y}(k+1)|\underline{x}(k+1), Y^k)p(\underline{x}(k+1)|Y^k)}{p(\underline{y}(k+1)|Y^k)}$$

where the prediction is given by the integration:

$$p(\underline{x}(k+1)|Y^k) = \qquad (10)$$

$$\int_{\underline{x}(k)} p(\underline{x}(k+1)|\underline{x}(k))p(\underline{x}(k)|Y^k)$$

and $p(\underline{y}(k+1)|Y^k)$ is a normalising constant. Thus, in order to form the posterior density at $k+1$, expressions are required for the prior and prediction densities of the state at k. The linear, Gaussian case is one instance where an analytical solution to these equations exists, leading to the Kalman filter. For nonlinear/non-Gaussian problems, however, there are usually no closed form solutions and therefore numerical approximations must be applied. Generally these approaches attempt to approximate the shape of the probability density functions.

The bootstrap filter(Gordon et al., 1993) is a recently proposed approach which exploits the duality between a density, $p(\underline{x}(k)|Y^k)$, and samples generated from that density, $\{\underline{X}_l(k|k) : l = 1, ..., N_s\}$. Rather than attempting to approximate the shape of the state prior and posterior density functions, samples from the prior density are propagated and updated to represent the posterior density as the set of samples, $\{\underline{X}_l(k+1|k+1) : l = 1, ..., N_s\}$.

Samples from the state prediction density are formed by passing each prior sample through the state transition function:

$$\underline{X}_l(k+1|k) = f(\underline{X}_l(k|k), \underline{W}_l(k)) \qquad (11)$$

where $\underline{W}_l(k)$ is a sample from the process noise

density. The posterior density samples are then chosen using the weighted bootstrap resampling technique(Smith and Gelfand, 1992) by resampling N_s times from the prediction set such that $\underline{X}_l(k+1|k+1) = \underline{X}_m(k+1|k)$ if

$$\sum_{i=1}^{m-1} \bar{q}_i < s_l \le \sum_{i=1}^{m} \bar{q}_i \qquad (12)$$

where s_l is a uniformly distributed random number in the range $(0, 1]$ and $\bar{q}_i$ is the normalised likelihood of $\underline{X}_i(k+1|k)$:

$$\bar{q}_i = \frac{p(\underline{y}(k+1)|\underline{X}_i(k+1|k))}{\sum_{l=1}^{N_s} p(\underline{y}(k+1)|\underline{X}_l(k+1|k))} \qquad (13)$$

The justification for the bootstrap filter is given in (Gordon et al., 1993). As shown in §2 the optimal solution to the multiple model problem requires either the formation of non-Gaussian densities or an exponentially growing number of Kalman filters. The next section describes the extension of the bootstrap filter to this problem.

4. EXTENSION TO MULTIPLE MODELS

The multiple model bootstrap filter augments the posterior density samples by an index vector, $A(k|k)$, representing the model parameter, $\alpha(k)$. The model conditioned state posterior densities, $p(\underline{x}(k)|\alpha(k) = \alpha_j, Y^k)$, are then approximated by the N subsets of these samples:

$$\{\underline{X}(k|k), \underline{A}(k|k) = j\}, j = 1, ..., N \qquad (14)$$

The posterior model probabilities, $p(\alpha(k) = \alpha_j|Y^k)$, are approximately equal to the proportion of samples in the index vector from each model:

$$p(\alpha(k) = \alpha_j|Y^k) \approx \frac{n(A(k|k) = j)}{N_s} \qquad (15)$$

where $n(A(k|k) = j)$ denotes the number of samples in $A(k|k)$ indexing model j.

The model branching is approximated by considering that, if $\alpha(k) = \alpha_j$, the probability of $\alpha(k+1) = \alpha_i$ is h_{ij}. Thus a further vector, $A(k+1|k)$, is added to the sample set which indexes the model at $k+1$ and, if $A_l(k|k) = j$, $A_l(k+1|k)$ is set to i with probability $\bar{h}_{ij}$. The actual values of $A(k+1|k)$ are selected in a similar manner to the likelihood sampling step of the bootstrap filter. Thus for $\underline{A}_l(k|k) = j$, $\underline{A}_l(k+1|k)$ is set to m where:

$$\sum_{i=1}^{m-1} h_{ij} < s_l \le \sum_{i=1}^{m} h_{ij} \qquad (16)$$

and s_l is a uniformly distributed random number in the range $(0, 1]$. The branched state densities (6) are then approximated as the subset:

$$\{\underline{X}(k|k), A(k|k) = j, \underline{A}(k+1|k) = i\}, \qquad (17)$$
$$i, j = 1, ..., N$$

and the merged model conditioned prior state densities (7) are approximated by:

$$\{\underline{X}(k|k), \underline{A}(k+1|k) = i\}, i = 1, ..., N \qquad (18)$$

For illustration the merging approaches of the IMM and bootstrap methods are compared for a scalar, two model case. The state prior densities are Gaussian:

$$p(x(k)|\alpha(k) = \alpha_1, Y^k) = N(-2, 1)$$
$$p(x(k)|\alpha(k) = \alpha_2, Y^k) = N(2, 2) \qquad (19)$$

The prior model probabilities are given as $p(\alpha(k) = \alpha_1) = 0.3$ and $p(\alpha(k) = \alpha_2) = 0.7$ and the Markovian probability transition matrix is:

$$H = \begin{bmatrix} 0.8 & 0.2 \\ 0.2 & 0.8 \end{bmatrix} \qquad (20)$$

The IMM approximation to the merged densities are determined as Gaussians with the same mean and variance as the true density:

$$p(x(k)|\alpha(k+1) = \alpha_1, Y^k) \approx N(-0.53, 5.09)$$
$$p(x(k)|\alpha(k+1) = \alpha_2, Y^k) \approx N(1.61, 3.30) \qquad (21)$$

Using the multiple model bootstrap approach with 5000 samples, $0.3 \times 5000 = 1500$ and $0.7 \times 5000 = 3500$ samples were used to represent the two posterior state densities. Upon merging, as described above, the means were determined as -0.52, 1.59 and the variances as 5.33, 3.42 respectively. Clearly the bootstrap approximation captures the moments of the merged density fairly accurately. Figure 1 plots the bootstrap and IMM approximations to the true merged densities. The bootstrap density is formed from the samples using Kernel Density Estimation techniques, (Silverman, 1986). In this case the bootstrap method also captures the general shape of the densities whereas the IMM approach does not.

For the multiple model, linear, Gaussian system (1) the multiple model bootstrap filter proceeds as in the single model case, but with the index vector, $\underline{A}(k+1|k)$, determining which model is applied to

produce the set of state prediction samples, $\{\underline{X}(k+1|k), \underline{A}(k+1|k)\}$.

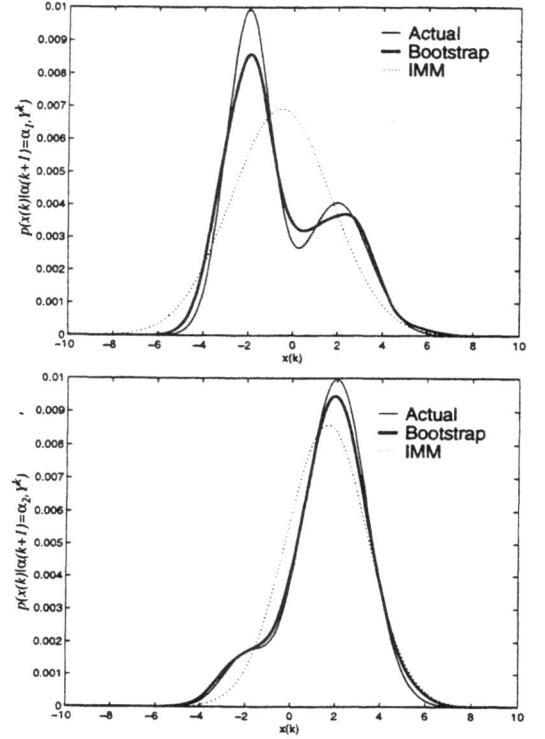

Fig. 1. Bootstrap and IMM approximations to merged densities, $p(x(k)|\alpha(k+1) = \alpha_i, Y^k)$

Clearly, just as the single model bootstrap filter is applicable to the general nonlinear system (8), so the multiple model extension can also be applied to the general nonlinear multiple model case:

$$\underline{x}(k+1) = f_\alpha(\underline{x}(k), \underline{u}_\alpha(k), \underline{w}_\alpha(k)) \qquad (22)$$
$$\underline{y}(k+1) = h_\alpha(\underline{x}(k+1), \underline{u}_\alpha(k+1), \underline{v}_\alpha(k+1))$$

The posterior density set is formed by choosing:

$$\{\underline{X}_l(k+1|k+1), A_l(k+1|k+1)\} = \qquad (23)$$
$$\{\underline{X}_m(k+1|k), A_m(k+1|k)\}$$

using the resampling procedure given in (12) and (13). In this way the posterior samples are weighted not just towards samples of high likelihood but also towards models with a high proportion of samples in the prediction density.

The computational load of the bootstrap filter is a function only of the number of samples, N_s, therefore it is possible to add more models to the multiple model bootstrap filter without incurring an increase in computations, although the accuracy of the sampled approximations will be affected as less samples per model are available.

5. APPLICATION

The IMM and bootstrap algorithms were compared using a target manoeuvre modelled as a step change in acceleration along the Cartesian x and y axes(Bar-Shalom and Birmiwal, 1982). The target begins at an initial position of $[2000, 10000]^T$ m in the x-y plane. It moves at a constant velocity of $[0.0, -15.0]^T$ ms^{-1} for 400 seconds when it begins a slow $90°$ turn of magnitude $U_x = U_y = 0.075 \ ms^{-2}$. The turn lasts 200 seconds after which the target reverts to moving at a constant velocity. After 10 seconds of this motion, a fast $90°$ turn is performed of magnitude $U_x = U_y = -0.3 \ ms^{-2}$ lasting 50 seconds, after which the target again moves with constant velocity for the remainder of its motion.

Throughout the motion the target has an additional zero-mean, Gaussian random acceleration noise on both axes of variance $10^{-6} \ (ms^{-2})^2$. The noiseless target trajectory is shown in figure 2. For simulations, a sample time of 10s was used, with measurements taken of position corrupted by zero-mean, Gaussian white noise of variance $10^4 \ m^2$. Measurements, $\{z^x(k), z^y(k)\}$ were assumed uncorrelated on each axis allowing separate multiple model filters to be used.

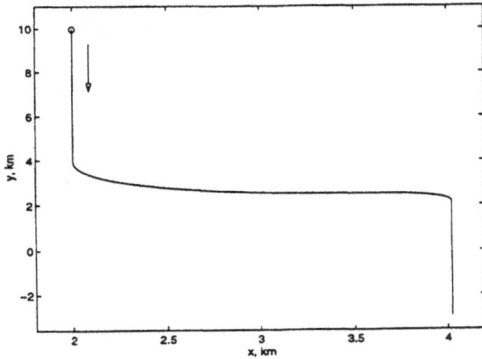

Fig. 2. Target Trajectory

Both algorithms used the same multiple model configurations. Each model assumed constant velocity motion, which for one axis is:

$$\ddot{x}(t) = u(t) + w(t) \qquad (24)$$

where $\ddot{x}(t)$ is the target acceleration, $u(t)$ is the deterministic target manoeuvre input and $w(t)$ is zero-mean, Gaussian white noise of variance q. This gives rise to the discrete time model:

$$x(k) = A\underline{x}(k-1) + Bu(k) + \underline{w}(k) \qquad (25)$$

where, for one axis, $A = \begin{bmatrix} 1 & T \\ 0 & 1 \end{bmatrix}$, $B = \begin{bmatrix} T^2/2 \\ T \end{bmatrix}$ and

$\underline{w}(k)$ is a zero-mean, Gaussian white noise sequence of covariance:

$$Q = q \begin{bmatrix} T^3/3 & T^2/2 \\ T^2/2 & T \end{bmatrix} \qquad (26)$$

Five models were used, each assigned to a different input $u(k)$ chosen from the set:

$$u(k) = [-0.3, -0.15, 0.0, 0.15, 0.3] \ ms^{-2} \qquad (27)$$

and each assumed $q = 10^{-6}(ms^{-2})^2$.

The model transition probability matrices were assigned as:

$$h_{ij} = \begin{array}{l} 0.95, i = j \\ 0.0125, i \neq j \end{array} \qquad (28)$$

The non-manoeuvring model, i.e. $u(k) = 0.0$, was initialised with a probability of $p(0) = 0.6$. The other models were initialised with equal probability: $p(0) = 0.1$.

The bootstrap filter was implemented using a sample size of 5000, giving a mean execution time on a Pentium 90 PC running MATLAB v4.2c of 1.1s for each iteration.

The Normalised Position Error(NPE) is defined as the ratio of the mean-square position estimation error to the mean-square measurement error over N simulations:

$$NPE = \sqrt{\frac{\sum_{i=1}^{N} [(x_i(k) - \hat{x}_i(k))^2 + (y_i(k) - \hat{y}_i(k))^2]}{\sum_{i=1}^{N} [(x_i(k) - z_i^x(k))^2 + (y_i(k) - z_i^y(k))^2]}}$$

$$(29)$$

Figure 3 compares the time variation in NPE over 50 simulations using the IMM algorithm and the multiple model bootstrap filter. Figure 4 illustrates the variation in the mean posterior model probability for the non-manoeuvring model on the x-axis. During the non-manoeuvring period of motion the model weightings of both algorithms are almost identical, but, observing figure 3, during these periods the multiple model bootstrap filter reaches a lower level of NPE than the IMM algorithm, suggesting that the bootstrap approximation to the posterior state density is more accurate than the IMM moment matched density. When the first manoeuvre, M_1, begins, the bootstrap filter exhibits a slightly higher peak error than the IMM algorithm even though the posterior model probability of the non-manoeuvring model falls more rapidly, indicating faster adaption. In this case the samples are switching from the non-manoeuvring to a manoeuvring model. The peak illustrates the delay in the number of distinct samples in the bootstrap approximation becoming sufficient to accurately represent the manoeuvring model density. This effect also occurs in the single model bootstrap filter when the likelihood of the samples is low.

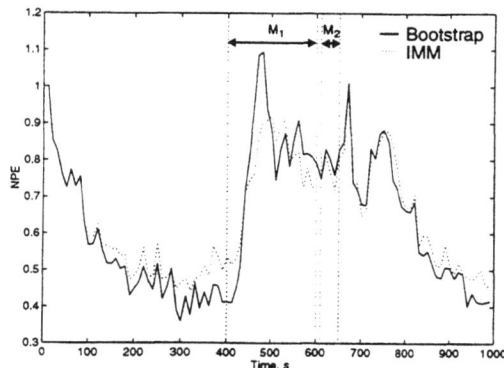

Fig. 3. Variation in NPE with time

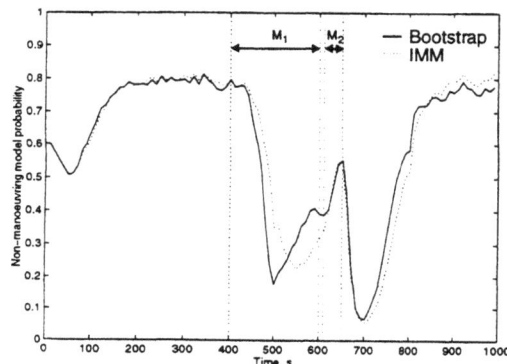

Fig. 4. Variation in posterior probability of non-ma-noeuvring model with time

During the first manoeuvre, M_1, there is no model matching the actual target acceleration. In this case, after the bootstrap filter has settled, both algorithms give similar errors. The same is true during the second manoeuvre, M_2, although the bootstrap filter appears to react more quickly to the end of the manoeuvre.

6. CONCLUSIONS

The bootstrap filter is an approximation to the recursive Bayesian estimator using samples to represent the required posterior densities. The multiple model bootstrap filter has been presented as an extension of the bootstrap filter to the solution of the optimal multiple model problem. This uses the Markovian transition probability matrix to approximate the model branching of the optimal solution and therefore keep the number of estimators

constant.

Simulations have shown very promising performance of the filter when compared to the IMM algorithm on a standard manoeuvring target scenario. Further work is necessary to determine approaches which avoid the fall in the number of distinct samples when a manoeuvre occurs. The bootstrap filter, and also the new multiple model extension, are not restricted to linear, Gaussian systems and future work will investigate the performance of the filter on nonlinear and non-Gaussian multiple model problems.

7. REFERENCES

Sorensen, H.W., (1974). On The Development Of Practical Nonlinear Filters, *Information Sciences*, **Vol. 7**, 253-270

Bar-Shalom, Y., Chang, K.C. and Blom, H.A.P., (1989). Tracking A Maneuvering Target Using Input Estimation Versus The Interacting Multiple Model Algorithm, *IEEE Transactions on Aerospace and Electronic Systems*, **Vol. AES-25, No. 2**, 296-300

Munir, A. and Atherton, D.P., (1995). Adaptive Interacting Multiple Model Algorithm for Tracking a Manoeuvring Target, *IEE Proceedings - F*, **Vol. 142, No. 1**, 11-17

McGinnity, S. and Irwin, G., (1997). A Fuzzy Multiple Model Approach to Manoeuvring Target Tracking, *Preprints 4th IFAC workshop on Algorithms and Architectures in Real Time Control*, Portugal, 264-269

Gordon, N., Salmond, D.J. and Smith, A.F.M., (1993). Novel Approach to Nonlinear/Non-Gaussian Bayesian State Estimation, *IEE Proceedings - F*, **Vol. 140, No. 2**, 107-113

Smith, A.F.M. and Gelfand, A.E., (1992). Bayesian Statistics Without Tears - A Sampling-resampling Perspective, *The American Statistician*, **Vol. 46, No. 2**, 84-88

Silverman, B.W., (1986). Density estimation for statistics and data analysis, Chapman and Hall

Bar-Shalom, Y. and Birmiwal, K. (1982). Variable Dimension Filter for Maneuvering Target Tracking, *IEEE Transactions on Aerospace and Electronic Systems*, **Vol. AES-18, No. 5**, 621-629

8. ACKNOWLEDGEMENT

Shaun McGinnity would like to acknowledge the financial support of the Department of Education for Northern Ireland and SMS Ltd. in the form of a CAST studentship.

RESPONSE TIME DRIVEN SCHEDULING FOR REAL-TIME PROGRAMMABLE LOGIC CONTROLLERS WITH NETWORK-BASED I/O SYSTEMS

Seungkweon Jeong* Naehyuck Chang** Wook Hyun Kwon*

**School of Electrical Engr., **Dept. of Computer Engr.
Seoul National University, Seoul, 151-742, Korea
*{jsk, whkwon}@cisl.snu.ac.kr, **naehyuck@comp.snu.ac.kr*

Abstract: To reduce the response time determinately, this paper proposes a scheduling method for a programmable logic controller(PLC) with network-based I/O systems, assuming the multi-tasking facilities. A generalized architectural and behavioral model is developed to yield precise timing specifications, by which the proposed method schedules program execution and data transmission. A synchronization scheme, which enables the generated schedule to be realized on different processing elements with minimal hardware resources of PLC, is also suggested. *Copyright © 1998 IFAC*

Keywords: programmable logic controller, network-based I/O, response time, scheduling algorithm

1. INTRODUCTION

PLCs were developed as sequential control devices to replace electro-mechanical relay-based hard-wired controllers. Now, it is widely used for automation of FMSs, chemical processes, transportation systems, etc(Simpson, 1994; Warnock, 1988). As factory automation advances to higher levels, the tasks of a PLC are getting larger and more complex. Generally, many small scale PLCs are used and each is assigned to the small portions of the whole control task. In this situation, the overhead for individual programming and monitoring is heavy. Thus, large scale PLCs are preferred nowadays. The large scale PLC controls a vast plant, where sensors and actuators are distributed. Since it costs much to wire all sensors and actuators, sensors and actuators are grouped and wired to I/O subsystems according to the location and I/O subsystems are interconnected over a network(Electric, 1994; Automation, 1992; Systems, 1994). Until now, most PLCs executes a whole sequence control program repeatedly at one rate. These PLCs are inefficient since some part

of the sequence control program can be executed with a lower rate. Thus, the multi-tasking facilities, which enable different parts to be executed at different rates, are currently adopted in the large scale PLC.

The important criterion of the PLC performance is the computational speed. It is commonly represented by the response time, that is the latency from the variation of states in the controlled system to the consequent reaction of the sequence control program(Bonfatti *et al.*, 1997). In the large scale PLC, the PLC job is divided into a series of small jobs. It is executed by several processing elements to increase the computational speed. In such a multi-processing architecture, the divided jobs are cascaded and the common resources such as common memory, system bus, network line are contended. The response time may be lengthened in the large scale PLC due to discontinuous execution of the cascaded jobs and contention of common resources(Jeong *et al.*, 1997). To reduce the response time when the common resources are contended in multi-processor architecture, some

Table 1. Nomenclature

ST	scan time
SD	duration of state in controlled system
$WCRT$	the worst case response time
$WCCD$	the worst case controller delay
PD	plant delay
CI	control interval
τ_n	nth task
ST_n	scan time for τ_n
$EX_n(k)$	program execution for τ_n in kth scan
$IN_n(k)$	input data transmission for τ_n in kth scan
$OUT_n(k)$	output data transmission for τ_n in kth scan
$WCRT_n$	the worst case response time for τ_n
$WCCD_n$	the worst case controller delay for τ_n
$WCPD_n$	the worst case plant delay τ_n
MCI_n	the minimum control interval for τ_n
e_n	the worst case execution time of τ_n
i_n	input data transmission time for τ_n
o_n	output data transmission time for τ_n
$s_n^e(k)$	initiation time of $EX_n(k)$
$s_n^i(k)$	initiation time of $IN_n(k)$
$s_n^o(k)$	initiation time of $OUT_n(k)$
$c_n^e(k)$	completion time of $EX_n(k)$
$c_n^i(k)$	completion time of $IN_n(k)$
$c_n^o(k)$	completion time of $OUT_n(k)$
I_n^e	time phase of EX_n
$I_{n,s}^i$	time phase of IN_n from sth I/O subsystem
$I_{n,s}^o$	time phase of OUT_n to sth I/O subsystem
M_T	number of task
M_S	number of I/O subsystem
$a \prec b$	b refers the results of a

methods were proposed about data transfer on system bus(Park *et al.*, 1993; Kwon *et al.*, 1994) and data communication over a network(Park *et al.*, 1997). However, these methods improve the performance partially without considering the end-to-end response time. Other method to reduce the response time could be scheduling with a synchronization scheme.

Lots of scheduling algorithms were proposed for general application. The objective of general scheduling is to find out the initiation time of the task with given deadline and period. However, the objective of PLC scheduling is to find out both the initiation time and the period. The response time is the sum of the period and the deadline, since it takes one period for PLC to sense the variation of controlled system and the processing time is bounded by the deadline. Thus, the period and the deadline should be determined firstly and then the initiation time should be found out to reduce the response time determinately in the PLC. There are trade-off between schedulability and efficiency depending on how to determine the period and the deadline. A schedule with long period and short deadline is efficient in the sense that the scheduled tasks occupy a small portion of resources. However, a feasible schedule may not exist for the tight deadline. Thus, a rule for determination of period and deadline is required. General scheduling algorithm may yield schedule jitter. Since jitter lengthens the response time in PLC, jitter should not be allowed or should be considered in determining the period and the deadline.

In order that tasks are executed at the scheduled time on different processing elements, each pro-

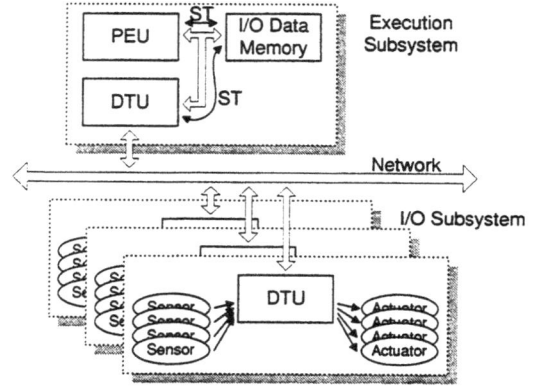

Fig. 1. Architecture of a PLC with remote I/Os

cessing element should be synchronized. Since the PLC is a dedicated system, it has minimal hardware resources. It is difficult to implement general synchronization schemes without additional hardware support. It is observed that the synchronization is required mainly between program execution and data transmission in PLC. Thus, a specific synchronization scheme which can be implemented with minimal hardware resources is required.

This paper suggests a general architectural model of a large scale PLC, and its scheduling method that reduces the response time. The generated schedule includes the period and the initiation time of program execution and data transmission for each task. The schedule is achieved by task-specific timing parameters derived from the end-to-end constraints and task precedence relations. The suggested method searches a schedule feasible with the largest period, thus, generates the most efficient solution. A synchronization scheme between program execution and data transmission is also proposed. It enables the generated schedule to be implemented easily on the PLC.

This paper consists of six sections. Section 2 analyzes characteristics of a large scale PLC and states the problems. Section 3 describes a scheduling method to reduce the response time determinately. Section 4 evaluates the performance of the proposed scheduling method. Section 5 suggests a synchronization scheme and demonstrates an implemented example. This paper concludes in Section 6.

2. PROBLEM STATEMENTS

2.1 *PLC architecture*

The primary functions of PLCs are reading sensors, executing sequence control programs and updating actuators. PLCs are generally composed of ES(execution subsystem) and IOS(I/O subsystem)s. While IOSs are interconnected over

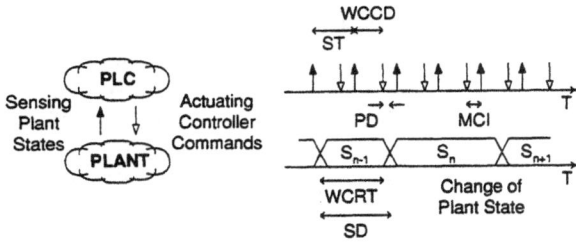

Fig. 2. Definition of parameters in a PLC

a local system bus in small scale PLCs, they are interconnected over a network in large scale PLCs. Fig. 1 shows a typical architecture of PLCs with network-based IOSs. The ES consists of PEU(program execution unit) and DTU(data transmission unit), which have separate processors. The PEU in the ES takes data of the IODM(I/O data memory) for the actual value of the sensors and the actuators, and executes sequence control programs. The DTU in the ES keeps the data of IODM coherent with the values of sensors and the actuators in IOSs. Both of them have the same processing period, which is called scan time in PLC. Because the DTU usually updates data by burst transfer, the order of the I/O instructions of the sequence control programs is not kept in I/O data transmission. Data transfer between DTU of ES and DTU of IOS should have bounded delay. Recently, field bus networks are suitable candidates for this purpose.

2.2 Performance indices

Since PLCs are synchronous machines that should handle asynchronous events from plants, the PLC with the higher sampling frequency(the shorter ST) affords to control the higher-speed plants, thus, the condition of $ST < SD$ is required(the symbols are defined in Table 1). Although the allowable minimum scan time has been accounted for the measure of the performance of PLCs, the response time indicates the performance more determinately. In terms of a plant, the response time is the delay between the change of its state and the change of corresponding controller action. The pessimistic value of the response time, $WCRT$, should be taken into account in view of hard real-time systems.

The state change of the plant is caused by two factors: a control update or an internal event. When the state of the plant changes by the control update, it goes without saying that $WCRT \leq SD$. $WCRT \leq SD$ is required as well, in order that the plant gets back the proper control update after change of its state, before leaving the current state by other internal events. Because the control update is generated after sensing and computation, i.e. $WCRT = ST + WCCD$, the

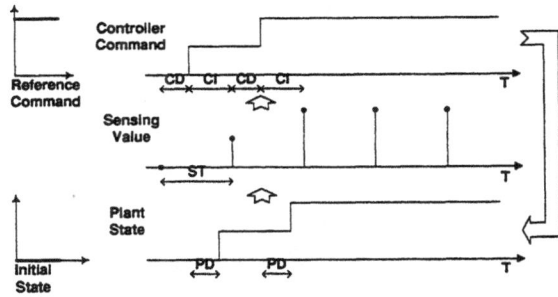

Fig. 3. Effect of CI(CI > PD)

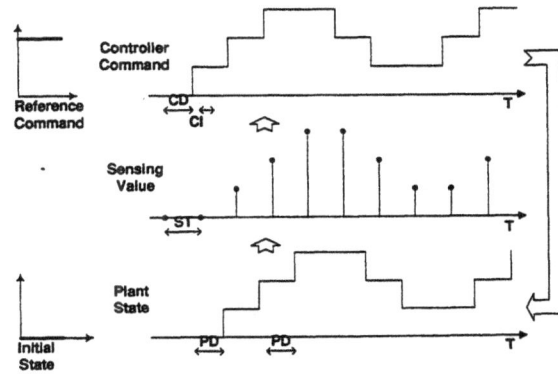

Fig. 4. Effect of CI(CI < PD)

requirement of $WCRT \leq SD$ includes the requirement of $ST < SD$. Long ST is desirable to maintain low occupation of resources as long as the $WCRT$ is bounded by a required specification. $WCCD$ depends on the hardware performance such as processor power, network bandwidth, delay of I/O devices. $WCCD$ additionally depends on task scheduling and synchronization between PEU and DTU for PLCs having network-based IOS and multi-tasking facilities.

PD is defined as the delay between control update and state change. CI is defined as the delay between the control update and the next input sensing(Fig. 2). To accomplish the proper feedback control, it is required that $CI > PD$. The significance of CI is demonstrated in Fig. 3 and 4, where the control rule is 1) increase the output if input is less than the reference, and 2) decrease the output if input is greater than the reference. Proper feedback control is showed in Fig. 3. In Fig. 4, ST is reduced in order to achieve fast control regardless of the condition $CI > PD$. Consequently, even though ST is shorter in Fig. 4, control becomes unstable since control update is computed with the old values of the plant state.

3. TASK SCHEDULING

Let us assume the following conditions. First, there is no sporadic task. Second, e_n, i_n and o_n are bounded and known for all τ_n. The e_n is able to be estimated as a rough bound(Koo and Kwon,

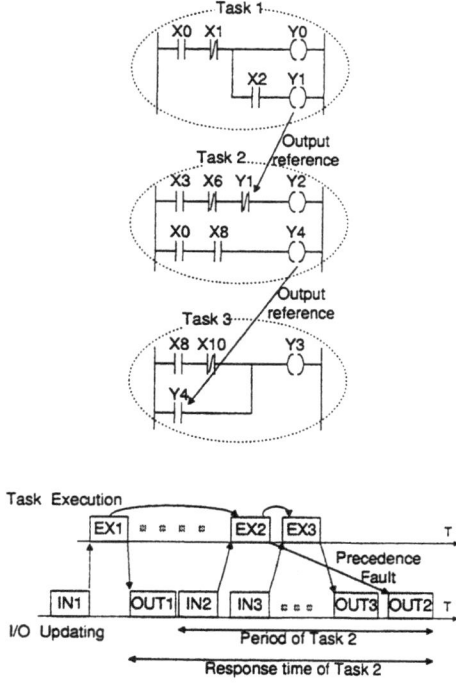

Fig. 5. Inter-task relations

1996). The size of the I/O data can be extracted by static measurement of sequence programs and networks can guarantee the transmission time for the bounded amount of data.

The utilization of the PEU and the DTU should be bounded to 1, and Eq. (1) should be met though program execution and data transmission are processed in parallel.

$$\forall n, \ e_n + i_n + o_n < ST_n \qquad (1)$$

Unless Eq. (1) holds, τ_n starts before its input data is ready in the IODM, and/or the output data is transmitted over the network before τ_n finishes. The former yields useless computation of τ_n and the latter yields useless data transmission. Though both of them may not cause system failure, they double or triple $WCRT$.

3.1 Scheduling constraints

The objective of scheduling establishes the following task-specific timing constraints.

(a) $c_n^o(k) - s_n^i(k) < WCRT_n - ST_n$
(b) $c_n^i(k) < s_n^e(k)$
(c) $c_n^e(k) < s_n^o(k)$
(d) $s_n^i(k+1) - c_n^o(k) > MCI_n$.
(e) If $\tau_n \prec \tau_m$, the sequence such as $c_n^e < c_m^e < c_m^o < c_n^o$ should be avoided,

The requirements in Subsection 2.2 yields the conditions (a) to (d). (e) is necessary to avoid the data coherent problem between the data in IODM and the values of sensors or actuators in

IOSs. It may happen that τ_m reads the data which is computed by τ_n from the IODM and updates its output data to the IODM, however, the DTU transmits the output data of τ_m in the IODM to the actuators, prior to the output data of τ_n. Fig. 5 shows a precedence fault where $Y3$ is caused by $Y4$ and $Y3$ is effective earlier than $Y4$ on actuators.

3.2 Scheduling algorithm

Generally, a pre-run time scheduling algorithm is suitable when there are hard real-time deadlines, precedence relations(Locke, 1992). Although lots of scheduling algorithms were proposed for general models, they cannot be used since PLCs with network-based IOS are multi-processor systems and require to meet many specific constraints. We suggest the scheduling algorithm that arranges task execution on the PEU and data transmission on the DTU, without violating five scheduling constraints, (a) to (e). The algorithm is a two-step process.

The first step is to determine the ST_n. $WCCD_n$ can be reduced to $e_n + i_n + o_n$. In this case, ST_n is achieved as the largest value. Although resource occupation is reduced by lengthening ST_n, small $WCCD_n$ makes deadline of task execution and data transmission tight with reducing schedulability. Fig. 6 shows an example that low resource occupation does not always enhance the schedulability. While a feasible schedule cannot be found with the smallest $WCCD_0$, $WCCD_1$, and $WCCD_2$, a valid schedule guaranteeing all $WCRT$ can be found with increased $WCCD_1$, $WCCD_2$ and reduced ST_1 and ST_2. The suggested algorithm selects the ST iteratively, from the largest value to the smallest value. This continues until a feasible solution is found or the utilization bounds are exceeded.

The second step is to generate I_n^e, $I_{n,s}^i$ and $I_{n,s}^o$ in the least common multiple of $ST_1, ..., ST_{M_T}$. This step constructs a tree whose nodes represent all possible combination of I_n^e, $I_{n,s}^i$ and $I_{n,s}^o$. Let B_T be the time-tick unit of the scheduler. I_n^e, $I_{n,s}^i$ and $I_{n,s}^o$ are integer multiple of B_T and their range is ST_n. This step traverses the tree until it find the node whose I_n^e, $I_{n,s}^i$ and $I_{n,s}^o$ meet all scheduling constraints.

4. PERFORMANCE EVALUATION

$WCCD$ can be three times as large as ST when (b) and (c) conditions are violated(Jeong et al., 1997). In this case, ST should be determined as $\frac{1}{4}WCRT$ to guarantee the WCRT. Under the condition of Eq. (1), the proposed scheduling algorithm is able to increase ST as follows:

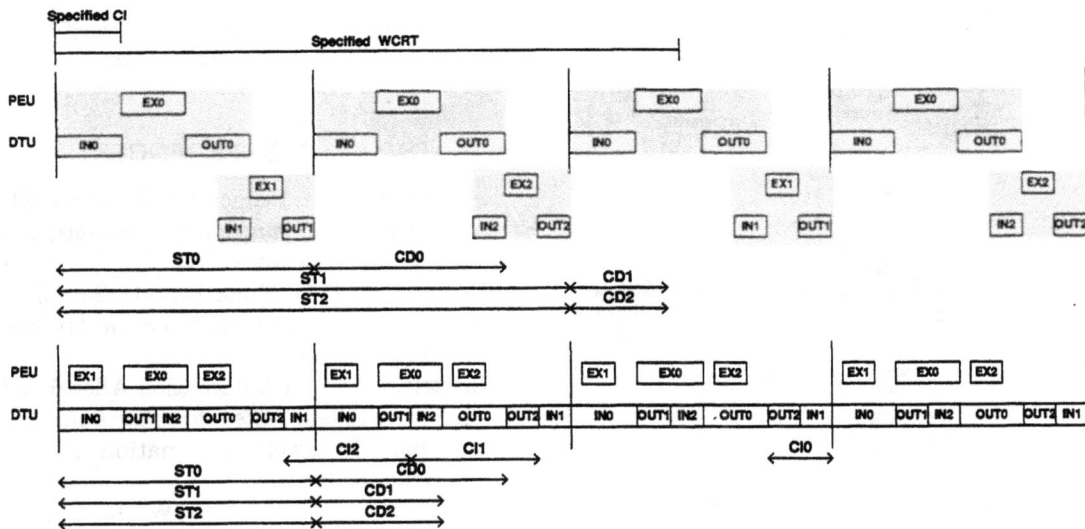

Fig. 6. Scan time and schedulability

$$\frac{1}{2}WCRT_n \leq ST_n \leq WCRT_n - e_n - i_n - o_n.$$

The feasible schedule generated from first iteration is the most efficient schedule since it guarantees $WCRT$ and MCI with the lowest resource occupation. As shown in Table 2 and Table 3, some schedules become feasible by reducing the ST. These demonstrate that schedulability may increase while the resource occupation increases.

From the comparison between Table 2 and Table 3, it is observed that a feasible schedules is hard to be found with the minimized $WCCD$ in case of the higher resource occupation. Increasing $WCCD$ is useful to find a feasible schedule in such a case.

Table 2. WCCD vs. schedulability (1)

1st IT (0.84)	2nd IT (0.91)	3rd IT (0.96)	% of ratio for all task sets
o	-	-	83.3
×	o	-	13.8
×	×	o	2.9

o: schedule can be found x: schedule cannot be found
1st IT(Iteration): WCCDs are minimized
2nd IT, 3rd IT: WCCDs are lengthened to some degree
() shows the average utilization

Table 3. WCCD vs. schedulability (2)

1st IT (0.67)	2nd IT (0.78)	3rd IT (0.88)	4th IT (0.95)	% of ratio for all task sets
o	-	-	-	92.3
×	o	-	-	0.8
×	×	o	-	6.4
×	×	×	o	0.5

o: schedule can be found   x: schedule cannot be found
1st IT(Iteration): WCCDs are minimized
2nd IT, 3rd IT, 4th IT: WCCDs are lengthened to some degree
() shows the average utilization

5. IMPLEMENTATION

5.1 Network for I/O data transmission

There are many kinds of networks, however, fieldbus could be taken into account to guarantee the worst case transmission delay. Among fieldbuses, FIP(Factory Instrumentation Protocol) is adopted in the implemented PLC. In FIP, each real-time channel has its slot reserved by pre-run time scheduling. There is a centralized controller that manages the packet scheduling. FIP guarantees deterministic data transfer for cyclic traffics. It is suitable for PLCs because program execution and data transmission are served with the period called ST.

5.2 Synchronization in the execution subsystem

Although FIP standard is said to guarantee the worst case transmission delay, it is only effective for the physical layer and the data link layer. Thus, the implementation should solve two problems. One is synchronization between the application layer and the data link layer. The other is bounding delay of the application layer. While it is not difficult to bound the delay of the application layer, the synchronization requires additional hardware support because the application layer runs on the PEU and the data link layer runs on the DTU. FULLFIP2, an FIP controller, supports "indication" that generates hardware interrupt to application processors when tagged data has been transmitted(WorldFIP, 1995). As shown in Fig. 7, the scheduler generates a list of phase of task execution and I/O data transmission. The code generator builds a list of indication variables(list of tagged data) having the same phases as the

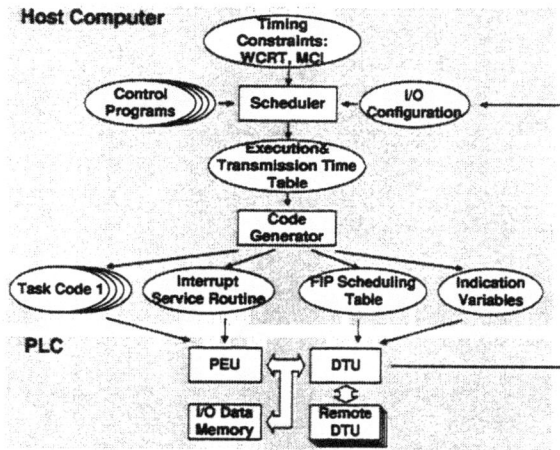

Fig. 7. Implementation of the scheduler

tasks. When the DTU generates an interrupt request to the PEU, PEU invokes the task having the same phase as the indication variable.

5.3 *Synchronization in the I/O subsystems*

One I/O subsystem consists of an FIP slave interface and I/O devices. The FIP slaves transmits data packet by packet. The order of packets is not in accordance with the order of I/O device access. However, the access interval is relatively negligible in comparison with the *ST* in ES. Our implementation does not care about the order of packets in the I/O subsystems and the synchronization in IOS.

6. CONCLUSIONS AND FUTURE WORKS

A scheduling method is proposed to reduce the response time determinately in a PLC which has network-based I/O systems. By analysis of specific behavior of PLCs, a precise timing model is built. From the timing model, five task-specific scheduling constraints are extracted. These constraints contain all correlations and individual characteristics of the plant and the controller such as response time, input to output delay of PLCs, plant delay and precedence relation among sequence programs. Under the derived constraints, task execution and data transmission are scheduled. Since a feasible schedule cannot always be achieved under low resource occupation as shown in computer simulation, the proposed scheduling method adjusts period as well as initiation time, thus, it enhances the schedulability. Implementation technique is described with the synchronization scheme between the execution processor and the network processor. The implemented PLC adopts FIP fieldbus network and resolves the synchronization with minimal hardware resources. Future works are reducing the complexity

of the scheduling algorithm and verifying the performance by experimental results.

7. REFERENCES

Automation, GE Fanuc (1992). *Series 90-30 Programmable Controller Installation Manual.* GE Fanuc Automation.

Bonfatti, F., P. D. Monari and U. Sampieri (1997). *IEC1131-3 Programming Methodology.* CJ International.

Electric, Rockwell Automation Allen-Bradley Reliance (1994). *Allen-Bradley Automation Systems.* Rockwell Automation Allen-Bradley Reliance Electric.

Jeong, S., N. Chang and W. H. Kwon (1997). Scheduling algorithm for programmable logic controllers with remote i/os. *Proceedings of RTCSA* pp. 87–94.

Koo, K. and W. H. Kwon (1996). Worst-case timing prediction of relay ladder logic by constraint analysis. *Proceedings of RTCSA* pp. 180–186.

Kwon, W. H., J. Park and N. Chang (1994). Real-time bus for multiprocessor-based programmable controller. *Proceedings of RTCSA* pp. 38–42.

Locke, C.D. (1992). Software architecture for hard real-time applications: Cyclic executives vs. fixed priority executives. *The Journal of Real-Time Systems* 4, 37–53.

Park, J., N. Chang, G. Rho and W. H. Kwon (1993). Implementation of a parallel algorithm for event driven programmable controllers. *Control Eng. Practice* 1, 663–670.

Park, J. W., H. G. Park and W. H. Kwon (1997). A contention-resolving algorithm for real-time communication between distributed programmable controllers. *Proceedings of DCCS* pp. 55–60.

Simpson, C.D. (1994). *Programmable Controllers.* Prentice Hall.

Systems, Siemens Automation Group Industrial Automation (1994). *SIMATIC S5 S5-135U, S5-155U and S5-155H Programmable Controllers Catalog ST 54.1.* Siemens Automation Group Industrial Automation Systems.

Warnock, I. (1988). *Programmable Controllers - Operation and Application.* Prentice Hall.

WorldFIP (1995). *General Purpose Field Communication System, prEN 50170.* WorldFIP.

A GRAPH-BASED METHOD TO TRANSLATE RELAY LADDER LOGIC DIAGRAMS

J. L. Azevedo, J. P. Estima de Oliveira

Department of Electronics and Telecommunications
University of Aveiro
3810 Aveiro, Portugal
Tlf. +351.34.370500
E-mail: jla@inesca.pt, jeo@inesca.pt

Abstract: The Relay Ladder Logic (RLL) graphic programming language is widely used to program Programmable Logic Controllers. A method to convert RLL diagrams into a set of Boolean expressions is presented. This method is based on the representation of each RLL network in terms of a directed graph. The original graph is reduced, in an iterative way, detecting and detaching parallel circuits by replacing groups of vertices by a single one. A the same time, each parallel circuit is independently translated. The logic expression for the reduced graph is then generated, and the final step is the recursive replacement of every reference to a parallel circuit by the sub-expression generated at reducing time. *Copyright © 1998 IFAC*

Keywords: Programmable Logic Controllers, Ladder Algorithms, Programming Languages, Graphs

1. INTRODUCTION

Factory automation and process control tasks are today mainly assigned to Programmable Logic Controllers (PLC). Although there are some other languages available (IEC1131-3, 1993), the most common language used to program these kind of devices is the graphical language normally called Relay Ladder Logic (RLL). A RLL diagram (Pessen, 1989) is usually translated to a literal language (either a Boolean based language or a stack based language), before being downloaded into a PLC.

The main goal of a translation method should be the generation of the least possible number of instructions which, in terms of PLC programming (Michel, 1990), has two main advantages: a better usage of the program memory and the reduction of program execution time (also called scan time). This is specially important for low range PLCs, that are usually low cost solutions where limitations in memory capacity and CPU execution time are more obvious.

In this paper a method to translate RLL graphic diagrams to a set of Boolean expressions is presented. A RLL diagram is a graphic representation of Boolean expressions, delimited by two vertical lines designated by power rails. As shown in Fig. 1, the basic elements of the RLL language are contacts (input and/or output contacts, normally open and/or normally closed), coils, and connecting lines. The left power rail can be connected to contact elements and/or connection lines; to the right power rail only coil elements can be connected. Coil elements may represent physical outputs, internal variables, timers or counters.

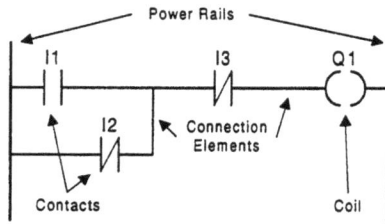

Fig. 1. Basic elements of the RLL graphic language.

In terms of logic flow, the left power rail represents the Boolean state TRUE. A RLL diagram is evaluated from left to right and from top to bottom.

2. DEFINITIONS AND METHODOLOGY

A graph G (Bondy, 1976) (Carré, 1979) can be defined as a pair (V, E), where V is a non-empty finite set, and E is a finite set of unordered pairs of elements of V. The elements of $V=V(G)$ are usually designated by vertices, and the elements of $E=E(G)$ by edges of the graph G. An edge $e=\{u,v\}$, with u, $v \in V$, is defined by two vertices (endpoints of an edge), and is said that e connects u and v, or that e is incident with u and v. Two vertices connected by an edge are said to be adjacent. A path P_n of G is a sequence of n distinct vertices v_1, v_2, ..., v_n, where v_i and v_{i+1}, with $i=1, 2, ..., n-1$, are adjacent. The length of a path P_n is the number of edges in P_n.

A directed graph (or simply digraph) is defined as a pair (V, E), where V is a non-empty finite set, and E is a finite set of ordered pairs of elements of V. This way, an edge connects a vertex to another vertex and it is called an arc (u, v), where u is the source vertex and v the destination vertex. Graphically, the direction of an edge is indicated by an arrow. A path P_n in a directed graph is a sequence of n distinct vertices v_1, v_2, ..., v_n, where v_i and v_{i+1}, with $i=1, 2, ..., n-1$, are connected by the arc (v_i, v_{i+1}).

In the method being presented, each RLL network is represented by a directed graph (referred later on simply as graph) which is a particular case without symmetric ordered pairs. The number of vertices in the graph is exactly the number of RLL elements in the network and, therefore, each vertex in the graph represents either a contact or a coil. An arc from vertex i to vertex j means that the RLL element i is connected to the RLL element j.

As stated before, the left power rail in a RLL diagram is the entry point to analyse/calculate each network. It is, therefore, the starting point for the logic flow in that network, and the graph should represent left to right connections. Since RLL elements have two sides, the graph is built verifying if the right hand side of a given element is connected to the left hand side of any element in the network, as shown in Fig. 2

(the left and right hand sides are represented by a cross and a circle, in this order).

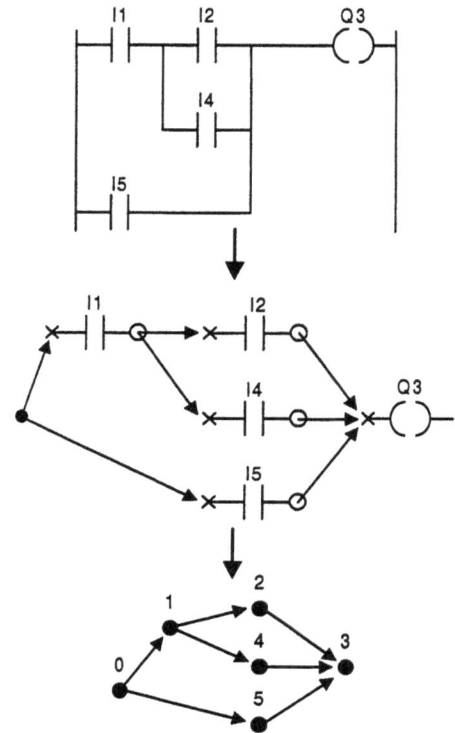

Fig. 2. Construction procedure of the RLL equivalent graph.

Each RLL element is assigned a number (1 to N), being N the total number of graphic elements in the RLL network. The left power rail is also represented by a vertex, which is assigned the number 0, and the right power rail is ignored. Vertex 0 is the graph entry point, and there is a path from vertex 0 to every vertex in the graph. A vertex representing a coil is easily distinguished from all others, since there aren't any edges starting from it.

The information symbolised by a graph is easily processed by a computer if it is represented in terms of the corresponding adjacency matrix. Let G be a graph and $V=\{v_1, v_2, ..., v_n\}$ the set of all vertices. The adjacency matrix of G, $A(G)=[a_{ij}]$, is a square matrix, where $a_{ij}=1$ if v_i and v_j are adjacent, and $a_{ij}=0$ otherwise. In a directed graph v_i and v_j are adjacent vertices if there is an arc from vertex i to vertex j. It is, therefore, a matrix representing every oriented arc in the graph, where rows are initial vertices and columns final vertices. A row filled with zeros means that there is no arc starting from the corresponding vertex and, from the RLL point of view, it is an output (coil). A column filled with zeros means that the given vertex is not destination of any directed arc (this situation is only allowed for vertex 0).

3. TRANSLATION METHOD

The translating method presented in this paper is composed of four main actions which evolve according to the following algorithm:

```
Detect short-circuit
IF NOT short-circuit THEN
    DO
        Reduce graph and do partial coding
    WHILE can reduce
    IF NOT completely reduced THEN
        Assign temporary variables
    ENDIF
    Generate final code
ENDIF
```

These four actions will be discussed in the following sections.

3.1 Short-circuit detection

As the graph is being built, it is easy to detect that parameters associated with a given RLL element, such as identification or timing values, are missing. By analysing the graph it is also easy to detect elements that are not connected (partially or totally disconnected). There is a third kind of error that can affect the RLL behaviour, designated here by short-circuit.

The short-circuit situation is characterised in Fig. 3a), where the action of vertex k is suppressed by the arc between vertices i and j.

The following two conditions together are necessary conditions for a short-circuit to exist:
1. There is a vertex i source of two or more arcs.
2. There is a vertex j adjacent to i, destination of two or more arcs.

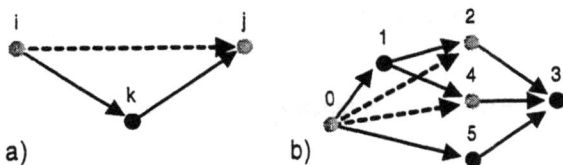

Fig. 3. a) Short-circuit characterisation. b) short suppressing the action of contact "I1" in the example of Fig. 2.

An effective short-circuit situation comes out when the two previous conditions are satisfied, and there is a path between vertices i and j of length equal to or greater than 2. In other words, a short-circuit exists if there are two or more paths between vertices i and j and the distance (length of the smallest path) between them is one.

3.2 Graph reduction and partial coding

In a RLL diagram the logic function carried out by a given set of contacts results from its relative position, that is established by graphic connections between them. Except for coils (which may have the function of output, timer or counter), there is no function pre-assigned to any graphic element. The translation is then obtained from context analysis of each contact in the RLL diagram.

The translation process starts with the detection of groups, which can be defined as RLL network parts that can be detached from the whole. Every time a group is detected it is immediately translated and so, instead of analysing the whole diagram, the operation performed by a given contact is determined by context analysis inside the group. This new entity, that can be thought as a macro, replaces the set of contacts forming the group, and interacts with the rest of the diagram as if it was a single contact.

The set of operations aiming group isolation and group translation, performed on the RLL equivalent graph, are designated here by "graph reduction and partial coding". The main goal of this process is to transform the original graph in a simplified one without forks, which means a graph with a single path between the initial vertex (vertex 0) and the vertex corresponding to the output element.

The reduction process is shown in Fig. 4. A group is represented by a circle surrounding a vertex. An underlined vertex number denotes the number which will be assigned to the group that is being processed.

The RLL contacts can be connected in a serial or a parallel fashion, corresponding to the logic operations AND or OR, or in a mixture of serial/parallel groupings. The RLL network parts to be isolated are composed of contacts in parallel or, more generically, composed of parallel paths of contacts in series.

In terms of graph analysis, the following three conditions characterize a group:
1. There is a vertex i, destination of zero or one arc and source of two or more arcs; such a vertex can be called a fork or a start vertex.
2. There is a vertex j, accessible from i, destination of two or more arcs and source of zero or one directed arcs; this type of vertex is named here a joint or a stop vertex.
3. There are two or more valid paths between vertices i and j, whose length is great than or equal to 2.

A valid path for condition 3, is a path without forks, that is, a path where every vertex has degree 2 (except for start and stop vertices) as shown in the example of Fig. 5a).

37

This is the RLL equivalent graph. The first group that can be detected is composed of vertices 2 and 4, and is labelled Group G2. The group is then translated to the literal form.

Group G2

G2 = I2 + I4

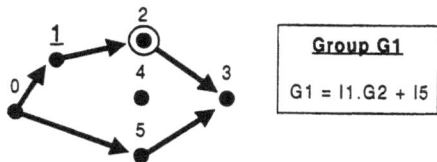

When a group is detected, all the vertices forming that group are replaced by a single vertex that preserves all the "border" connections (vertex 2 in the example). A new group (G1) is then formed, composed of vertices 1, 2 and 5.

Group G1

G1 = I1.G2 + I5

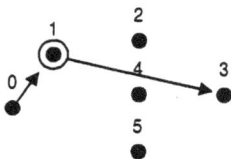

The resulting graph has a configuration where it is not possible to detect any new group. All vertices in the graph are of degree 2 (except the entry vertex 0, and the exit vertex 3). The reduction process terminates.

Fig. 4. Graph reduction and code generation for each detected group, taking the RLL example of Fig. 2.

In the RLL equivalent graph, each vertex represents a contact or a coil. There is no support for connection lines, that are implicitly represented by directed arcs. Therefore, a vertical connection line, shunting both sides of two or more contacts, is represented by multiple arcs between vertices, as shown in Fig. 5b). This way, logic commonality between contacts is masked in the resulting graph and, as a consequence, the notion of valid path must be redefined. Hence, a valid path between vertices i and j must have length equal to or greater than two and fulfil one of the three following conditions:

1. Every intermediate vertex between i and j have degree 2.
2. The vertex preceding a joint vertex has degree three or more, but it is a fork vertex.
3. The vertex following a fork vertex has degree three or more, but it is a joint vertex.

Assuming this definition of valid path, in Fig. 5b) two different groups can be identified: the first composed of vertices 1 and 4 (parallel circuit "0,1,2" / "0,4,2" or "0,1,5" / "0,4,5"), and the second composed of

vertices 2 and 5 (parallel circuit "1,2,3" / "1,5,3" or "4,2,3" / "4,5,3").

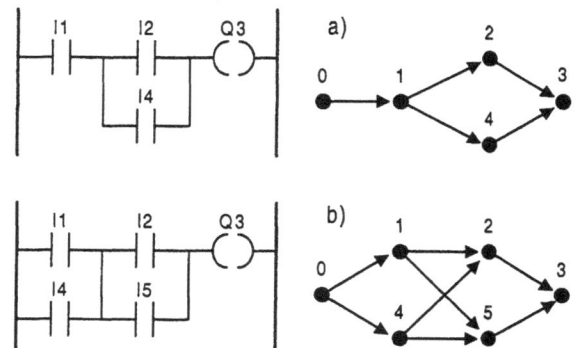

Fig. 5. Definition of a group.

The parameters that identify a group are: the initial and final vertices, the first path, and the number of consecutive paths (that correspond to the number of branches in the parallel circuit). Once identified, a group is translated according to the following rules:

- all vertices in the same path are combined with the logic operation AND. In terms of code, $n-1$ AND instructions must be generated, where n represents the number of vertices in the path;
- all paths in the same group are combined with the logic operation OR. The code must exhibit $m-1$ OR instructions, where m represents the number of paths in the group.

After being translated, the whole group is replaced by a single vertex (the first vertex in the first path as in Fig. 4) which takes on the group representation in the graph by preserving its border connections. All remaining vertices are neutralized by deleting all incident and adjacent arcs.

The graph reduction and group translating process continues repeatedly, trying to detect new groups possibly made up of others previously processed. This action terminates whenever it is not possible to identify any new group in a complete iteration.

3.3 Assignment of temporary variables

The original configuration may prevent the graph to be completely reduced. This means that in the reduced graph there is more than one path between vertex 0 and the vertex corresponding to the network output, as shown in the example of Fig. 6. Under these circumstances, it is not possible to operate all inputs in the RLL network without having temporary variables to store intermediate results. Therefore, this step of the translation method is done only if after the reduction process the graph has one or more forks.

A temporary variable is assigned to each remaining fork in the reduced graph (a fork starting in vertex 0 doesn't need to have a temporary variable, since its logic value is always TRUE). Such variable allows the logic result in a fork to be saved, so that it can be used as the initial value in the processing of every path starting from it. The graph parts preceding a fork are therefore codified only once.

During the final code generation process (see next section), a given temporary variable is written once and read every time a new path (starting from the same fork vertex) is processed. In order to easily recognize write time from read time, temporary variables (that are represented by numbers) are assigned according to the following rules:
1. The variable associated with a fork is in fact assigned to every adjacent vertex.
2. The first fork path is distinguished from all others, by setting the most significant bit of the temporary variable number.

The assignment process is represented in Fig. 6.

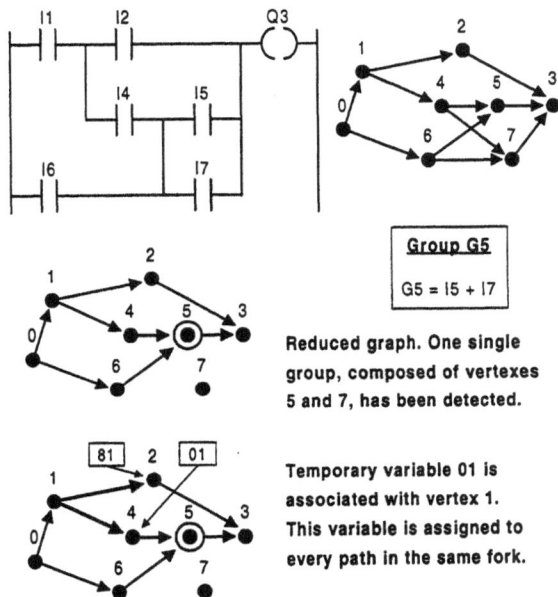

Group G5

G5 = I5 + I7

Reduced graph. One single group, composed of vertexes 5 and 7, has been detected.

Temporary variable 01 is associated with vertex 1. This variable is assigned to every path in the same fork.

Fig. 6. Example of a RLL diagram whose graph can not be completely reduced. In such case it is necessary to assign temporary variables to each remaining fork.

3.4 Final code generation

In the first step of the translation process the graph is reduced and, each time a reduction takes, place the code for each detected group is emitted. After that, if the graph doesn't become completely reduced, temporary variables are assigned to each remaining fork. These two steps prepare the graph for the last operation called "final code generation", which consists of two main actions:
1. Code generation taking the reduced graph as an input.

2. Recursive replacement of all references to groups by their respective contents, taking the result of the previous operation.

The logic result of a RLL network is the sum (logic OR) of the contributions of each path leading to the same output element. This means that, to produce the equivalent code, each network path must be traversed starting from the left power rail and ending at the output element.

The reduced graph may be thought as a tree, whose entry point (root) is the coil equivalent vertex. The leaves of that tree represent either a contact connected to the left power rail or a contact connected to a fork. In terms of logic flow, a contact connected to the left power rail has to be intersected (AND operation) with the logic value TRUE (and so is ignored), while a contact connected to a fork must be intersected with the logic value stored in its own temporary variable. Fig. 7 shows the code generation technique applied to the two examples previously analysed, presenting the reduced graphs as trees. Temporary variables assigned to vertices are referenced by "Tn".

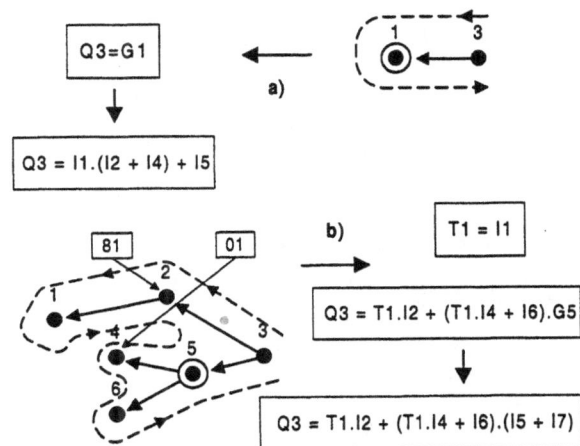

Q3 = G1

Q3 = I1.(I2 + I4) + I5

a)

b)

T1 = I1

Q3 = T1.I2 + (T1.I4 + I6).G5

Q3 = T1.I2 + (T1.I4 + I6).(I5 + I7)

Fig. 7. Code generation technique applied to the two examples previously analysed: a) tree with a single leaf which represents a graph without forks. b) tree with three leaves, one of them (node 4) representing a fork connection.

As mentioned before, each path leading to the same output element must be scanned, in order to process all logic contributions. Looking the reduced graph as a tree makes this task easy, since it ensures that:
• the hierarchical structure of the RLL diagram is preserved;
• all vertices belonging to the tree are visited in the correct order;
• the code generation process is done following the common procedure, that is, from left to right, and from top to bottom.

The technique used to traverse the tree is the normal one, which can be expressed in the following way (see Fig. 7): 1) start from the root; 2) if v_i (the vertex currently being visited) has successors not yet visited then visit them, respecting a top to bottom order. Otherwise return back to predecessor of v_i, or terminate if the predecessor is the root.

4. CONCLUSIONS

RLL graphic programming language is widely used in industry to program PLCs. In order to use PLC memory space in an effective way, the conversion process must generate the minimum possible number of instructions. Having that in mind this paper presents a method to translate RLL diagrams into a set of Boolean expressions. Although this output is not adequate to directly program PLCs, it can be easily translated (Estima de Oliveira, 92) to a Boolean type or stack based PLC language.

5. REFERENCES

Bondy, J. A. and U. S. R. Murty (1976). *Graph Theory with applications*. Macmillan Press, London.

Carré, Bernard (1979). *Graphs and networks*. Clarendon Press, Oxford.

Estima de Oliveira, J. P., Azevedo, J. L., *et al* (1992). Software development for Programmable Logic Controllers - a methodology and a system. In: *Proceedings of the IFAC CIM in Process and Manufacturing Industries*, 233-238.

IEC1131-3 (1993). Programmable Controllers - Part 3: Programming Languages. *International Electrotechnic Commission*.

Michel, Gilles (1990). *Programmable logic controllers: architecture and applications*. John Wiley & Sons, Chichester.

Pessen, D. W. (1989). Ladder diagram design for programmable controllers. *Automatica*, **25**, 407-412.

PID_ATC: A REAL-TIME TOOL FOR PID CONTROL AND AUTO-TUNING

J. Portillo, M. Marcos, D. Orive, F. López, F. Pérez

Escuela Superior de Ingenieros de Bilbao (University of the Basque Country)
Alameda Urquijo s/n 48013 Bilbao-SPAIN
e-mail: jtapobej@biha01.bi.ehu.es, jtpmamum@bi.ehu.es

Abstract: This work deals with the design of a real-time tool that implements a set of PID-type controllers using different algorithms and different structures. The tool not only performs the real-time control of a process but also performs on-line monitoring. It also offers an auto-tuning option that includes the evaluation of the dynamic characteristics of the process (via step response or a frequency domain model) and the computation of controller parameters using different methods. Once the auto-tuner has computed the PID parameters and these settings have been tested on the system, the auto-tuner sends the settings to the commercial PID. *Copyright © 1998 IFAC*

Keywords: PID control, autotuners, real-time systems, Ada tasking programs.

1. INTRODUCTION

The well-known PID controllers are found in large numbers in all types of industry. They are often combined with logic, sequential machines, selectors and simple function blocks to build complex automation systems.

Thanks to the great advances in technology, these controllers are currently going through an interesting phase of development. The improvements in microelectronics have given cheap microprocessors whose computational power is continuously increasing. This fact allows manufacturers to offer equipment with features like: parameter adjustment aids (*automatic tuning* or *pre-tuning*), automatic adjustment procedures (*auto-tuning*), *gain scheduling* or *adaptation*. All these innovations attempt to achieve a higher degree of automation by adjusting parameters of the controller in an ordinary feedback loop (Åström, *et al.*, 1993; Åström and Hägglund 1995; Åström 1996).

Even so, recent studies (Bialkowski 1993; Ender 1993) show that a 30% of the installed process controllers operate manually, that 20% of the loops use default parameters set by the controller

manufacturer and that 30% of the loops function poorly due to instrumentation problems.

The present work focuses on the auto-tuning technique (also called automatic tuning, tuning on demand or one shot tuning) to tune automatically a controller on demand from a user. The tuning procedure starts when the operator suplies the initial PID settings, or to modifies them to adapt to a changing situation. The auto-tuning tool proceeds to obtain the optimal settings automatically. This man-machine interactive scheme differs from the self-adaptive controller that continuously estimates the process parameters and adjusts the settings of the controller.

An automatic tuning procedure consists of three steps:

- Generation of a process disturbance
- Evaluation of the disturbance response
- Calculation of controller parameters

The first step allows the process dynamics to be determined. The evaluation of the disturbance response may include a determination of a process model or a simple characterisation of the response.

Commercial PID controllers with automatic tuning facilities have only been available since the beginning of the eighties. Automatic tuning can also be performed using external equipment. These products are connected to the control loop only during the tuning phase. When the tuning experiment is finished, the products suggest controller parameters (Åström, et al., 1993). Since they are designed to work together with controllers from different manufacturers, they must be provided with information about the controller. This information should include the controller algorithm (interactive, non-interactive or parallel), controller structure (PID, PI-D or I-PD), sampling rate, filter time constants and units of the different controller parameters, as well as the estimation and tuning methods to be used.

If this external equipment is an application running in a PC, it is straightforward to add more control features to the software tool, apart from the tuning, e.g. one could control the process (once tuned) in order to test the new controller settings, monitor the process variables and store them into a file for later processing.

This paper presents the design and development of a real-time tool that offers such characteristics. After the test of the controller settings, it is also able to communicate with an industrial PID, sending the controller settings using the appropriate serial protocol.

The paper is laid out as follows Section 2 presents previous work in this field. In Section 3 the different algorithms and structures of the PID controller are described. Section 4 presents the auto-tuning methods that have been implemented. Finally, Sections 5 and 6 discuss the general requirements and the detailed design of the software, respectively.

2. RELATED WORK

There are several commercial products with automatic tuning facilities. The Foxboro EXACT (760/761) uses step response analysis for automatic tuning, and pattern recognition technique and heuristic rules for its adaptation; the Alfa Laval Automation ECA400 controller uses relay auto-tuning and model-based adaptation. The Honeywell UDC 6000 controller uses step response analysis for automatic tuning and a rule base for adaptation. Yokogawa SLPC 181 and 281 use step response analysis for auto tuning and a model based adaptation (Åström 1995; Åström, et al., 1993).

There are also commercial computer-based products available. For instance, Intelligent Tuner and Gain Scheduling is a software package used in distributed control system by Fisher-Rosemount. Looptune is a tuning program package to be used within the DCS system Honeywell TDC 3000. DCS Tuner is a

software package for controller tuning in the ABB Master system and Techmation Protuner is a process analyser, that is only connected to the control loop during the tuning and analysing phase. A description of these industrial packages can be found in Åström and Hägglund (1995).

Within the applied research field, several authors describe the design and implementation of auto-tuners. For instance, in Cox, et al., (1997) a CAD software package that implements PI and pPI controller to be used in water industry is described. In addition, in Schrönberger and Poli (1997) an auto-tuner based on the phase margin of the compensated system is presented. This system implements a finite automaton.

The work described in this paper deals with the design of the software structure that allows the integration of the auto-tuning methods described in the literature. Some of them, presented in section 4, are implemented here but the design allows an easy integration of new ones.

3. PID CONTROL: ALGORITHMS AND STRUCTURES

The effect of the PID parameter settings on the system response depends on the type of algorithm used by the controller. Most of the commercial PIDs implement one of the three algorithms known as: interactive, non-interactive and parallel (Gerry 1987). Table 1 presents the ideal equations for these algorithms.

Table 1. Ideal PID algorithms.

NON-INTERACTIVE:

$$u(t) = Kp \cdot \left[e(t) + \frac{1}{Ti} \cdot \int e(t) \cdot dt + Td \cdot \frac{de(t)}{dt} \right]$$

PARALLEL:

$$u(t) = Kp' \cdot e(t) + \frac{1}{Ti'} \cdot \int e(t) \cdot dt + Td' \cdot \frac{de(t)}{dt}$$

INTERACTIVE:

$$u(t) = Kp'' \cdot \left[e(t) + \frac{1}{Ti''} \cdot \int e(t) \cdot dt \right] \cdot \left[1 + Td'' \cdot \frac{de(t)}{dt} \right]$$

Each algorithm requires different tuning parameters so it is necessary to know which one the controller implements before trying to tune it. Moreover, there are formulas to convert parameter values from one algorithm into another.

Several structures are also available for PID controllers: PID, PI-D and I-PD. Each structure will provoke different responses to reference changes (Morilla 1990).

Seven PID controllers can be designed as combination of algorithms and structures, apart from PI or P controllers (interacting algorithm does not admit P-ID or I-PD structures).

4. AUTO-TUNING METHODS

Depending on the estimation procedure, auto-tuning methods can be classified on those that use Time Domain models and those that use Frequency Domain models.

4.1 Estimation with Time Domain Models

Typical time-domain features are static gain (K), dominant time constant (T_p) and dominant dead-time (T_0). These parameters lead to a first order plus dead time model for the process as given in equation (1) (see Smith and Corripio 1997; Åström and Hägglund 1995).

$$G_P(s) = \frac{K}{1 + sT_P} \cdot e^{-sT_0} \qquad (1)$$

This model can be determined from the response of the process to step input (for this reason estimation methods with time domain models are also called open loop procedures).

The analysis of the step response to obtain the first order model parameters can be done by measuring the times in which the response reaches a percentage of the final value (see Figure 1) or by determining certain areas (see Figure 2).The first procedure is called the "slopes method" (Smith and Corripio 1997) and the second one the "areas method" (Nishikawa, et al., 1984).

After estimating the process as a first-order model, different formulas can be applied to obtain the PID parameters (K_p, T_i, T_d).

Irrespective of the method used for estimating the process dynamics, there are several tuning tables by different authors that provide the PID parameters.

Tuning tables take the model estimated and give different values for PID parameters depending on (1) the selected tuning criteria (2) the PID algorithm and (3) whether the tuning is for reference or load changes. Typical tuning criteria are Quarter Decay Ratio Response, MIAE, MISE and MITAE (Smith and Corripio 1997).

The selected tuning tables implemented in this work are: Ziegler-Nichols (Ziegler and Nichols 1942), López-Murrill-Smith (López, et al., 1967), Rovira-Murrill-Smith (Rovira, et al., 1969) and Kaya-Scheib (Kaya and Scheib 1988).

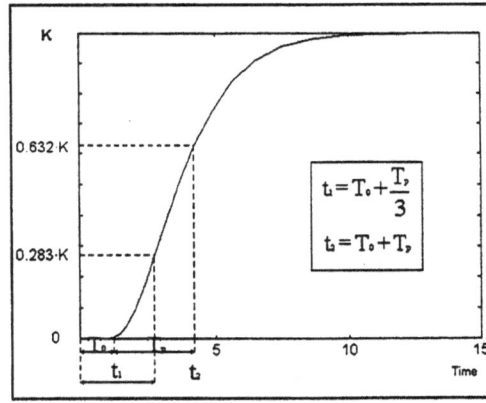

Figure 1. Estimation of a model can be performed using the "slopes method".

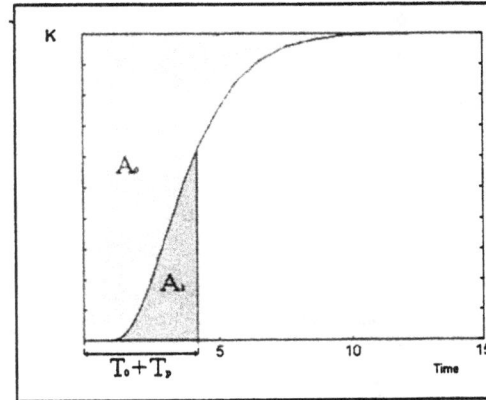

Figure 2. Estimation of a model can be performed using the "areas method".

4.2 Estimation with Frequency Domain Models

Typical Frequency Domain characteristics are ultimate gain (K_c) and ultimate period (T_c). These quantities are related to simple properties of the Nyquist curves.

A way to estimate the process model consists in using a relay feedback to get an oscillation with a period close to the ultimate frequency. It is the "Relay method" (Åström 1995) and it is a closed loop procedure because the loop must be closed to get the desired oscillation.

There are tables by Ziegler-Nichols Åström, et al., 1993) and Åström (Åström 1982) to obtain the PID parameters from K_c and T_c.

5. PID_ATC: GENERAL REQUIREMENTS

5.1 Functional requirements

The program whose design is studied in this work should integrate other functionalities besides auto-tuning functions in order to be a **comprehensive application**. These extra features should include monitoring every involved signal, saving the results of trials into files and controlling the process via computer (as the way to test the quality of the results before using them in an industrial PID controller).

Most commercial controllers include tuning features but they are designed to work with proprietary equipment and normally implement only one tuning method. It is desirable to have a generic software structure, which is able to tune any kind of PID algorithm or structure and offers several tuning procedures. The designed application should be **non-specific** in this sense, as well as the functional sense.

Another strength of the application should be **automatic operation**. After launching the tuning process, the application should be able to take the output to the selected set point, identify the process and tune the parameters without any external indication.

5.2 Operational Requirements

Before tuning, **initial information** has to be provided by the operator: the PID algorithm and structure, the estimation method and the tuning criterion to be selected. Online help for the user would be advisable.

Sometimes, it is necessary to take the process to a **set point**. *Manual* and *automatic set point* should be implemented. The calculation of the excitement signal value in every sampling time and the ability to decide whether the output has reached the stationary state must be implemented by a specific algorithm, in order to achieve the automatic set point mode.

In order to **identify** the system dynamics both time domain and frequency domain models should be available to suit any kind of process.

During the execution, **monitoring** and **storage** functionalities would allow the evolution of every signal (control action, process output, reference and load signals) to be followed on the screen and to be stored into a file.

The resulting **parameters** for the PID should be **tested** from the auto-tuner. Operator could change reference and load from keyboard and he could see on screen if process output is satisfactory.

The program has to offer a **user interface** to modify the working conditions or to stop the execution. The user must be able to change the visualisation scale, activate or inhibit the storage, switch between manual or automatic control, change reference and load, launch the auto-tuning and select different tuning tables or methods.

5.3 Real time requirements

The fact that the application should run in real-time adds an important constraint. Critical activities must be finished on time, assuming they meet their deadlines. Therefore, critical activities must be distinguished from interactive and background activities so that the scheduling method can assure system stability in the sense that hard-real-time tasks are guarantied even in system worst-case behaviour.

6. DETAILED DESIGN

The described application has been developed using Ada'83 as programming language and a PC (with acquisition board for A/D and D/A conversions) under DOS as target platform. This hardware platform has been selected because of its widespread use and to its favourable cost-performance ratio. Ada'83 offers concurrency (a Run Time Kernel allows concurrency even for a non-multiple tasking OS), easy access to hardware and special mechanisms for developing real-time applications.

6.1 Application timing

It is important to guarantee that the actions (represented in figure 3): (1) *acquiring data from the process*, (2) *computing the control algorithm* and (3) *sending the control action* are performed periodically (at the sampling instants) and in sequence.

To achieve this, the acquisition board includes a programmable timer that can be programmed in different operation modes. One of them uses two counters to trigger periodic A/D conversions followed by an interrupt signal. In such a mode, the application software is interrupted at the sampling frequency. The interrupt processing ensures that the three actions are executed in sequence.

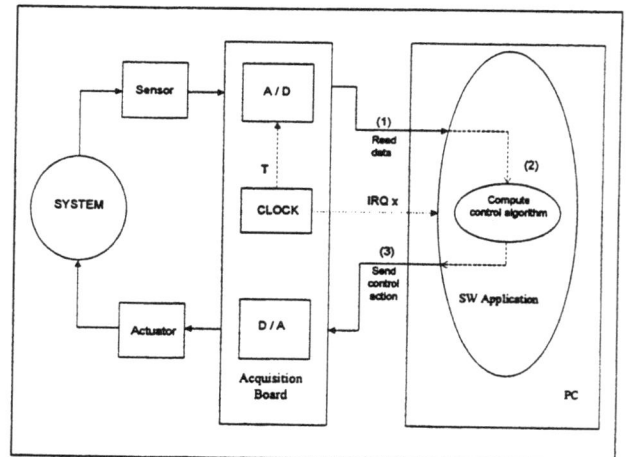

Figure 3. Data acquisition and control action performed every T msec (in instants ordered by the clock).

44

The following tasks have been identified to integrate the structure of the software application.

Data acquisition is the name of the interrupt handler used to read values from the process. This is a periodical task whose activation depends on the associated hardware (it should be modified for a different A/D board).

Engine receives the input data from the interrupt handler and depending on the functionality being carried out (taking the process to a set point, performing tuning or controlling the process) calls different procedures.

In the design, these two tasks have been considered as hard real time tasks.

Monitor (presents data on screen) and *Storage* (saves data into a file) are background tasks that implement soft real time activities.

User interface is an interactive task that allows the user to modify the operational conditions.

Synchronisation assures mutual exclusion access for shared data, and it is implemented as a pure server task. It lets other tasks read or write data in an inner structure (process variables). This task does not add any functionality by itself, it is a server task that encapsulates the structure containing information about the overall system (process output, control action, reference signal...) at the current sampling period. Other tasks access the process variables through the procedures exported by *Synchronisation*.

Figure 4 represents the design of the application using Buhr's notation (Buhr 1984). As it is shown in the figure, only *Engine* can write process variables and other tasks (*Monitor* and *Storage*) always read shared data in *Synchronisation* task. The Ada rendezvous mechanism (DoD 1983) has been used to implement the access to the shared data.

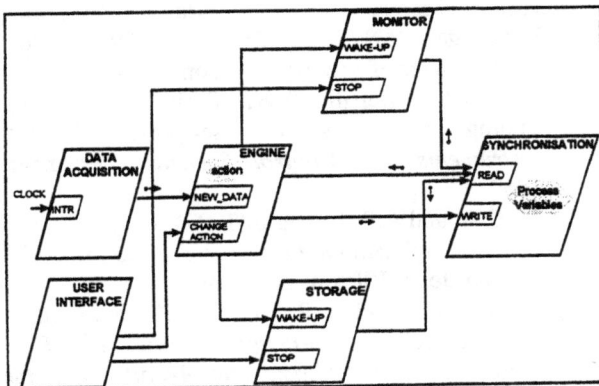

Figure 4. Exchange of data between tasks every sampling period.

Note that, in the control field, the use of priorities in a concurrent system has to be used to break the concurrency in order to assure that some actions (acquiring data, computing the control algorithm and sending the control action) execute sequentially.

In this case, the highest priority is assigned to *Data acquisition* (the critical activity) through the interrupt level (every hardware interrupt has higher priority than any software task). The highest software priority is assigned to *Engine*, then *Synchronisation* and the lowest priority to *Monitor*, *Storage* and *User interface*, which are background and interactive tasks, respectively. These priorities result in the sequential execution of figure 5.

It is important to point out that independently of the failure of background or interactive tasks during execution, the rest of them can continue working properly and thus the system remains under control.

7. APPLICATION EXAMPLE

As an example of its application, the auto-tuner has been applied to tune and control a second order process simulated in an analogue computer and a scale model called Ball & Hoop apparatus (Wellstead 1983). This system consists of a hoop that is free to rotate under the action of a DC servomotor, the hoop constitutes a simple inertial load on the motor axis and the control goal is the hoop speed.

The evolution of every signal involved is shown in figure 6 for the simulated model and figure 7 for the Ball & Hoop. First of all, the system is taken to a set-point (automatically in the first case and manually in the second). If automatic set-point is being performed, an algorithm decides when the process response has reached the operating point (phase 1 in figure 6). Then auto-tuning process starts (phase 2). Open loop identification can be seen in figure 6 and closed loop in figure 7. As soon as the selected tuning procedure achieves the PID parameters, the auto-tuner uses them in order to control the process (phase 3). Both figures represent the system responses to step changes in the set-point (changes in the load are also possible).

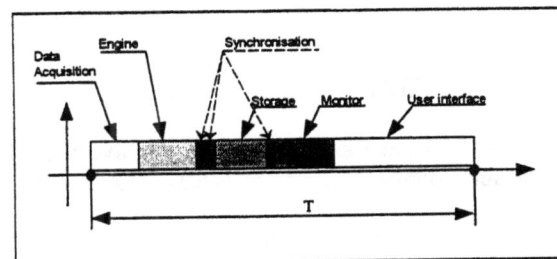

Figure 5. Sequential order of the execution.

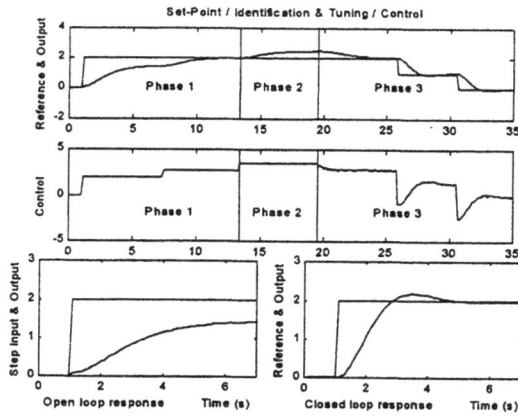

Figure 6. Tuning (Rovira-Murrill-Smith tables and MITAE criteria, open loop identification with slopes method) and control (PI-D non interactive) of a second order process simulated in an analogue computer.

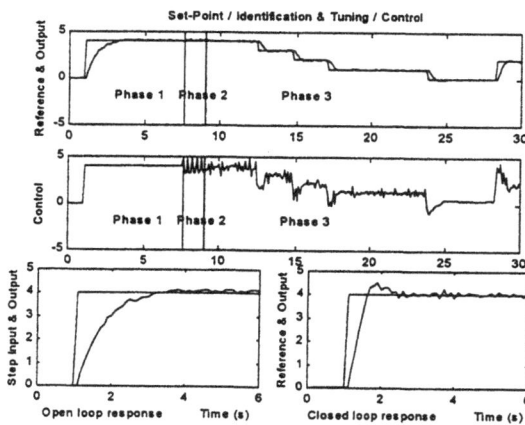

Figure 7. Tuning (Åström tables and rele identification method) and control (I-PD non interactive) of the Ball & Hoop scale model.

CONCLUSIONS

The designed real-time software implements auto-tuning techniques and different algorithms and structures of a PID. The software also allows testing the parameters of the PID on the process before sending them to the PID controllers. The combination of tuning methods and control techniques in the same tool seems to be an interesting strategy for the global tuning and testing of PID controllers.

ACKNOWLEDGEMENTS

This work has been supported by the University of the Basque Country under grant 146.345-TA 003/97.

REFERENCES

Åström, K.J. (1982). *Ziegler-Nichols Auto-Tuners.* Report from Lund Institute of Technology in Sweden (department of Automatic Control). Coden: Lutfd2/(tfrt-3167)1-025/(1982).

Åström, K.J., T. Hägglund, C.C. Hang and W.K. Ho (1993). Automatic Tuning and Adaptation for PID controllers. A survey. *Control Engineering Practice,* **1**, n° 4, pp. 699-714.

Åström, K.J. and T. Hägglund (1995). *PID Controllers: Theory, Design and Tuning.* ISA.

Åström, K.J. (1996). Tuning and Adaptation. *Plenary Paper in Pre-prints of the 13th world congress of IFAC.*

Bialkowski, W.L. (1993). Dreams versus reality: A view from both sites of the gap. *Pulp and Paper.* Canada, **24**, n° 11.

Buhr, R.J.A. (1984). *System Design with Ada.* Prentice Hall, Englewood Cliffs, NJ.

Cox, C.S., P.R. Daniel and A. Lowden (1997). Quick tune: A reliable automatic strategy for determining PI and PPI controller parameters using a FOPDT model. *Pre-prints of the 4th IFAC Workshop on Algorithms and Architectures for Real-Time Control.* Vilamoura (Portugal).

Ender, D.B. (1993). Process control performance: Not as good as you think. *Control Engineering,* **40**, n° 10, pp. 180-190.

Gerry, J.P. (1987). A Comparison of PID Control Algorithms. *Control Engineering,* March 1987, pp. 102-105.

Kaya, A., Scheib, T.J. (1988). Tuning of PID controls of different structures. *Control Engineering,* July 1988, pp. 62-65.

López, A.M., Miller, J.A., Murrill, P.W., Smith, C.L. (1967). Tuning controllers with error-integral criteria. *Instrumentation Technology,* **14**, n° 11, 1967.

Morilla, F. (1990). Controladores PID: algoritmos y estructuras. *Automática e Instrumentación,* July 1990, n° 204, pp. 131-136.

Nishikawa, Y., Sannomiya, N., Ohta, T., Tanaka, H. (1984). A method for Auto-tuning of PID Control Parameters. *Automatica,* **20**, n° 3, pp. 321-332, 1984.

Reference Manual for the Ada Programming Language. *ANSI/MIL-STD-1815A* (1983). DoD United States of America. Silicon Press.

Rovira, A.A., Murrill, P.W., Smith, C.L.(1969). Tuning controllers for set-point changes. *Instruments and Control Systems.* December, 1969.

Smith, C.A. and A.B. Corripio (1997). *Principles and Practice of Automatic Process Control.* 2nd Edition. John Wiley & Sons, Inc.

Schrönberger, F. and E. Poli (1997). An auto-tuning PID controller. *Pre-prints of the 4th IFAC Workshop on Algorithms and Architectures for Real-Time Control,* pp. 105-110.

Wellstead P.E. (1983). The ball and hoop system. *Automatica.* Vol 19. 401-406.

Ziegler, J.G., Nichols, N.B. (1942). *Optimum setting for automatic controllers.* Trans. ASME, 64, pp. 759-768, 1942.

AN OBJECT LANGUAGE WITH STATES: SOL

TINE Houari, and
COMMERÇON Jean-Claude

Laboratoire d'Ingénierie de l'Informatique Industrielle (L3I),
INSA, Bat 502, 20 Avenue A. Einstein, 69621 Villeurbanne Cedex, FRANCE.

Abstract: Synchronous reactive systems continually react with their environment. Based on formal hypothesis and complete semantics, this approach brings rigor and flexibility. The aim of some research is to specify a common code which can be used as an intermediate code or gateway for the present synchronous languages; this paper is consistent with this spirit. SOL (State Object Language) is proposed and its concepts are presented. From an object code of synchronous language, the equivalent SOL program is generated. The result is a clean program with a high abstraction level; this allows the application evolution and maintenance in an incremental way. *Copyright © 1998 IFAC*

Keywords: object-oriented programming, synchronous theory, finite automata, states, statecharts.

1. INTRODUCTION

Object analysis methods, object design methods and object programming languages, are more benefit than the classical ones thanks to the abstraction, encapsulation, reusability, maintainability, hierarchy... They are used more in several areas: management, networks, real-time systems, DBMS, etc. But the state notion is not well defined yet. For example, water is a substance that may be a liquid, a solid or a gas. The liquid —solid or gas— water state is a proper object. Thus, the state is modelled by an object —this is the main difference with similar approaches such as the *modes* (Taivalsaari, 1993) or the *states* (Lecoeuche, 1994). This involves introducing a new relation called *state relation* allowing one to connect an object-with-states to its state-objects. The new state relation cohabits and completes the inheritance relation, so it is possible to model some problems without losing the polymorphism.

A state can be defined as some information which sums up all the useful past of the object, that is to say, this information is sufficient to assure the correct object behaviour in the future. Rumbaugh (1991) gives this definition : "A state is an abstraction of the attribute values and links of an object". So, *ordinary* objects, as encountered in Smalltalk-80 (Goldberg and Robson, 1983) for example, can be seen as objects with states, but with only one state, that is to say, they have only one behavior. Common object oriented programming languages, such as Smalltalk-80, Eiffel (Meyer, 1988), C++ (Stroustrup, 1994), etc., own no special features to handle (multiple) state-object properly. CCS (Milner, 1989) and CSP (Hoare, 1985), focus on the events and the processes, so an object is the *a posteriori* gathering of some processes, and the object notion does not appear clearly; moreover, there is no possible hierarchy notion allowing good design. Lotos (Iso, 1988) shares many things with CCS or CSP, furthermore it extends Finite State Machine (FSM) declarations, but it does not allow neither object hierarchy nor state hierarchy. Petri nets with inhibitor arcs have the Turing machine

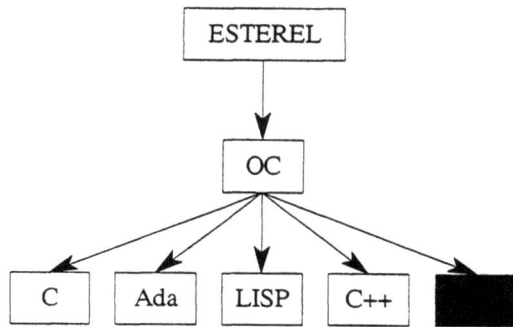

Fig. 1. Process of code generataion.

power (Peterson, 1977). The objects are partitions of the graph, this means that an object is a set of places; the object events are the labels of the object inner transitions and the labels of the transitions with an arc leading to the object; the state is the ordered collection of the token numbers in object places; so it is difficult to consider a state as an object distinct of the object the places represent; also, the state behaviour does not appear clearly because we must look for the enabled transition in a particular state.

For objects with dynamic behaviour —possibly computed— the actor model (Agha, 1986) can be used, but this approach lacks widespread commercial products and also, these kinds of objects can be forgotten at first glance if engineering is the main concern. The synchronous approach (Benveniste and Berry, 1991) —represented by the language such as: ESTEREL, LUSTRE, SIGNAL, etc. — is used for programming the reactive systems[1]. The synchronism hypothesis considers that the system response is instantaneous because reaction time is null. This approach allows one to treat the time with rigor and to combine parallelism and determinism.

ESTEREL is an imperative language enriched by concepts allowing the specification and the programming of real-time systems (Berry, *et al.*, 1987). It does not have the features of the data structuring and computing (Phan, 1992). The compilation of an ESTEREL program generates the object code called *OC* which is not directly executable. This code defines a state machine that expresses the reactive behaviour of the application. This machine is completely deterministic, so the logic correctness of the application can be checked.

The transition from OC to an executable code requires two steps, see figure 1 (André, *et al.*, 1991):

---

[1] The reactive systems can be programmed with an FSM. The advantages are: determinism and good performance at the execution. However, when the system is complex, the FSM size can become huge. Moreover, small modifications in the system specification can lead to total change the machine structure.

- The translation from OC to a host language (C, C++, Ada, LISP, etc.),
- The construction of an execution machine over this automata in order to establish some links between the synchronous core and the environment.

The object concepts are introduced from OC. The original concepts of the SOL language are located in the black box of figure 1 and are presented in the following section.

2. OBJECT MODEL BASED ON STATES

It is a general approach which is based on the explicit definition of the states according to the domain created by the *state-attributes* of a class. Unlike the modes (Taivalsaari, 1993) and the states (Lecoeuche, 1994), the states are considered as proper objects, see (Tine and Commerçon, 1997b, 1998b) for motivations. Therefore there are three kinds of objects: the *object-with-states*, the *state-objects* and the *conventional objects*. The conventional objects have no *states-attributes* (only *non-states-attributes*). The object-with-states have some state-attributes (and some non-states-attributes, eventually) on which a division of the domains they are forming is operated; the domains are matched with an *Invariant* (Meyer, 1988). The leaves of the state-relation hierarchy are the state-objects. There is only one version of object-with-states attributes which are shared by all their state-objects. The invariant is a function which maps the state-attributes of its object-with-states (direct or indirect) to the domains. The invariant defines the domain in which it is allowed to evolve. Object-with-states and state-object are linked by the state-relation. For example the notation *Water↓ (Liquid, Solid, Gas)* describes the object-with-states *Water* with *Liquid*, *Solid* and *Gas* as state-objets. The *state-relation* (denoted by "↓") includes the inheritance relation semantics (denoted by "↑"). An object can be an object-with-states and a state-object: it is a superstate in the statecharts terminology. This allows one to build up a hierarchy of states to specialise behaviours (Tine and Commerçon, 1997a).

The *modifiers* are the methods which alter at least one state-attributes. The *accessors* are the methods which never modify the state-attributes, but which may modify the non-state-attributes. The modifiers always hold the *become* feature contrary to accessors.

The *become* feature allows to achieve the transition, it must be executed, automatically, after a modifier. It has at least the semantics encountered in Smalltalk, in the actors of Agha (1986) or in the modes. The current state-object will be transformed into another object (among the list of objects indicated in parameter of *become*). If the successor object is explicitly specified then it will say *closed modifier*. If the *modifier* has a *become(NULL)* feature, then it will say *open*; and will need to be dynamically *closed*. This process is a major one for generic components since it allows the manipulation of state machines with dynamic structures which can evolve during execution (Tine and Commerçon, 1998b). The state management is implicit (become must indicate a valid successor), and explicit (the object-with-states can find the successor, state-object, thanks to the invariants and the state-attributes). If the indicated successor does not correspond to state attributes then an exception is triggered.

Other state approaches (Kafura and Lee, 1989; Taivalsaari, 1993; Matsuoka, *et al.*, 1993; Gangopadhyay and Mitra, 1993; Lecoeuche, 1994; Paech and Rumpe, 1994) cannot model some problems without losing the state polymorphism.

For example, triangle is an object with caracteristics such as: isosceles, right-angled, right-angled-isosceles, equilateral or any other types. This example cannot be modelled with the previous approaches, because the link between (right-angled, right-angled-isosceles), (isosceles, right-angled-isosceles) or (isosceles, equilateral) is broken. Here is the solution thanks to state relation.

```
TRIANGLE↓(REC, ISO, ISO-REC, EQUI, ANY)
ISO-REC↑(ISO, REC)
EQUI↑(ISO)
```

The first relation denotes the machine and its states. The two other relations allow to keep the inheritance links, hence the state polymorphism. This solution is allowed due to: 1) the introduction of the state relation and 2) the promotion of the states to the status of objects. The following relation: $ISO\downarrow(ISO_{12}, ISO_{13}, ISO_{23})$, can specialise the state-object ISO, so the object-with-states TRIANGLE can be used on different level of ramifications (in ISO_{ij}, vertex i = vertex j). For details concerning the state polymorphism, see (Tine and Commerçon, 1998a). The semantics errors such

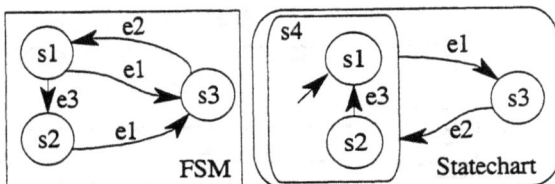

Fig. 2. Correspondance FSM-Statechart.

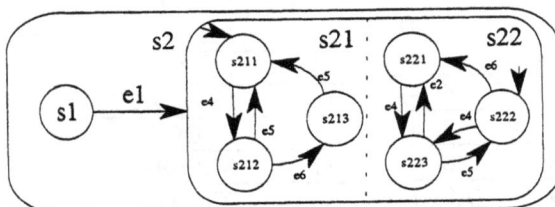

Fig. 3. Decomposition with an *AND* function semantics.

as: TRIANGLE↓(REC, ISO, ANY) must be detected because a right-angled and isosceles triangle would be in the two states, ISO and REC. This is due to the intersection of the right-angled domains and the isosceles domains which is not empty. Thus the state-object ISO-REC must be added. These concepts are implemented in the SOL language.

3. STATECHARTS

Statecharts (Harel, 1987), popular for modelling system behavior in the structural analysis paradigm, are used as formalism to represent the behavioural semantics of SOL. Statecharts can manage the explosion problem of FSM based system description into *OR* and *AND* function composition of smaller FSMs.

Statecharts extend finite state machines in order to take account into three notions: hierarchy, concurrency (orthogonality) and communication by broadcasting. They form the core of many works, UML, Statemate (Harel and Naamad, 1996), Rhapsody (Harel and Gery, 1997), ObjChart (Gangopadhyay and Mitra, 1993). SOL can be seen as an integration of this idea. The concepts of statecharts are briefly summarised in this section.

3.1 Superstate and default entry

Figure 2 represents the correspondence between classical FSMs and statecharts. So, *s4* is a *superstate* and *substate s1* is its *default entry*. That is, it moves transitionally from state *s3* to state *s1* in the event *e2*.

3.2 Refinement

The process of decomposing a state into substates is called *refinement*. The semantics of that operation is an *OR* function. For example, in figure 2, when the statechart is in state *s4* at the higher level of abstraction, it is really in either *s1* or *s2* at the lower level of abstraction.

3.3 Orthogonal decomposition

The semantics of this kind of a decomposition is an *AND* function. It is represented by splitting box using dashed lines. The example of figure 3 is borrowed from Davis (1988); the state s2 is decomposed into s21 and s22 by an *AND* function semantics. Hence, when the statechart is in state s2 then it is in states s21 *AND* s22. And since s21 and s22 are machines then it is really in their default entry, namely s211 and s222. On receipt of event e1, in state s1, it enter both states s211 and s222 simultaneously. If event e4 now arrives, the statechart transition from states s211 and s222 simultaneously into states s212 and s223. Now, if e6 arrives, it transition from states s212 and s223 simultaneously into states s213 and s223, only the s21 machine changes.

4. MODELING

The semantics of an object-with-states is a statechart, and the leave state-objects are states. An object with a double status, object-with-states and state-object is a superstate. The specification of initial/final states is given by the state relation. Orthogonal decomposition expresses the multiple inheritance. Each statechart corresponding to an ESTEREL module will be modelled by an object-with-states and each state by a state-object. The method name corresponds to an event, and its body to an action triggered by that event. The methods of one class are inputs of the statechart. A method called inside an action corresponds to an output of statechart which must be sent to another machine. The accessors can be executed simultaneously.

The inheritance between object-with-states is mainly based on the simple rules given for statecharts. The combination of these rule gives a complex object-with-states but this complexity is encapsulated into simpler object-with-states. For details concerning inheritance, object semantics, relation semantics, etc. see (Tine and Commerçon, 1997a).

The different (synchronous) object-with-states are connected through the conventional statements. Thus, the assignment (A.x:=B.y) means that the output signal y of B is connected to an input signal x of A —y can be a function but not a procedure. The type system of compiler checks the signal (type) compatibility. The entire application is composed by a set of object-with-states, each one having one and only one active state-object; the state of the system at

time t is the aggregation of these state-objects. Only active state-objects are sensitive on events. An event is defined by a sender, a receiver, a name and a list of parameters. If the receiver is not specified then the event is broadcasted to all active state-object.

5. EXAMPLE

The compilation of an ESTEREL program generates a file containing a set of tables, a finite state machine and actions; this machine is completely deterministic. SOL can also model non-deterministic machines. The following example come from (Phan and Lecompte, 1994), it corresponds to an *Interface de Transmission Non Permanente* (ITNP) which is a sub-set of an environment simulator, see appendix 1 for ESTEREL program. The information of OC code generated from this program are given in table 1, the actions are given in table 2.

Translating the previous OC code in SOL is automatic (cf. appendix 2). This is the passage from event-based to object-based programming. The different states are modelled by classes and linked with the state relation "↓". So, IDLE, WAIT_CONNECT, WAIT_INIT, INIT and RUN are the bricks which built the object-with-states ITNP through the state relation that plays the role of the glue. The (Inv$_i$(sa)) expression after a state-object is an invariant of these classes. A method not belonging to the class interface triggers an exception; then it is not necessary to specify the error case in table 1. The actions common to several state-objects are implemented in object-with-states. State changing is done by the language that verifies the matching between the successor state in the become primitive and the state-attributes. Notice that the PO_SEND_Req and the SU_SEND_Req events trigger the same action. So, to optimize the code, a solution could be: the event PO_SEND_Req triggers the SU_SEND_Req event. This is possible in this case because SU_SEND_Req is an accessor.

Table 1 Transitions. From a given state, an event triggers the couple (execution of an action, moving to a new state). (—, —) corresponds to an unrecognised event.

Events \ States	Idle (0)	Wait_Connect (1)	Wait_Init (2)	Init (3)	Run (4)
SU_INIT_Req	(A1, 1)	(—, —)	(—, —)	(—, —)	(—, —)
T_CONNECT	(—, —)	(A2, 2)	(—, —)	(—, —)	(—, —)
INIT_I_Dem	(—, —)	(—, —)	(A3, 3)	(A3, 3)	(A3, 3)
INIT_I_Ack	(—, —)	(—, —)	(—, —)	(A4, 4)	(—, —)
PO_SEND_Req	(—, —)	(—, —)	(—, —)	(—, —)	(A5, 4)
SU_SEND_Req	(—, —)	(—, —)	(—, —)	(—, —)	(A5, 4)
MSG_PAOCR	(—, —)	(—, —)	(—, —)	(—, —)	(A6, 4)
MSG_SUPER	(—, —)	(—, —)	(—, —)	(—, —)	(A7, 4)
SU_DECX_Req	(—, —)	(A8, 0)	(A8, 0)	(A8, 0)	(A8, 0)
T_DISCONNECT	(—, —)	(A9, 1)	(A9, 1)	(A9, 1)	(A9, 1)
T_LISTEN	(—, —)	(A2, 2)	(—, —)	(—, —)	(—, —)

Table 2 Actions. The corresponding names of actions used in Table 1.

Actions	Names
A1	INIT_LIAISON
A2	IT_CNX_IND
A3	INIT_I_Cnf_RTC + INIT_I_Cnf_RADIO
A4	IT_INITt_IND
A5	ITNP_ACTION
A6	PAOCR
A7	SUPERCOM
A8	XTI_DECX
A9	IT_DECX_Ind

The evolutivity and the maintainability of applications can be do from generated code. For example, the class RUN can be written RUN↓(RUN1, RUN2), in which RUN1 and RUN2 extend the application to take into account two types of new services. This way, it is not necessary to rewrite and to recompile the ESTEREL program leading to a new global machine, because modifications are local. The evolutivity and the maintainability of the applications are facilitated by the inheritance, the polymorphism, the overloading, the redefinition and the ramification of states, all available in SOL. The comprehensibility of the code and its high abstraction level is obvious. The C code of the ESTEREL module (ITNP_STRL) of appendix 1 generated by "AGEL" (Phan and Lecompte, 1994) for example is very difficult to read, and to maintain. Furthermore, no extensions are possible.

6. CONCLUSION

SOL is not a language implementing statecharts but it is a general language based on the object paradigm that can model some objects through the state notion allowing to have multiple interfaces and giving a kind of object-oriented FSMs. Unlike other techniques that generate a low level code where states, transitions and structures of machines are hidden in details of programs and where the result machine is enormous because it is a Cartesian product of machines. The code generated saves the semantics of machines and where each one stands a part of the complexity. This way can be more advantageously used in lieu of OC code. It allows to maintain the applications without changing the initial structures. In this manner the migration of ESTEREL applications (or other state-based programs) into object-oriented one can be done smoothly and elegantly. Finally, about the reproach sometimes given about the explicit states approaches, say, the inheritance anomalies (Kafura and Lee, 1989; Matsuoka, et al., 1993) does not exist in SOL because the coding of transitions are not put in the code of modifiers, but by typing the states.

REFERENCES

Agha, A.G. (1986). *Actors: A Model of Concurrent Computation in Distributed Systems.* MIT Press.

André, C., J.P. Marmorat and J.P. Paris (1991). *An execution machine for ESTEREL.* ECC'91, Grenoble, France, July, Vol. 2, HERMES Edition, pp. 1672-1677.

Benveniste, A. and G. Berry (1991). *The synchrounos Approach to Reactive and Real-Time systems.* Proceedings of the IEEE, vol. 79, n° 9, September.

Berry, G., P. Couronne and G. Gonthier (1987). *Synchronous programming of reactive systems:* *An introduction to ESTEREL.* Research Report n° 647, INRIA-Sphia Antipolis, France.

Davis, A.M. (1988). *A comparison of techniques for the specification of external system behavior.* Comm. of the ACM, 31, 9, pp. 1098-1115.

Gangopadhyay, D. and S. Mitra (1993). *ObjChart: Tangible Specification of Reactive Object Behaviour.* Proceedings of ECOOPS'93, Kaiserslautern, Germany, July, pp. 433-457.

Goldberg, A. and J. Robson. (1983). *Smalltalk-80: The language and its Implementation.* Addison Wesly.

Harel, D. (1987). Statecharts: a visual formalism for complex system. *Science of computer programming,* 8, pp. 231-274.

Harel, D, A. Naamad, (1996). The STATEMATE Semantics of Statecharts. *ACM Transaction on Software Engineering Methodology,* Oct. pp.293-333.

Harèl, D. and E. Gery (1997) Executable Object Modeling with Statecharts. *Computer,* Jully.

Hoare, C.A.R. (1985). *Communicating sequential processes.* Prentice Hall.

Iso, ISO/IEC (1988). *LOTOS, A Formal Description Technique Based on the Temporal Ordering of Observational Behaviour.* International Standard 8807, International organisation for standardisation -Information Processing Systems - Open System Interconnection, Genève.

Kafura, D.G. and K.H. Lee (1989). *Inheritance in Actor based concurrent object-oriented languages.* Proceedings of ECOOP, pp. 131-175. Cambridge University Press.

Lecoeuche, H. (1994). *Intégration des aspects dynamiques dans le modèle objet.* Thèse de doctorat, INSA de Lyon, France.

Matsuoka, S., K. Taura and A. Yonezawa (1993). *Highly Efficient Re-use of Synchronisation Code.* Proceedings of OOPSLA, Washington, DC, USA, 26 Sep. 1 Oct., pp.109-126.

Meyer, B. (1988). *Object Oriented Software Construction.* Prentice Hall.

Milner, P. (1989). *Communication and Concurrency.* Prentice Hall.

Paech, B. and B. Rumpe, (1994). *A New Concept of Refinement used for Behaviour Modelling with Automata.* Formal Methods Europe, LNCS 873, pp. 154-175, und Bericht TUM-I9413.

Peterson, J.L. (1977). Petri nets. *ACM Computation Survey,* 9, 3, pp. 223-252.

Phan, S. (1992). *Programming Reactive Systems with Synchronous Languages : A survey latest developments.* EDF Technical Report, 92 NI J 0007.

Phan, S. and V. Lecompte (1994). *Synchronous languages: Study of a Concrete case with AGEL/ESTEREL.* EDF Technical Report, 94 NJ 00011.

Rumbaugh, J., M. Blaha, W. Premerlani, F. Eddy and W. Lorensen (1991). *Object-Oriented Modeling and Design.* Prentice Hall.

Stroustrup, B. (1994). *The C++ programming language*. 2nd edition, Addison Wesly.

Taivalsaari, A. (1993). Object-Oriented Programming with Modes. *Journal of Object Oriented Programming*, June, Vol. 7, pp. 25-32.

Tine, H. and J.C. Commerçon (1997a). *Objets à Etats*, Internal Report.

Tine, H. and J.C. Commerçon (1997b). *Un Modèle à Etats Orienté Objet*. LMO conference, session posters, Brest, France.

Tine, H. and J.C. Commerçon (1998a). *A State Model Ensuring the State Polymorphism*. to appear in IDPT'98.

Tine, H. and J.C. Commerçon (1998b). *Object with States in the Object paradigm*. Paper submitted to OOIS.

APPENDICES

Appendix 1

```
module ITNP_STRL:
input
    SU_INIT_Req, INIT_I_Ack, INIT_I_Dem,
    T_CONNECT, SU_DECX_Req, T_DISCONNECT,
    PO_SEND_Req, SU_SEND_Req, T_LISTEN
    MSG_PAOCR(integer), MSG_SUPER(integer);
relation
    SU_DECX_Req # SU_INIT_Req # INIT_I_Ack #
    INIT_I_Dem  # T_CONNECT # T_DISCONNECT #
    PO_SEND_Req # SU_SEND_Req #
    MSG_PAOCR # MSG_SUPER;
output
    INIT_LIAISON, IT_CNX_Ind, INIT_I_Cnf_RTC,
    ITNP_ACTION, INIT_I_Cnf_RADIO,
    PAOCR(integer), SUPERCOM(integer),
    XTI_DECX, IT_DECX_Ind, IT_INIT_Ind;
loop
    await SU_INIT_Req do
        emit INIT_LIAISON
    end;
    do loop do
        await [T_CONNECT or T_LISTEN] do
            emit IT_CNX_Ind
        end
        every INIT_I_Dem do
            emit INIT_I_Cnf_RTC;
            emit INIT_I_Cnf_RADIO;
            await INIT_I_Ack do
                emit IT_INIT_Ind
            end;
            [every [PO_SEND_Req or SU_SEND_Req] do
                emit ITNP_ACTION
             end
            ||
             every MSG_PAOCR do
                emit PAOCR(?MSGPAOCR)
             end
            ||
             every MSG_SUPER do
                emit SUPERCOM(?MSGSUPER)
             end
            ]
        end every
        upto T_DISCONNECT;
        emit IT_DECX_Ind;
    end loop
    upto SU_DECX_Req;
    emit XTI_DECX;
end loop

end module
```

Appendix 2

```
class ITNP {
    void INIT_I_Cnf_RTC(void)   {...}
    void INIT_I_Cnf_RADIO(void) {...}
    void XTI_DECX(void)         {...}
    void IT_DECX_Ind(void)      {...}

    public:
        state-attributes: sa1, sa2, ...;
        modifiers:
        T_CONNECT(); SU_DECX_Req();T_LISTEN();
        SU_INIT_Req(); T_DISCONNECT();
        INIT_I_Dem(); INIT_I_Ack();
        accessors:
        PO_SEND_Req(); SU_SEND_Req();
        MSG_PAOCR(integer p);
        MSG_SUPER(integer s);
} (Inv(sa))
```

```
class INIT {
    IT_INIT_Ind() {...}
    public:
    INIT_I_Dem()->become(INIT) {
      · INIT_I_Cnf_RTC();
        INIT_I_Cnf_RADIO();}
    INIT_I_Ack()->become(RUN) {IT_INIT_Ind();}
    SU_DECX_Req()->become(IDLE) {XTI_DECX();}
    T_DISCONNECT()->become(WAIT_CONNECT)
        {IT_DECX_Ind();}
} (Inv3(sa));
```

```
class RUN {
    ITNP_ACTION()       {...}
    PAOCR(integer)      {...}
    SUPERCOM(integer) {...}
    public:
    INIT_I_Dem()->become(INIT) {
        INIT_I_Cnf_RTC();
        INIT_I_Cnf_RADIO();}
    PO_SEND_Req() {ITNP_ACTION();}
    SU_SEND_Req() {ITNP_ACTION();}
    MSG_PAOCR(integer p) {PAOCR(p);}
    MSG_SUPER(integer s) {SUPERCOM(s);}
    SU_DECX_Req()->become(IDLE) {XTI_DECX();}
    T_DISCONNECT()->become(WAIT_CONNECT)
        {IT_DECX_Ind();}
} (Inv4(sa));
```

```
class IDLE {
    INIT_LIAISON() {...}
    public:
    SU_INIT_Req()->become(WAIT_CONNECT)
        {INIT_LIAISON();}
} (Inv0(sa));
```

```
class WAIT_INIT {
    public:
    INIT_I_Dem()->become(INIT) {
        INIT_I_Cnf_RTC();
        INIT_I_Cnf_RADIO();}
    SU_DECX_Req()->become(IDLE) {XTI_DECX();}
    T_DISCONNECT()->become(WAIT_CONNECT)
        {IT_DECX_Ind();}
} (Inv2(sa));
```

```
class WAIT_CONNECT {
    IT_CNX_Ind() {...}
    public:
    T_CONNECT()->become(WAIT_INIT)
        {IT_CNX_Ind();}
    SU_DECX_Req()->become(IDLE) {XTI_DECX();}
    T_DISCONNECT()->become(WAIT_CONNECT)
        {IT_DECX_Ind();}
    T_LISTEN()->become(WAIT_INIT)
        {IT_CNX_Ind();}
} (Inv1(sa));
```

```
ITNP↓(initial final IDLE, INIT, WAIT_INIT,
        WAIT_CONNECT,RUN)
```

PROGRESSIVE DOMAIN FOCALIZATION IN INTELLIGENT CONTROL SYSTEMS

Ricardo Sanz, Idoia Alarcón[†], Miguel Segarra*,
Jose A. Clavijo* and Angel de Antonio**

Universidad Politécnica de Madrid *
Instituto de Ingeniería del Conocimiento [†]
Madrid, SPAIN
e-mail: sanz@disam.upm.es

Abstract: Advanced information processing technologies are providing advanced controllers with capabilities to tackle control problems of intrinsic complexity. Methods to cope with the software-intensive control system development are needed, because quality issues in control software are of extreme importance. *Architecture based development* provides a clear pathway to quality; focusing on component based control system development. In the case of intelligent controllers, advanced control modules (expert, fuzzy, neural, etc.) can be built to be reusable. In the ideal situation, application construction would be a matter of plugging in the controllers and then making a customization work of them. *Copyright © 1998 IFAC.*

Keywords: Intelligent Control, Industrial process control, Architecture, Software Engineering, Distributed Computer Control Systems.

1. INTRODUCTION

Complete autonomy is the final –quite far- target for most of us. Autonomy means, for a system, to be capable of performing the assigned tasks without recurring to external help sources. Our approach to autonomy is constructive. This means we are trying to achieve a method to build up the required degrees of intelligence to perform a specified collection of tasks using a modular-functional approach (Sanz et al., 1994).

1.1 The Concept of Autonomy and Intelligence

The characterization of what is *autonomy* or what is *intelligence* has been a matter of disagreement during decades. For us the terms have simple engineering meanings, because we are aware that these terms refer to attributes that are not classic -boolean valued- concepts. The concepts of *autonomy* and *intelligence* are pure *fuzzy* concepts. They refer to the degree of presence of some property in a system, always bearing in mind that these properties should be analyzed from the point of view that they have only sense in relation with task accomplishment of the system.

This means that *autonomy* and *intelligence* are not intrinsic properties but relations among systems that perform activities; and they are not absolutely measurable properties of an isolated system. What we can conclude is that one system behaves more intelligently or more autonomously than other. Intelligence is related to the capability of solving complex tasks. This means that intelligence is considered greater in those systems than can solve more complex tasks. Autonomy is related to the capability of performing the tasks in an independent, helpless way.

We conclude that intelligence relations should be established as a function of quality of results for specific tasks. So we should talk about intelligence in performing specific tasks (finding objects, controlling arms, ordering boxes, perceiving targets, etc.) and not as global properties of systems.

The same type of analysis can be done for autonomy. We must say that A is more autonomous than B performing task T and not that A is an autonomous system[1].

When we use the expression *intelligent autonomous systems* we mean systems that solve more complex tasks than others, that obtain better results and that do

---

[1] Obviously, in ideal conditions, there will be *absolutely* autonomous and intelligent systems.

need less help from external entities to achieve their results.

In accordance with some trends in the artificial intelligence community regarding the embodiment need for true intelligence, we believe that the true search of artificial intelligence is the search done by control engineers. Control systems are minds for

2. WHAT IS ARCHITECTURE ?

There are lots of definitions of software architecture. Almost all of them agree in the basic idea that it is a design of something. The disagreement appears when trying to identify that *something*.

As a thought experiment, think in the design of a

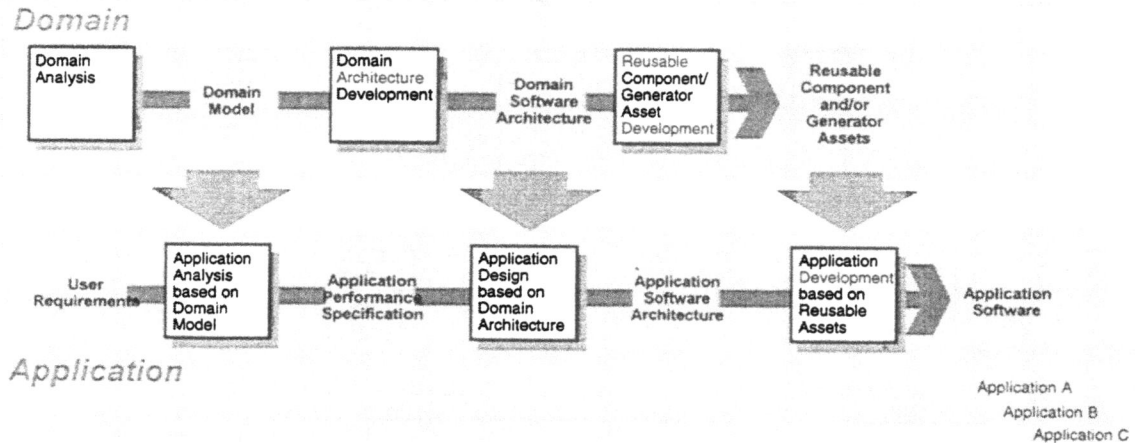

Fig. 1: The ARPA STARS architecture based development process.

artificial bodies; minds that are being built –this is the *artificial* part- to perform well in the tasks assigned to the entity –this is the *intelligence* part.

1.2 Intelligent Control Systems

The basis for autonomous intelligent behavior is the control system of the autonomous entity[2]. The control system of an intelligent autonomous machine is an artificial mind in the sense used by Franklin (1995).

The term *Intelligent Control Systems,* used by most researchers in the automatic control community, groups an heterogeneous collection of control systems whose complexity ranges from the simplest fuzzy controller to an integrated complete control system for a whole plant. For many authors intelligent control is related to the type of technology employed in its construction (usually expert, fuzzy or neural systems).

For us, it means capability to perform tasks achieving better results. Normally, depending on the complexity of the task, the soft computing systems attain better results than conventional controllers. However, in many cases, conventional control systems (i.e. mathematical controllers) perform better. In these situations, we will say that these controllers are more intelligent than their *"intelligent"* counterparts.

complex system: it is -almost ever- a hierarchical description. It has a basic depiction at the root node and progressively clarifying descriptions at progressively deep echelons. It is the same for complex software systems. For us, the architecture of such a system is some subtree rooted in the root of the hierarchy. The actual extent is a matter of preferences. For some, even the last leaf of the tree is part of the architecture. For others, only the first echelon of nodes.

2.1 Architecture Based Development

The software engineering community, and in particular, the autonomous systems community, is becoming progressively aware of the importance of software architecture. Architecture based development is seen as the "good way" to achieve high levels of software quality. Especially those referred to non-functional requirements for the software systems.

As Clements (1996) say, architecture based development offers promising perspectives that we are trying to achieve for autonomous control systems:

- Systems can be built in a rapid, cost-effective manner by importing (or generating) large externally developed components.
- It is possible to predict qualities analyzing the architecture
- Developing product lines sharing architectural design
- Separation of interface and implementation at the component level
- Advantages of design variability restrictions

---

[2] Apart from the intrinsic physical capabilities to perform the tasks assigned to it.

Based on *domain analysis*, generic architectures are proposed to address a whole bunch of applications in a specific domain. See in Figure 1 to the architecture based process proposed by the ARPA STARS project. Development is divided in two separate phases:

- *Domain engineering*: One effort shared by a complete collection of instances of a product line. Its final product is threefold: a domain model, a generic architecture and some reusable components.
- *Application engineering*: The process of getting the real applications (the products in the product line).

A great level of effort has been put in the last years in addressing the architectural problem for all types of software systems. In the case of autonomous systems, many generic architectures have been proposed and components developed. Stunningly, most of them use a layered approach based on three levels of control.

2.2 The magic of three

The layering concept is quite natural from the engineering point of view. What is not so immediate is that in many cases the proposed number of - relevant- layers is three. **3** seems to be a magic number for layered intelligent control, why?

Some examples at hand of these 3 tier layered architectures are:

HINT: Heterogeneous Integration Architecture (HINT, 1994). The three intelligent levels specified for the HINT blackboard are:

- *Operational*: In charge of continuous state maintenance.
- *Tactical*: In charge of task level related activities (by example problem solving).
- *Strategic*: In charge of attaining high level objectives: production optimization, quality, safety, etc.

The reasoning and deciding component (RDC) of the COSY agent architecture proposed by Haddadi (1996) is decomposed in three layers:

- *Strategic*: What to do.
- *Tactical*: How to do it.
- *Executional*: When to do it.

3T is a three tier (so the name) architecture for intelligent robots (Bonasso et al., 1996). The layers are:

- A dynamically reprogramable set of reactive skills coordinated by a skill manager
- A sequencing capability to activate or deactivate skills to accomplish specific tasks.
- A deliberative planning capability with deep reasoning.

Fischer (1996) proposes with INTERRAP an agent architecture with three layers:

- A behavior based layer
- A local planning layer
- A cooperative planning layer

The theory of Hierarchically Intelligent Control systems proposes the following layers:

- An execution level: hardware and software controllers.
- A coordination level
- A knowledge based organization level

The contents of the layers seem quite similar in most cases but without a clear relation between them. The triple layering seems as an externalization of our modes of behavior:

- · Reaction: A reaction follows immediately a perception.
- Task: A sequence (possibly a tree) of terns perception→action→follow-up
- Plan: Neither real perception nor real action. Making task trees for the future.

2.3 Still some problems

In the case of the layered architectures mentioned, when going down to real applications in the industry, the layering concept is still maintained as a principle. But in the final implementation, the layering is not as clear as it should be. It is not easy to say neither the number of layers nor their relations. In many real implementations, engineers solve real problems patching the clean layered design, implementing ad hoc solutions to specific sections of the system under development.

Component reusability is not as high as it should be if we give credit to the claims of the architectural method connoisseurs. The final niches where these systems are applied are too narrow to reuse generic components thoroughly. Great efforts in adaptation and construction of specific elements are needed to be able to reuse even the generic architectural design.

3. THE COMPLEXITY OF INTELLIGENT CONTROL SYSTEMS

The conventional approach to intelligent control system is to employ an off-the-self architecture, normally based on the type of tool employed (For example BB if using HINT or distributed if employing RT-Works).

But there are lots of matters to solve:

- *Feature issues*: Real-time, distributed, heterogeneous, intelligent, etc.
- *Development issues*: analysis, design, implementation, integration, testing, maintenance, etc.

55

- *Support issues*: Operating systems, protocols, artificial intelligence tools, etc.

New trends in software architecture for complex controller implementation are appearing to tackle with these problems.

4. A CONSTRUCTIVE APPROACH TO ARTIFICIAL MINDS

Our research is based in a simple hypothesis: that foundations of intelligent behavior can be independent of the task at hand. This leads to a component based, modular approach to artificial mind construction (Alarcon et al., 1994). We think that there are mechanisms for minds that can be reused in several tasks. The level of task independence can vary from mechanism to mechanism, but we believe that they can be constructed with modularity and composability properties (van der Linden and Muller, 1995).

Whole autonomous systems, elementary mechanisms of mind or basic implementation technologies, can be managed with a design space (Garlan, 1995) approach in which different orthogonal properties are used to classify mind designs. Some of the dimensions of this design space are:

- **Analysis vs design**: Some artificial mind designs are based on models of biological systems; they are built using an analytical approach. From the other side, the type of minds we want to address are designed minds; models of minds that do not resemble known models of biological systems, but that have been designed for a specific purpose.

- **Top-down vs bottom-up**: Some minds are built using high-level cognitive approaches –for example expert systems- and others are built using subcognitive approaches –for example neural networks. In the last case, some form of learning or adaptation is necessary to achieve specific functionality.

- **Simplicity vs complexity**: It is not the same a mind for a welding robot that a mind for a submarine exploration robot. Complexity, as mentioned before, is the main issue of intelligence.

- **Constructive vs evolutive**: Minds can be built from their parts producing a somewhat static final structure or can be grown up from stupid stuff using some form of stuff evolution.

We want to address the issue of constructiveness of complex, design based minds. We want to use a globally top-down designed mind architecture, using a constructive approach to the mind component integration problem. We believe that the constructive way is the most effective way to achieve really complex artificial minds, in terms of money spent in their construction.

This does not mean that we neglect the use of evolutionary or bottom-up models. We think that these models are extremely useful –and even better in some cases- to build specific mind components, but their usefulness decreases when the global mind complexity increases; obviously due to task complexity.

Our proposed method to autonomous intelligent system construction is based on an almost classical sequence of steps:

- Specification of complex tasks
- Analysis and decomposition in canonical tasks
- Design of task solving components

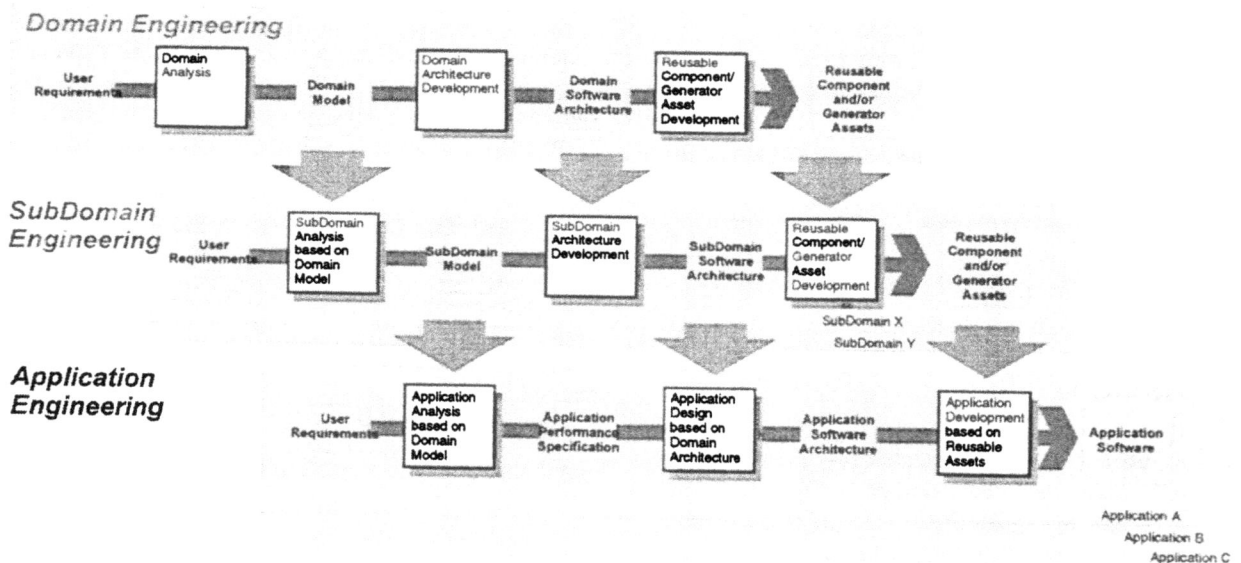

Fig. 2: Progressive domain focalization. A way to deep reuse and domain generalization.

- Design of integration components
- Construction of task components
- Construction of integration components
- Component integration

Our approach is near to what Stein calls modular functionalism, because we think that evolutionary or not modular approaches do not scale well from the toy problems in which they are usually employed.

5. A REUSE-BASED CONSTRUCTION MODEL

The model we propose for artificial minds is based in the agency concept. We have selected this model not because it resembles the human mind but because it offers a whole bunch of engineering alternatives, based on its intrinsic flexibility.

When searching for a technology to implement autonomous systems, we must identify the requirements posed to that technology. Requirements are so varying that no simple technology can comply with this. Flexibility, extensibility and scalability are desirable properties for any type of software. In the case of background software for autonomy, they are not desirable but indispensable properties.

5.1 Progressive Domain Focalization

The way we have envisioned to support this constructive approach is to provide a framework for intelligent component reuse where flexibility (adaptability and extensibility) are main concerns for component developers.

This method is based in what we call *progressive domain focalization*. The concept isn't new at all, it is a classical object-oriented concept. The main idea is that the separation into domain and application proposed by most architecture base development methods is valid only in crisp domains. Autonomous systems domain is intrinsically fuzzy. Domain analysis cannot be performed adequately if the domain is unclear.

The basic idea is that what you can do is to progressively focalize domains until final, deployable applications are reached. But components can be produced at any subdomain engineering level. In Figure 2, a development method derived from the STARS model is presented with one intermediate subdomain. The labeling of layers as *domain*, *subdomain* and *application* is done to match it against the original STARS model. In reality, they should be labeled always as domain activities, being application engineering another term to refer to the narrower domain engineering (or the deployment domain).

6. ICA: A CORBA BASED SCALABLE APPROACH

The approach we have taken to achieve this component development in progressively narrow domains, is the implementation of an extensible collection of agents upon an integrative middleware using the CORBA model of heterogeneous integration. The architectural methodology is based in the use of these agents to implement control systems *design patterns* that address specific problems in an autonomous controller environment.

In order to provide a coherent agent implementation framework we have developed ICa. It offers technology to adapt generic architectural design patterns to specific implementations, tailoring artificial intelligence based architectures to the concrete tasks posed to the autonomous system. The system can then be built by means of composition of functional modules atop the basic integration infrastructure. The modules are constructed by instantiation and adaptation of task oriented object frameworks to implement the collection of interacting agents that will constitute the system mind.

A sample collection of agents can be seen in figure 3. This is the inheritance tree of agents used to implement fuzzy systems for control and data validation purposes in the ESPRIT DIXIT project.

The agents shown in the figure are:

- **ICa Core Agent**: basic behavior for all ICa compliant agents.

- **ICa MultiThreading Agent**: Extensions to basic behavior to provide a MT behavior to ICa agents.

- **ICa Real-Time Agent**: Extensions to manage real-time issues: fixed priority request scheduling, QoS in transports, etc.

- **ICa Fault Tolerant Agent**: Extensions to provide FT behavior: replication, mobility, persistence, etc.

- **DIXIT Agent**: Basic behavior for the subdomain of strategical continuous process control. The target domain of DIXIT.

- **FL Agent**: A fuzzy inference engine and fuzzy knowledge base management system.

- **RiskMan FFV Agent**: Fuzzy technology based filtering and validation of data coming from plant (RiskMan is the name of a DIXIT demonstrator for chemical plant emergency management).

- **Fuzzy Control Agent**: A fuzzy control agent for domains more generic than DIXIT (Sanz et al. (1996).

7. CONCLUSIONS

Domain focalization is not linear nor tree structured. It is a directed graph of domain analysis nodes with focalization relations. This graph can be mapped to the implementation of generic agents, that are designed and implemented bearing in mind user requirements for a specific domain. With users we mean potential users of the component, not only final users of a deployed application.

ICa is still under development of basic components, in particular we are working in the implementation of RT an FT agents, addressing the problems of

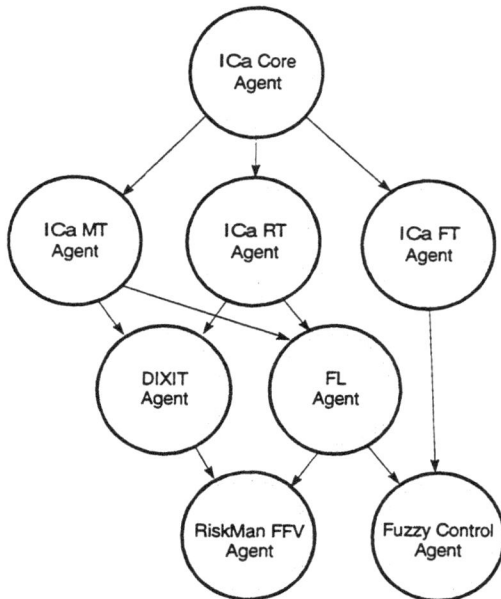

Fig. 3: ICa Generic Agent. Promoting component reuse in progressive domain focalization architecture based development.

functional composability of agents.

The DIXIT project under which this development has been done will finish in July 98.

8. REFERENCES

Alarcón, M.I., P. Rodríguez, , L.B. Almeida, R. Sanz, L. Fontaine, P. Gómez, X. Alamán, P. Nordin, H. Bejder and E. de Pablo (1994). Heterogeneous Integration Architecture for Intelligent Control.; Intelligent Systems Engineering, Autumn 1994.

Bonasso, R.P., D. Kortenkamp, D. P. Miller and M. Slak. (1996) Experiences with an Architecture for Intelligent, Reactive Agents. In *Intelligent Agents II*. M. Wooldridge, J.P. Müller and M. Tambe. Springer Verlag, pp. 187-202.

Clements, P.C. (1996) Coming Attraction in Software Architecture. CMU/SEI-96-TR-008.

Fischer, G. (1994) Domain-Oriented Design Environments. Fischer, G. *Automated Software Engineering, Vol. 1 No. 2*, pp. 177-203.

Franklin, S. (1995) *Artificial Minds*. Stan Franklin. MIT Press.

Garlan, D. (1995) Research Directions in Software Architecture. *ACM Computing Surveys*, Vol. 27, No. 2.

Haddadi, A (1996) *Communication and Cooperation in Agent Systems*. Afsaneh Haddadi. Springer.

HINT (1994) *HINT Manual for System Developers*. HINT Consortium, 1994.

Van der Linden, F.J. and J.K. Muller (1995) Creating Architectures with Building Blocks. *IEEE Software*, November 1995.

Sanz, R., R.Galán, A.Jiménez, F.Matía, J.Velasco and G.Martínez (1994) Computational Intelligence in Process Control. *ICNN'94, IEEE International Conference in Neural Networks*. Orlando, USA.

Sanz, R., F.Matía, R.Galán and A. Jiménez (1996) Integration of Fuzzy Technology in Complex Process Control Systems. *FLAMOC'96*. Sydney, Australia, 1996.

A PROPOSAL TO INCREASE THE FLEXIBILITY OF MANUFACTURING CELL CONTROLLERS

Jesús Sánchez, Maricela Quintana

Instituto Tecnológico y de Estudios Superiores de Monterrey
Campus Estado de México, Depto. de Ciencias Computacionales
{jsanchez,mquintan}@campus.cem.itesm.mx

Abstract: In this paper we describe the software and architecture components of a
parallel controller for flexible manufacturing cells. This controller, called PARDICO
(PArallel and Distributed Intelligent COntroller), is a multitransputer machine with
redundant RS232 interfaces and a reconfigurable topology. The software is composed
of three layers: physical, data link/network and application. Each one of these layers,
and their relationship with hardware, is described. The controller hardware and some
of the software, is fully functional and has been tested in real manufacturing cells. The
results show that parallel machines and IA techniques may greatly enhance flexibility
in manufacturing cells. *Copyright © 1998 IFAC*

Keywords: Parallel architectures, intelligent control, manufacturing systems.

1. INTRODUCTION

A *manufacturing cell* is a set of robots and machines used to build products in industrial environments. A *flexible* manufacturing cell must have
the following additional characteristics:

- **Planning flexibility:** the ability to recognize plans in several formats.
- **Reconfigurability:** the ability to change its
  configuration in order to manufacture new
  products.
- **Fault detection and isolation:** the ability
  to function properly after the occurrence of
  faults.

Several authors have proposed the use of parallel
machines to build flexible manufacturing cell controllers (Sánchez, 1996; Sinha *et al.*, 1996; Gray,
1996). In (Sánchez, 1996) we proposed the use of
a parallel architecture called PARDICO (PArallel
and Distributed Intelligent COntroller), to fulfill
this goal. In this paper we show how this controller

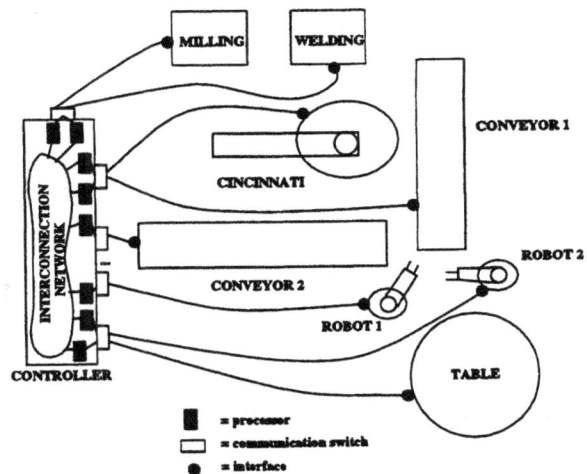

Fig. 1. A manufacturing cell controlled by PAR-
DICO

can be built, and how it achieves the reconfigurability and fault tolerance properties needed for
flexible manufacturing cells.

In order to achieve greater flexibility, we propose
a multi-agent architecture for the controller. In
the literature there are already several propos-

* *This work is supported by Conacyt-NSF Grant C016A.*

als to use multi-agent systems for manufacturing cells. In (Parunak, 1987) the authors use multi-agent communication techniques to coordinate a manufacturing cell. They have found that negotiation works well for systems with high *volatility*, a mesure of how likely a system changes during operation. In (Pancerella *et al.*, 1995; Balasubramanian and Norrie, 1997), the authors propose the use of an agent-based concurrent design environment to improve the part inspection and verification process in manufacturing cells. Finally, the AARIA (Autonomous Agents at Rock Island Arsenal) project (Parunak *et al.*, 1997) pursues the goal of designing a manufacturing system whose performance and functionality supersedes that in current centrally controlled manufacturing systems, by using a group of individually agented manufacturing system.

Our proposal has the advantage of using a parallel and distributed controller to support real-time operation with a multi-agent approach. We show in figure 1 a schematic view of the proposed controller. It is a *distributed system*, because of the physical distribution of the elements in manufacturing cells. Being a distributed system, the controller has to deal with problems such as synchronization, communication delays, time shifts, etc. (Sánchez, 1996). The controller is also *parallel:* it is composed of several processors controlling each of the elements of a manufacturing cell, as shown in figure 2. There are two reasons why we chose to design a parallel controller: to implement fault tolerance, and to be able to support complex operations in real time.

This article is organized as follows. In section 2 we show in detail how is built PARDICO, and how it is connected to the manufacturing cell. In section 3 we analyze its reconfiguration properties, and the control software running in this architecture is described partially in section 4. Finally, in section 5 we show some conclusions and we describe the planned future work.

2. CONTROLLER ARCHITECTURE

The basic element of PARDICO is the *transputer,* a RISC-processor containing a CPU, a FPU, local memory and four high-speed links (Michel *et al.*, 1995). Up to 10 transputers can be connected in an IMS B008 motherboard hosted by a PC, in a pipeline fashion as shown in figure 2. All transputers have two of their links connected via a reconfigurable 32×32 crossbar (IMS C004), which allows the establishment of redundant communication paths, as we will show in section 3. Several transputer boards can be put inside one, or several PCs if needed.

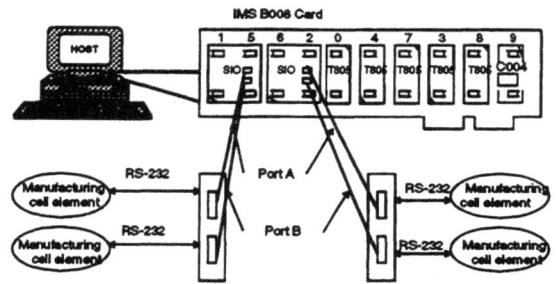

Fig. 2. PARDICO's structure

2.1 Controller-cell interface

Most robots in today manufacturing cells have three kinds of interfaces with the external world: *serial* (RS232), *parallel* (Centronics) or *MAP* (Token bus). We worked with robots having RS232 interfaces, so we had to develop a circuit (Alarcón *et al.*, 1996; Alarcón, 1997) to synchronize transputer link speed (20 megabits/s) with RS232 interface speed (20 kilobits/s).

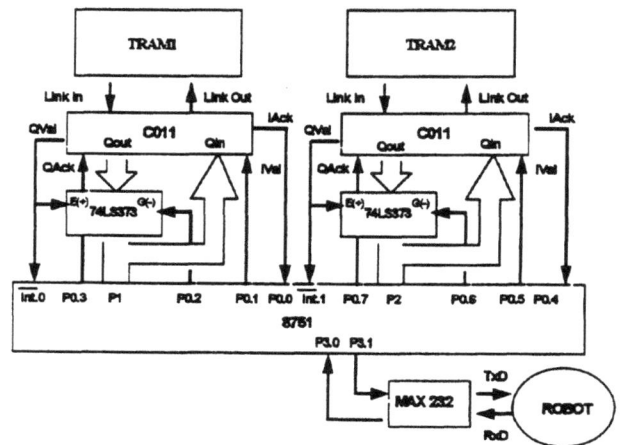

Fig. 3. Interface switch

This circuit, shown in figure 3, is composed of two chips C011 for serial-parallel conversion, and a microcontroller MC8751. It allows any of two transputers to communicate with a manufacturing cell element in a half-duplex fashion, using a handshake protocol in the physical layer (see section 4.1) to synchronize the elements with different speeds. The microcontroller software and hardware implement a 2×1 switch, so that we can have alternate paths between the manufacturing cell elements and the transputers.

Our interface occupies a half-sized PC board, and costs 10 times less than a similar circuit provided by the transputer manufacturer, the SIO-232, although this one implements a 2×2 switch. These interfaces are described more in detail in (Alarcón. 1997; Salmerón. 1997).

60

2.2 *Base topology*

The topology of a parallel architecture is an important factor affecting the performance of the applications running on it. We followed these criteria to choose PARDICO's topology:

- **Application needs.** The controller must communicate with every cell element and coordinate their actions. This implies some communication patterns that we will explain later (see section 4).
- **Hardware redundancy.** We must provide a framework which enables us to implement fault tolerance techniques.
- **Board optimization.** As a manufacturing cell has several elements and we need hardware redundancy, we must maximize the use of the space available in the circuit board.
- **Reconfigurability.** The controller must be able to change its configuration to execute other plans or avoid faulty elements (see section 3).

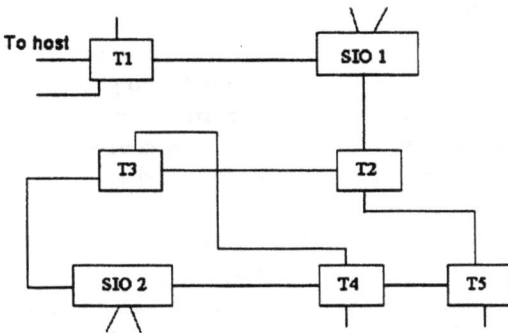

Fig. 4. Base topology

We show in figure 4 the base topology satisfying these criteria. It has two switches 2×2 built either using the interface described in section 2.1, or a SIO-232. These switches allow the transputers $T_1 \ldots T_4$ to control four elements of the manufacturing cell. There are two links from every interface to transputers, so it can be accessed by either two of them for hardware redundancy. This topology has also several redundant paths enabling the controller to change its configuration, as we will explain in next section.

T_5 is used as a monitor for the whole system, so it must be connected to every other transputer. This is not physically possible because the transputer has only four links, but it can be implemented by software, as we will see in section 4.2.

3. RECONFIGURATION PROPERTIES

Flexible manufacturing cell controllers need to be reconfigurable, in order to adapt to changes in planning and faults. *Fault tolerance* is the ability

of a system to function properly after the occurrence of faults. The basic fault-tolerant technique is *redundancy*, i.e. the multiplicity of hardware and software elements. We will talk of software redundancy in section 4.3, and we already described the *processor redundancy* found in our base topology (see section 2.2). In this section we will describe how we implement alternate paths between transputers and the elements of the manufacturing cell *(path redundancy)*.

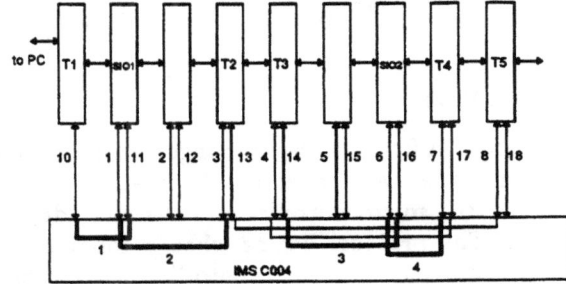

Fig. 5. Complete topology

The basis of our approach is to extend the base topology described in section 2.2, using the IMS C004 switch as shown in figure 5. Four alternate paths (bold lines) are established, so that every transputer has two links with its corresponding I/O interface (Salmerón, 1997). A diagram showing all connections of this extended topology is shown in figure 7.

4. CONTROLLER SOFTWARE

The controller software has been divided in three layers, following the OSI model for local networks: physical, data link/network, and application. The main purpose of these layers running in every transputer, is to establish and maintain communication between the elements in our parallel/distributed system (transputer and robots), in different abstraction levels. In the following sections we will explain in detail each of these layers.

4.1 *Physical layer*

The physical layer is in charge of the transmission of bytes over physical links. Transputer-transputer communication is implemented by hardware, whereas the microcontroller MC8751 in the controller-cell interface (see section 2.1) contains the protocol between robots and transputers. This handshake protocol was designed to synchronize robots with transputers (Alarcón, 1997; Mota, 1997), and was validated (together with the data link/network protocol described later) using the formal verification system PROMELA (Holzmann, 1991).

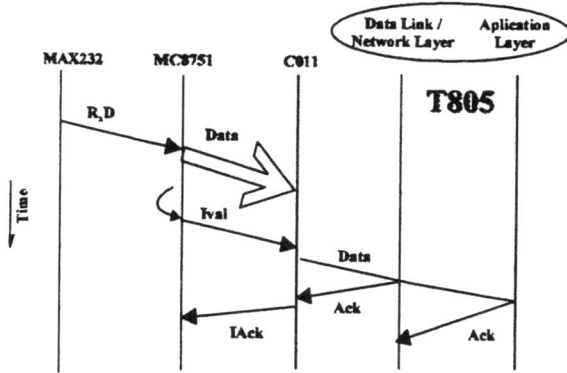

Fig. 6. Physical layer protocol

We show in figure 6 the exchange of information (in only one way for clarity) between an RS232 source, the microcontroller and the link adapter C011. The protocol begins when a robot sends data serially to the microcontroller. After 8 bits are received, the data is presented in parallel to the link adapter, and a signal Ival is sent. The link adapter sends data serially to a transputer, and when it receives an acknowledge, a signal Iack is sent to the microcontroller to finish communication. This protocol is very simple, allowing the transputer hardware and software to run at full speed (Alarcón, 1997).

4.2 Data Link/Network layer

The data link/network layer is responsible for the establishment and maintenance of communication between pairs of elements in the system, even if they are not directly connected.

The data exchange between robots and transputers is handled by a point-to-point protocol detailed in (Mota, 1997). The transputer and its development software support well our requirements for transputer-transputer communications, because it handles the notion of *virtual links*.

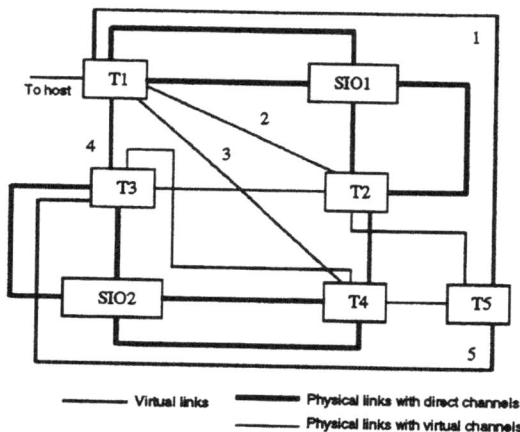

Fig. 7. Communication links

A virtual link is a connection between transputers that uses several physical links, and it is implemented by routing messages through the network. We show in figure 7 the five virtual links used in our controller. Two of them are needed to implement the monitor process in T_5, which needs to communicate with every other process in the controller. The rest of them are needed to implement a *logical crossbar* between processes in the controller. This full interconnection is needed because of the requirements of the application layer, as we will explain later.

4.3 Application layer

The application layer has two main functions: execution and supervision. The execution relies in a *multiagent system*, whereas the supervision is implemented by a *monitor* and *recovery blocks*. The main motivation for the use of an intelligent (multiagent) system to control the manufacturing cell, is to increase flexibility.

We show in figure 8 our agent's structure. Each agent is responsible of controlling one of the elements in the manufacturing cell: a robot, a machine, etc. The agent has a *control structure*, responsible of plan execution and monitoring, as well as communication with other agents. It has also a *knowledge base*, which is used to make local decisions during execution.

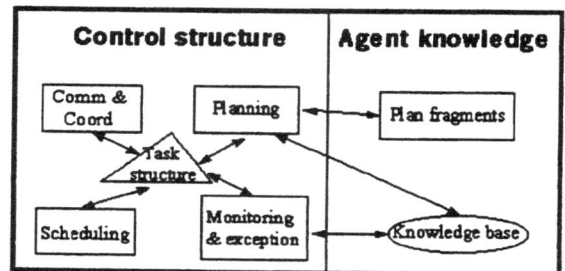

Fig. 8. Agent structure

The detailed description of this system is beyond the scope of this paper, and can be consulted elsewhere (Sánchez *et al.*, 1997; Castillo, 1996); we need only to know that all transputers must be able to communicate with every other one. This is implemented by the use of virtual links, as explained before.

The supervision system is responsible for the implementation of software fault tolerance. PARDICO, uses the following procedures to implement fault tolerance (Sánchez, 1996; Sánchez *et al.*, 1997):

- An active monitor, running in transputer T_5, tests periodically hardware and software with validation and acceptance tests.

62

- There are alternate (redundant) paths between processors and robots, as described in section 3.
- Each transputer runs several processes concurrently: if there is a faulty processor, another processor takes its job.

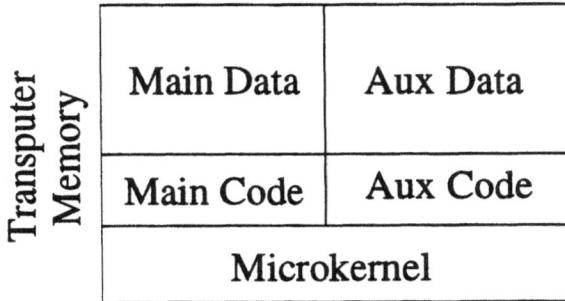

Main Data	Aux Data
Main Code	Aux Code
Microkernel	

(Transputer Memory)

Fig. 9. Recovery blocks in transputer memory

Recovery blocks (Mullender, 1993) allow the controller to implement this last procedure. Each transputer in PARDICO controls a robot with a main or primary code, and has an auxiliary area where lies the recovery block of another transputer, as shown in figure 9. The primary version of a code is always used, unless it fails to pass an acceptance test performed by the monitor, or a timeout is detected. In the case of a processor fault, the auxiliary code located in another processor takes control of the corresponding manufacturing cell element.

5. CONCLUSIONS AND FUTURE WORK

In this paper we have presented the hardware and part of the software of PARDICO, a parallel controller for flexible manufacturing cells. The hardware is fully functional and has been tested in real manufacturing cells at ITESM (Alarcón, 1997; Salmerón, 1997; Mota, 1997). Two layers of the controller software are already implemented, and work is in progress for the application layer. Our first experiences encourage us to continue using parallel architectures to enhance flexibility in manufacturing cells. Two areas are still under research: fault tolerance measurement in our architecture, and the design of the multiagent system.

6. REFERENCES

Alarcón, J., M. Rodríguez and S. Mota (1996). Diseño de una arquitectura reconfigurable basada en transputers. *2o. Congreso nacional sobre electrónica y comunicaciones. Colegio de Ingeniería en Comunicaciones y Electrónica. EXPO-CICE 96.*

Alarcón, Jaime (1997). Diseño e implementación de una interfaz de comunicación entre robots industriales y conmutadores a base de transputers. *Tesis de Maestría en Ciencias Computacionales, ITESM-CEM.*

Balasubramanian, S. and Douglas Norrie (1997). A multi-agent intelligent design system integrating manufacturing and shop-floor control. *http://ksi.cpsc.ucalgary.ca/DME/AnAgent.html, Department of Mechanical Engineering, University of Calgary.*

Castillo, Sergio Luis (1996). Diseño de un sistema multiagentes para controlar una celda flexible de manufactura. *Propuesta de Tesis de Maestría en Ciencias Computacionales, ITESM-CEM.*

Gray, P. (1996). A flexible manufacturing cell designed with petri nets with transputer control. *IEEE Publications on line.*

Holzmann, Gerard (1991). *Design and validation of computer protocols.* Prentice Hall Software Series.

Michel, Francisco, Alejandro Ramírez, Rafael San Vicente and Jesús Sánchez (1995). Integración de una celda de manufactura flexible utilizando tecnología de transputers. *Reporte técnico ITESM/CEM/DGI-9512-C02.*

Mota, Sara (1997). Protocolos de comunicación en el controlador de una celda flexible de manufactura. *Tesis de Maestría en Ciencias Computacionales, ITESM-CEM.*

Mullender, Sape (1993). *Distributed Systems.* Ed. Addison Wesley.

Pancerella, C., A.Hazelton and H.Frost (1995). An autonomous agent for on-machine acceptance of machined components. *Proceedings of modeling, simulation, and control technologies for manufacturing. SPIE's International Symposium on Intelligent Systems and Advanced Manufacturing* pp. 25–26.

Parunak, H. Van Dyke (1987). Manufacturing experience with th contract net. In: *Distributed Artificial Intelligence.* Morgan Kaufman Ed.

Parunak, H.V.D., A.D. Baker and S.J. Clark (1997). The aaria agent architecture:an example of requirements-driven agent-based system design. *Proceeding of the First International Conference on Autonomous Agents (ICAA'97).*

Salmerón, Mirna (1997). Diseño de una arquitectura reconfigurable basada en transputers. *Tesis de Maestría en Ciencias Computacionales, ITESM-CEM.*

Sánchez, Jesús (1996). A flexible manufacturing cell distributed controller. *4th International Simposium on Applied Corporate Computing* pp. 71–77.

Sánchez, Jesús, Isaac Rudomín, Carlos Rodríguez and F. Ramos (1997). Advanced communications technology in robotics and flexible manufacturing cells project: status as of november 1996. *Reporte técnico ITESM/CEM/DGI-9701-C04.*

Sinha, A., A. Chaudhuri and P.K.Das (1996). Transputer implementation of a fault tolerant distributed architecture for critical real-time application. *IEEE Publications on line.*

AN INTEGRATED SYSTEM FOR REAL-TIME SUPERVISION AND ECONOMIC OPTIMAL CONTROL WITH APPLICATION TO MINERAL PROCESSING PLANTS

A. Cipriano, J. Concha, E. Pajares and C. Muñoz

Faculty of Engineering, Catholic University of Chile
P.O. Box 306, Santiago 22, Chile
Phone: 56-2-6864284; fax: 56-2-5522563; e-mail: aciprian@ing.puc.cl

Abstract: In this article, a support system for the economic optimal control of industrial plants, is described. The system collects and processes, in real time, the data obtained from sensors and informs the operator about the optimal values of manipulated variables to obtain maximum economic benefits. An application to a grinding-flotation plant for copper mineral processing is also presented. *Copyright © 1998 IFAC*

Keywords: Process supervision, economic operation, model-based control, optimisation.

1. INTRODUCTION

The last technological advances and the decreasing cost of digital hardware have promoted the automation of industrial processes, based on distributed control systems or computer networks connected to PLC's. However, in most industrial plants, the control functions are still in a regulation level, with PID controllers, despite the modern methodologies like robust, fuzzy, expert and neural control.

The causes of the preceding may include the following: 1) all control strategies require reliable instrumentation; 2) modern control methodologies require knowledge usually not available to the end user; and 3) in many processes, an economic analysis of the benefits of automatic control is not performed.

This work intends to solve the third problem, designing control strategies that incorporate economic aspects, through the definition of an economic index of the plant operation, which should be optimised.

In this paper we present an *Integrated System for Supervision and Economic Optimal Control*, called SISCO in Spanish, that collects and processes, in real time, the data obtained from sensors, with the aim of informing the operator about the current economic performance of the plant. The tasks carried out by the system can be divided in three groups: a) it displays the current net dollar benefit obtained from plant operation; b) it answers operator questions about possible benefits to be obtained by changing process variables and set points; and c) it computes the optimal values of these variables in order maximize benefits while complying with the technical constraints.

The system uses simple static process models in order to represent the plant behavior and a sequential quadratic programming algorithm to solve the optimisation problem.

The system run over SCAUT-3G for Windows NT. SCAUT-3G is an automation and supervision software installed in several plants in our country.

A laboratory example and an application to a grinding-flotation plant for copper mineral processing are considered. It is shown that the system can increase the economic benefits obtained from the plant.

2. DESCRIPTION OF THE SYSTEM

SISCO is an information system for real-time supervision, control and optimisation, oriented to support the decision making that maximizes the economic benefits of industrial processes. Its main features are:
- On-line evaluation of economic indexes like income and costs.
- Model-based prediction of technical and economic variables, with automatic tuning of some parameters.
- Global economic optimisation of a plant subject to operational constraints.
- Configuration according to the typical characteristics of each process.
- Basic software structure and additional modules that expand the system capabilities.

The system consists of a set of programs that conform three modules, the operation interface and the communication system (see Fig. 1). The three modules are: Evaluation, Consulting and Optimisation. Each module is able to compute information and display it in spanish.

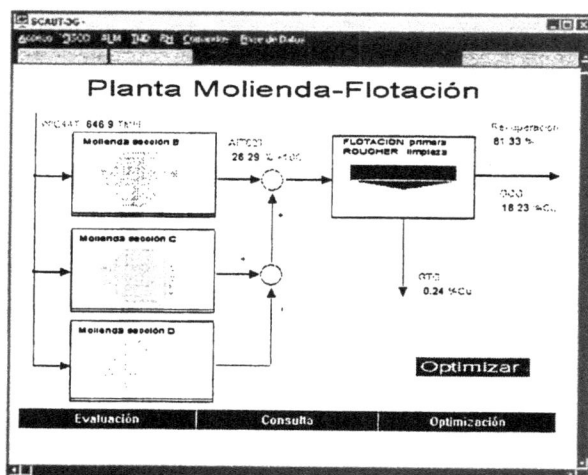

Fig. 1. SISCO in SCAUT-3G.

The Evaluation and Consulting modules are programmed in SCAUT-3G. The Optimisation module is a mixed C++ and FORTRAN program. Through the SCAUT-3G operation interface, the different modules can be accessed and the user can interact with the system.

The system has two possible modes of operation: Manual Mode and Optimising Mode. In Manual Mode, the values of the process manipulated variables are input manually by the operator. In Optimising Mode, the optimal values of these variables (as computed by the Optimisation Module) are used.

In the following sections, the three modules are described.

2.1 Evaluation module

This module presents information about user-defined indexes of the plant operation and performance. These indexes can be of technical and/or economic nature. The information presented includes current values, mean values and a trends graph for each index.

Fig. 2 presents the display of this module. The trends graph shows different indexes with different colors. The range for each variable can be changed with a *click* on the value of interest. The time range is in the upper left corner and can also change between 2 min., 4 min., 2 hr., 8 hr. and 24 hr. The graph allows the user to know instantaneous values by clicking on the point of interest. The lower part of the display shows the name of the indexes and their monthly mean value, current value, possible optimal value at that time (computed by the optimisation module), and the measure unit.

Finally, at the bottom of the display, there are buttons for accessing the other displays:

Planta:	plant flowsheet
Consulta:	consulting module
Optimización:	optimisation module
Restricciones:	display with the constraints of the process variables.

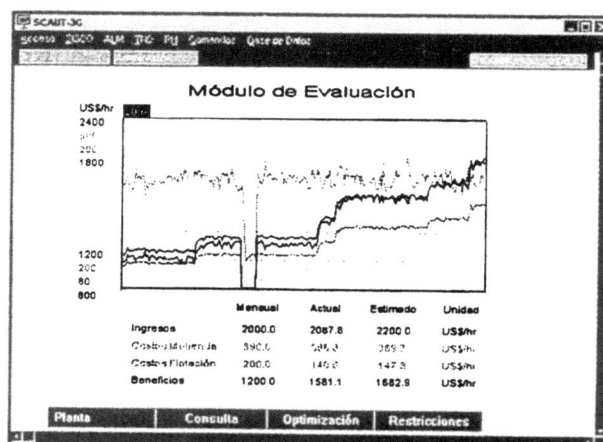

Fig. 2. Evaluation module display.

2.2 Consulting module

This module answers operator queries about the impact on the indexes of changes in the manipulated variables. This helps the operator to decide whether some specific change would be beneficial for the plant. This module requires mathematical models of the process. The module display is shown in Fig. 3.

The lower half of the display presents a list of the manipulated variables and three columns of values:

Manual:	current values of the manipulated variables. These can be changed

by the operator when Manual Mode is on.

Predecir: query values of the manipulated variables.

Optimizador: optimal values (computed by the Optimisation module) of the manipulated variables.

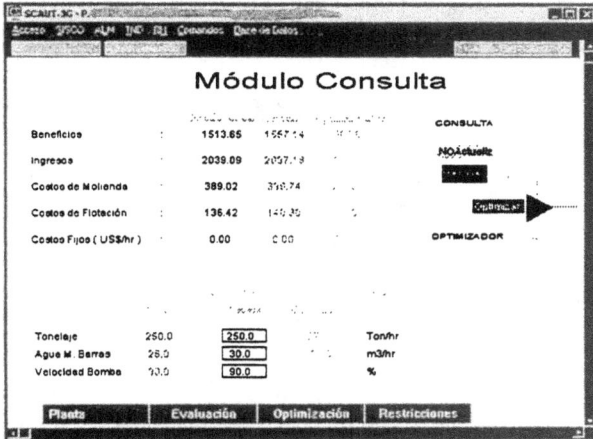

Fig. 3. Consulting module display.

The upper half of the display gives the result of the operator query. A list of the indexes is shown, along with the following three values for each index:

Predicción: predicted steady-state index value for the query values of the manipulated variables.

Actual: current index value, computed by the Evaluation module.

Optimizador: optimal steady-state index value. This value would be obtained if the optimal values of manipulated variables were used.

In the upper right area, schematic buttons allow the user to choose which values of the manipulated variables are applied to the process. The button Manual/Optimizar selects between Manual and Optimising Modes. The button Actualizar/NOActualiz applies the query values to the plant.

Finally, buttons for direct access to other displays are at the bottom of the display.

2.3 Optimisation module

This module calculates optimal values for the manipulated variables of the process with the objective of optimising a plant performance index, typically the net economic benefits obtained with the plant operation. The optimisation takes into account user-defined constraints, such as bounds on process variables.

Fig. 4 presents the display of the module. At the upper left area a trends graph shows the time evoultion of the optimal values of the economic indexes. At the right, the optimal and current values of the indexes are updated in real time. The button Manual/Optimizar toggles between Manual and Optimising Modes.

At the lower area of the display, the current and optimal values of the manipulated variables are shown. Buttons for direct access to other displays are located at the bottom of the display.

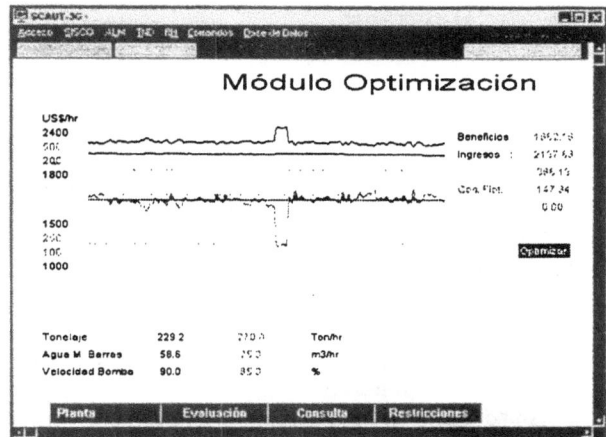

Fig. 4. Optimisation module display.

3. TECHNICAL CHARACTERISTICS

3.1 Software platform

The current version of SISCO is implemented on SCAUT-3G, for Windows NT. SCAUT-3G is a software for automation and supervision of industrial processes, consisting essentially of a real-time database containing process information. This database resides on RAM and is operated by a SCAUT-3G Manager. The system also has several modules for auxiliary tasks, including: historical records of process variables, trends records, computation of arithmetic expressions and warnings generation.

SCAUT-3G provides a C++ function library, that includes, for example, routines for database access. Thus, additional modules can be programmed by the user.

3.2 Configuration

The database of the system contains process variables and configuration variables. The user should configure the system, for the particular application, through the configuration display. It is possible to configure:

Performance indexes. The objective function is written as one or more arithmetic expressions, which usually include parameters that can be modified.

Constraints. These are defined by upper and lower bounds for process variables. The bounds can be modified by the user.

Models. The mathematical models used by the Consulting and Optimisation modules are static model and relate process variables and manipulated variables. These models depend on a set of numerical parameters, such as coefficients in arithmetic expressions, that can be modified by the user.

3.3 Optimisation algorithm

The Optimisation module utilizes a Sequential Quadratic Programming (SQP) method for performing the optimisation (Luenberger, 1994). This algorithm can be applied to non-linear problems with constraints.

SQP is an iterative algorithm and the search direction, in each iteration, is given by the solution of a quadratic programming subproblem. This subproblem is obtained with a second order approximation of the objective function around the current estimated solution.

3.4 Plant installation

The following 5 steps should be carried out when installing the system in a production plant:

1. Definition and configuration of the performance indexes and objective function.
2. Definition of the operational constraints.
3. Assignment of values to the involved parameters.
4. Adaptation of SISCO to the automation platform existing on the plant.
5. Definition and identification of the predictive models used by the Consulting and Optimisation modules.

4. LABORATORY EXAMPLE

The process considered is a grinding-flotation plant, with three mills and one flotation cell bank. This process can be modeled with a set of nonlinear differential and algebraic equations (Cipriano *et al*, 1997), based on which a Matlab simulator was developed. The simulator and the system were installed and run in two different computers in order to mimic a real application.

Steps 1 through 5 of the preceding section were followed in order to properly configure the system. The index to be optimised was defined to be the net

economic benefit obtained from plant operation, in US$/hr. Other indexes were: total income, grinding costs and flotation costs. Results showed that:

- The communication programs worked correctly.
- The predicted steady-state values of the indexes were correct with an error of less than 4%.
- The optimiser was able to take the plant to an operating point with a high value of the net economic benefit as shown in Fig. 5.

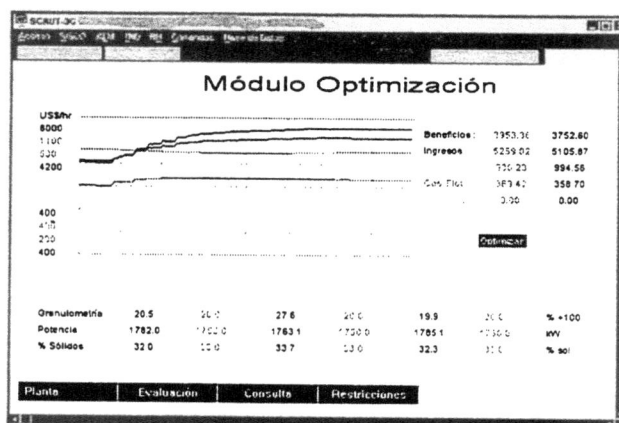

Fig. 5. Laboratory example.

5. APPLICATION TO A COPPER MINERAL PROCESSING PLANT

In order to evaluate the possibility of using the system in a real plant, operation data were obtained from a section of a real mineral processing plant in Chile. This section processes an average of 5500 tons of mineral per day.

As before, steps 1 through 5 of the system configuration procedure were performed. The indexes defined were the same as those in the laboratory example, namely: net economic benefit, income, grinding costs, and flotation costs. The models were developed using the data obtained from the plant.

The time evolution of the indexes are computed by the Evaluation Module, as shown in Fig. 2.

It was found that the economic benefit is highly variable, with a standard deviation greater than 20% of the mean. The use of the system would help the operator to keep the benefits at a more constant level with a higher average.

6. CONCLUSIONS

In this article, a support system for the economic optimal control of industrial plants, is described. The system collects and processes, in real time, the data obtained from sensors and performs three tasks:

1. It shows the current values, trends, and average values of user-defined performance indexes of the plant.
2. It answers operator queries about results of possible changes in manipulated variables.
3. It informs the operator about the values of manipulated variables that optimise a user-defined index, while complying with plant constraints.

It is shown with a laboratory example that the system, and particularly the optimiser, works correctly. Preliminary studies in a real plant show that the use of the system can increase economic benefits obtained from plant operation.

REFERENCES

Luenberger, D. (1984). *Linear and Nonlinear Programming*, Adison-Wesley Massachusetts.

Cipriano, A., M. Ramos and C. Muñoz (1997). Economic optimal control of a grinding-flotation plant using parallel processing (in Spanish). *Proceedings of the XII Chilean Congress of Electrical Engineering*, Temuco, November 3-8, pp. 497-502.

ACKNOWLEDGEMENTS

The authors wish to acknowledge the support of FONDEF, grant MI-17, and FONTEC, grant 96-0722.

SIMULATION OF DISTRIBUTED FAULT TOLERANT HETEROGENEOUS ARCHITECTURES FOR REAL-TIME CONTROL

H. Benitez-Perez, H.A. Thompson and P.J. Fleming,

Rolls-Royce University Technology Centre in Control and Systems Engineering,
Department of Automatic Control and Systems Engineering,
University of Sheffield, Mappin Street, Sheffield, S1 3JD, United Kingdom.

Synopsis

Increasingly, there is a move towards in-built intelligence for sensors and actuators to produce "smart" components leading to distributed heterogeneous systems. Of particular interest to safety-critical systems is utilisation of the local intelligence to provide health monitoring, fault detection and fault tolerance. However, crucial to system safety is the ability to predict time delays in the system and analyse their effect on control system performance. For real-time systems the delays associated with communicating diagnostic messages may result in safety deadlines being missed. At present, smart components are integrated in an ad-hoc way. This research is exploring how distributed processing capability can be exploited to improve controller performance and increase fault tolerant capability to create high availability systems for gas turbine engine control.
Copyright © 1998 IFAC

Keywords: Multi-Objective Optimisation, Distributed Systems

1. INTRODUCTION

Currently, Rolls-Royce utilise a centralised dual-lane High Integrity Computer configuration (HIC). This is a well proven concept with several million hours of operational experience. Although this is a tried and tested architecture, increasing miniaturisation through the use of advanced electronics such as Application Specific Integrated Circuits (ASICs) and Multi-Chip Modules (MCMs) has reduced electronics "real-estate" to such an extent that the controller box size and weight are now dictated by the number of connections to sensors and actuators. Cabling weight, flexibility to change and associated maintenance cost in a centralised system is also a concern with hard-wired connections to each dumb sensor or actuator. In order to remove these limitations and gain further significant weight and size reductions a radical change in architecture to a distributed system is required.

The emergence of smart sensor and smart actuator technology removes the need for centralised control with feedback loops to dumb peripheral actuators (a major source of wiring complexity and weight) replacing it with a databus connection. Potentially, this will give lower actuator installation and life-cycle costs and if two-way communication with the main computer is employed, the intelligence imparted to actuators can be used for local control and a number of other functions such as self-calibration, error detection and accommodation and health monitoring. This

would remove the tight coupling found in present systems with numerous dumb peripherals making alterations and maintenance easier.

2 DISTRIBUTED ARCHITECTURE

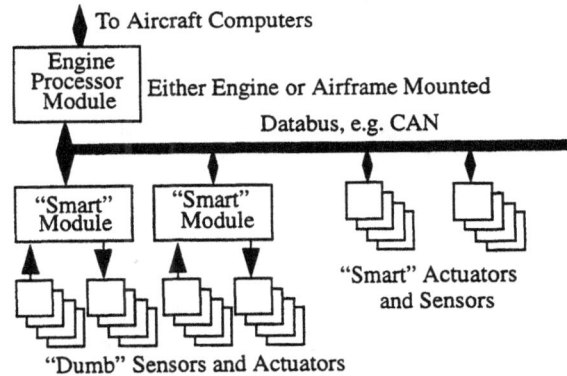

Fig. 1 Likely Future Architecture

In some cases smart actuators and sensors which meet requirements are already available, but in other cases further development is needed to meet the harsh environmental conditions. As a full set of smart sensors and actuators which meet all engine controller requirements is not available at present it is likely that a future architecture will use a mixture of smart sensors and actuators connected directly to a databus and smart modules that then interface to dumb sensors and actuators as shown in Fig. 1. The future architecture will thus be a distributed heterogeneous system. As building a distributed system is an expensive process the research is investigating the implications of using a databus on the aero-engine connected to a mixture of dumb and smart peripherals. The aim is to be able to demonstrate the advantages or disadvantages of adopting smart elements within the system and develop methodologies for exploiting the intelligence within the system for fault diagnosis and fault tolerance.

The first consideration is the impact of introducing a databus onto the engine. This will introduce delays into the control system. Safety requirements dictate that certain loops must be controlled within very tight timescales; typically in the order of a few milliseconds. For instance, an actuator can easily run away within a few milliseconds resulting in a catastrophic failure which may lead to large loss of life. It is thus important to have a means to assess the potential for this unsafe situation by means of modelling these delays within a closed-loop engine and controller simulation. The delays themselves are dependent upon the data traffic on the databus. This is composed of controller data input, controller data output, fault diagnosis and health monitoring messages. Although these can be defined for existing engines where centralised controllers operating cyclic schedules are used it is more difficult to anticipate the increase in data traffic from exploitation of smartness to improve levels of diagnostics and health monitoring. Already the control laws only account for 20% of the overall control task. Fault diagnosis and monitoring routines account for the remaining 80% of the control cycle. In addition it is also necessary to consider the effect of fault accommodation mechanisms to give degraded modes of operation. This will alter the delays experienced within the control system dependent on the detection, isolation and accommodation strategies used.

Within this work two alternative fault tolerant strategies are being considered, Hierarchical Fault Tolerance and Global Distributed Fault Tolerance. In the first case, one element in the system is responsible for fault diagnosis and reconfiguration. In the second approach, the diagnosis and the fault tolerance is handled across the intelligent resources with local reconfiguration mechanisms built-in to the sensors and actuators. In particular, tight timing to achieve consistency and synchronisation needs to be considered. Partial interaction between smart sensors or actuators is also possible to provide self-reconfiguration. The intention is to investigate reconfiguration strategies in both cases including analytical-based techniques (Benitez-Perez *et al.*). System synchronisation and maintaining data consistency for input parameter validation and inter-unit voting of controller outputs also needs to be addressed.

3 ON-ENGINE DATABUSES

The connection between the engine and the aircraft is defined by the airframe manufacturer. This is likely to be ARINC 429, ARINC 629 or MIL-STD 1553 (Reynolds, 1996). When selecting a databus standard to use on-engine the existing aerospace standards are not suitable due to their weight and cost. An exploration of Commercial Off-The-Shelf databuses from the fields of process and automotive control was performed. This highlighted that standardisation for process control is still some way off and the protocols used, e.g. Fieldbus, are more complex than are needed for aero-engine control (Pleinevaux and Decotignie, 1988). Automotive databuses, e.g. CAN (McLaughlin, 1997) are, however, simple, lightweight and cheap. Additionally, the appearance of cheap combined microcontroller/interface chips is likely to result in an explosion in the use of these databuses over the next few years. Many of the factors which make databuses attractive for cars are also true for aero-engines. In particular, they are retrofittable so that once a databus has been "designed-in" other features can be added at minimal hardware cost and change to the system architecture giving flexibility. There is a major improvement in reliability since a databus removes significant redundant wiring and reduces the risk of cut or shorted wires. Finally, diagnostics can be used to pinpoint the source of a fault. With traditional systems if one of the many wires in a wiring harness is cut or shorted figuring out which one is involved and where exactly the fault lies can take hours. Using a bus, fault isolation is far more efficient since the system can be configured so that each node on the bus receives a broadcast enquiry to its health. Overall, experience has shown that cars which use databuses are more reliable than conventional cars, easier to service and lighter in weight. As a result work is already underway for certification of CAN bus for aerospace applications (Thompson, 1996).

4 GENERIC NETWORK SIMULATION

From Fig. 1 it can be seen that there are two requirements: modelling the delay from the aircraft databus (ARINC 429, ARINC 629 or MIL-STD 1553) and also modelling the delay of the on-engine databus (e.g. CAN bus). This delay is dependent upon the topology of the interconnection architecture and the level of databus traffic. The intention in this work is to provide a set of simulation tools covering a range of databus types to enable future distributed control systems to be evaluated in closed loop including the impact of fault diagnosis and reconfiguration mechanisms on control performance.

Much of the work on modelling delays in hard real-time systems has concentrated on scheduling for processor tasks. For example, in Tindell *et al.*, 1994, the authors propose to build a system from a number of sporadic and periodic tasks each assigned static priorities and dispatched at run-time according to the static priority pre-emptive scheduling algorithm. For this, worst-case execution time and worst-case response time of a task can be defined.

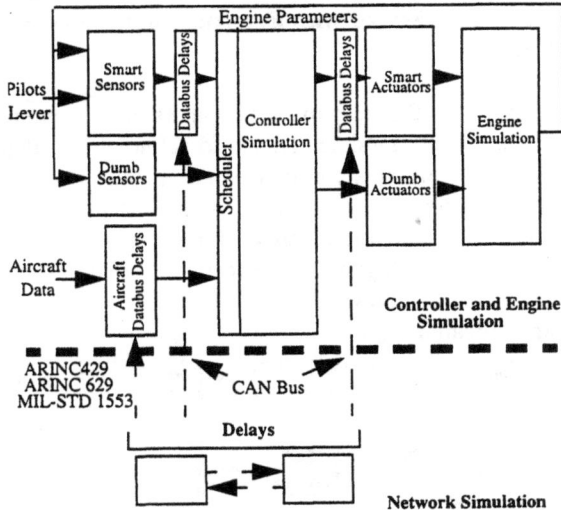

Fig. 2 System Concept

For multi-processor systems similar techniques are used but taking into account the message transmission delays between processors. For instance, Burns *et al.*, 1995, propose a computational model based upon distributed objects, asynchronous communication, and protected shared data areas. This views the application as a collection of tasks which are statically allocated to a primary resource in the system. Timing requirements are then mapped into the tasks. A problem with this application, as highlighted in Thompson, 1991, is that computation time varies dependent upon the flight envelope. This has been addressed in Scholefield *et al.*, 1994, using a "Temporal Agent Model" (TAM) of the system where different terms for real-time systems tasks, computation and minimum and maximum execution times are defined when the computation is non-deterministic. Another problem is that in this case we are particularly interested in dynamic allocation of tasks. The amount of communication will be dependent upon fault detection and reconfiguration actions being performed. Sporadic communication is addressed in Joseph and Pandya, 1986, through the use of system scheduling and in Barret and Couch, 1979, a concept for a highly dynamic and non-deterministic environment with adaptable tasks for fault tolerance is considered.

Since a number of databuses need to be evaluated the aim of this work was to produce a generic network simulation which could be easily tailored. A Finite State Machine (FSM) approach was adopted since within a FSM model, a digital (discrete) system is viewed as one that moves in discrete steps from one state to another (Gersting, 1982). A major advantage

of using this approach is that it is possible to represent the model in a general form without the need to define any detailed information. It is thus relatively simple to model a number of different databuses using the same simulation. The use of Petri Nets (Garg, 1985) was also considered but discounted because of complexity of the systems being considered. The problem of "explosion" with system complexity would have resulted in a large and complex model.

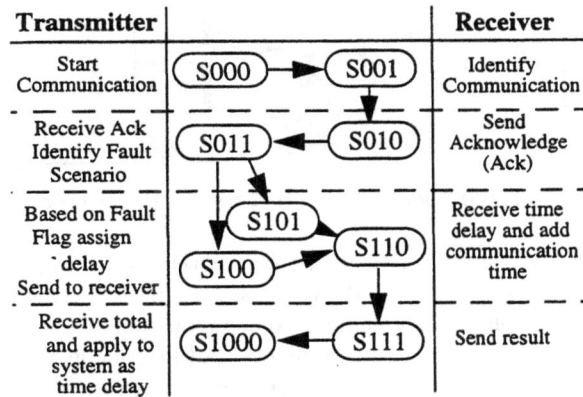

Fig. 3 FSM for Remote Terminal

A common Finite State Machine (FSM) can be defined for each Remote Terminal (RT) in the system as shown in Fig. 3. Note that a bus controller may also be needed, for databuses such as MIL-STD 1553. Of particular interest in this system is dynamic redundancy where reconfiguration is used to switch in spare elements or provide graceful degradation. Here the delays are variable dependent on the fault detection and accommodation technique used. In order to explore performance degradation it is necessary to define an overall FSM scheduler for the system. This module must determine when there is a fault and the fault location and duration. It also determines which fault tolerant strategy is adopted to accommodate the failure

5 GAS TURBINE ENGINE CONTROL EXAMPLE

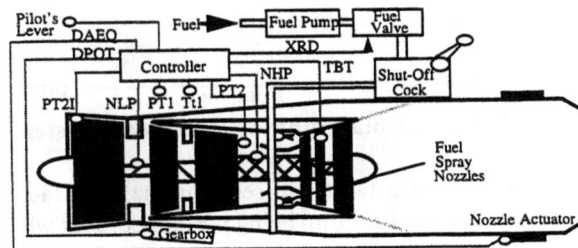

Fig. 4 Gas Turbine Engine

The system being explored is a smart distributed gas turbine engine controller as shown in Fig. 4. The controller utilises five different sample rates and to achieve this the major sample frame is divided into a series of minor cycles. The aim of the work is to simulate the distributed system in terms of the effects of time delays on system dynamics. In particular this is a safety-critical application where it is important to monitor actuators to catch runaway faults before a catastrophic failure occurs. This can require update times of a few milliseconds. Three different smart element approaches are considered, distributed, hierarchical and a voter where the voter can be distributed or hierarchical.

6 DISTRIBUTED FAULT TOLERANCE

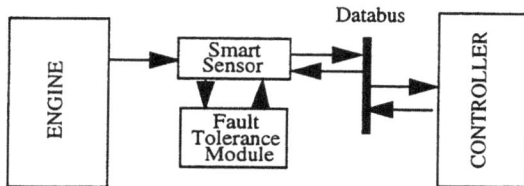

Fig. 5 Smart Sensor with In-Built Fault Tolerance

With Distributed Fault Tolerance (DFT) the smart element is integrated with its own in-built Fault Detection Isolation and Accommodation (FDIA) strategy (See Fig. 5.). In the event of local faults this element can mask the fault and inform the rest of the system about its condition.

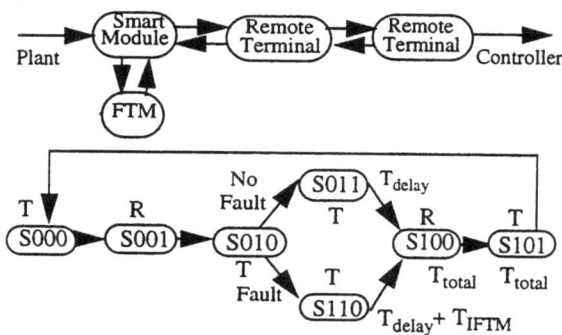

Fig. 6 FSM for Smart Element with Integral FTM

The Finite State Machine (FSM) for this type of element is shown in Fig. 6.

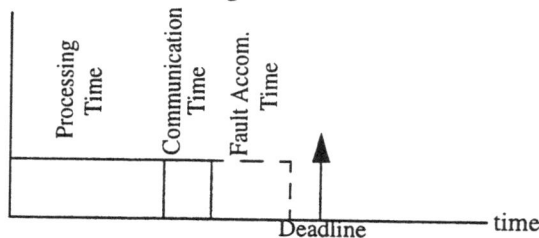

Fig. 7 Deadline Constraint for Distributed System

The presence of a fault will result in the local fault accommodation algorithm being performed. This will lead to the task time increasing as shown in Fig. 7. In this case it is only important to ascertain whether a critical deadline will be missed for that particular element as this could lead to the system being potentially unsafe if an actuator fault were to occur. If this is not the case it may be allowable to allow the system to operate with stale data in a degraded mode. An advantage of this approach is that only minor changes are needed to the scheduler to accommodate the addition of a new element. The approach is thus flexible.

7 HIERARCHICAL ARCHITECTURE

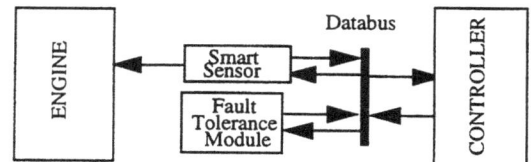

Fig. 8 Smart Sensor with External Fault Tolerant Mechanism

In Fig. 8 the FDIA is provided by an external element or fault tolerant module. This is to acknowledge that different levels of smartness exist. In this case the smart sensor may contain compensation and autocalibration functionality but may not be sufficiently intelligent to perform self-diagnostics. The sensor does, however, have a databus interface for communication.

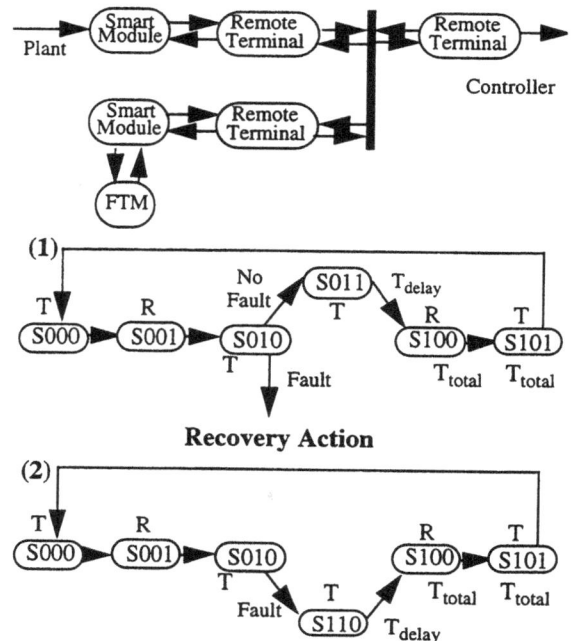

Fig. 9 FSM for Smart Elements with External FTM

In this case two communication procedures are needed (See Fig. 9). This is to represent the case when no fault exists but detection algorithms are performed and the case when a fault is detected and recovery action is needed from the external fault tolerance element. The recovery action is initiated by the scheduler. The scheduler must therefore be designed to perform the compensation action. In terms of scheduling this means that a new task must be scheduled as shown in Fig. 10. The delay from performing this task will have an effect on the overall schedule within the minor cycle. In order to avoid other tasks missing time critical deadlines then it is important to ensure that the timing issues of reconfiguration action are considered in the scheduler design. It can be seen that adding other elements to the system will have a large impact on the scheduling. In particular, if there are tight timing constraints it may be difficult to add further elements.

74

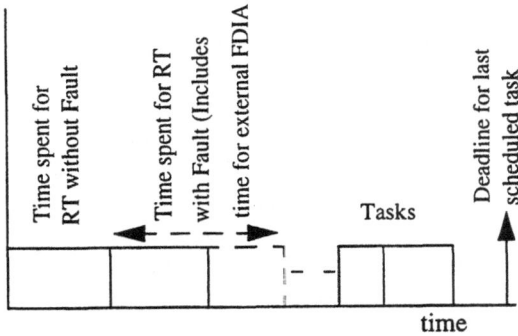

Fig. 10 Deadline Constraint for Distributed System

8 VOTER ELEMENT - FAULT MASKING

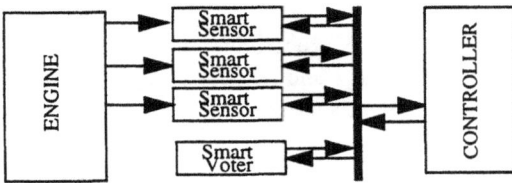

Fig.11 Group of Smart Sensors and External Voter

Fig. 11 shows grouping of smart sensors in a system and performing a vote on their output. This has the key advantage that it provides fault masking directly in the system. In this case an external voting element is shown (hierarchical FDIA), however, it is also possible to perform the voting in the sensors themselves (distributed FDIA). For a DFT approach it would still be necessary to consolidate the results of the voting in a single element.

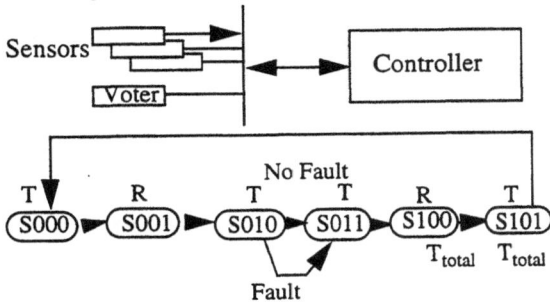

Fig. 12 FSM for Voter using Fault Masking

In the case of a voter (See Fig. 12) the fault tolerance algorithm will remain the same whether a fault exists or not. There are four communication messages; three between the 3 sensors and the voter and one with the controller. In this case there is no degradation in performance as the smart elements and the scheduler considers the fault masking strategy to be part of normal behaviour.

9 RESULTS

In reality, in the application being considered it is likely that a combination of these techniques will be utilised dependent on the criticality of signals. In order to ascertain the impact of these different approaches in an integrated system it is necessary to define the system architecture and scheduler.

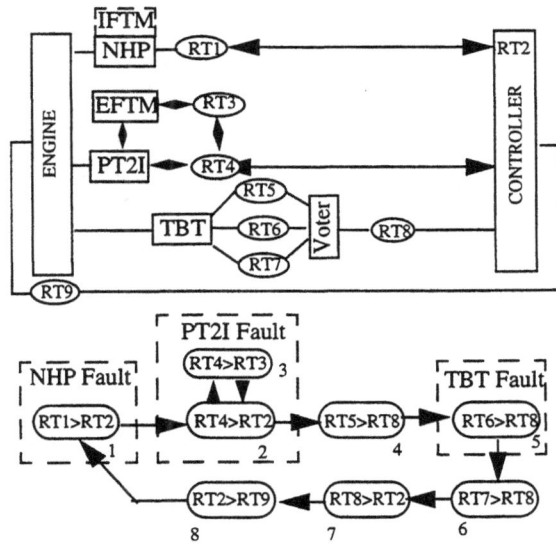

Fig. 13 Example Distributed Controller

In Fig. 13 a simple architecture is shown with the three different fault tolerance scenarios corresponding to the strategies shown in Figs. 5, 8 and 11 are considered. In this case the high pressure spool speed (NHP) sensor has local fault detection, isolation and accommodation (FDIA), the turbine inlet pressure sensor (PT2I) relies on an external FDIA mechanism and the turbine blade temperature (TBT) sensor utilises a triplex smart sensor arrangement with voting. The role of the scheduler is to inject faults into the simulation which will result in the following three different types of time delays:

t_{comi} => communication time

t_{delay} => processing time

t_{iftm} => fault accommodation strategy time

The delays incurred by NHP and PT2I can be calculated based upon the communication required to detect, isolate and accommodate a fault and the processing time. For NHP, however, fault masking is utilised and the effect of the failure in this case will not increase the time delay experienced by the system.

The specific format of communication depends on the databus that is being used and the mode selected. It is also necessary to define values for delays for system elements.

Fig. 14 shows the effect of the total time delay for a non-faulty case (1) and 5 different fault scenarios (cases 2-6) corresponding to:

1) Non Faulty Scenario
2) Faulty Scenraio1 (PT2I)
3) Faulty Scenario 2 (NHP
4) Faulty Scenario 3 (TBT)
5) Faulty Scenario 4 (Actuator)
6) Multiple Fault Scenario

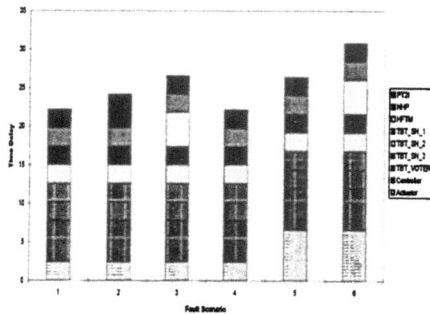

Fig. 14 Time Delays Introduced by Various Fault Cases

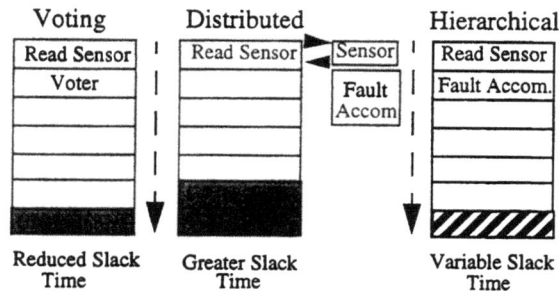

Fig. 15 Comparison of Approaches for One Minor Cycle

In reality systems may contain 200 or more sensor inputs and actuator outputs and in a complex distributed system there will be a large number of fault tolerant modules. The work shows that from the point of view of timing analysis adopting a triplex I/O grouping with a voter minimises analysis effort but this is at the expense of available time in which to schedule as shown in Fig. 15. If this approach is not possible due to cost/weight restrictions then the in-built FDIA approach is advantageous from the point of view of system flexibility since additional elements can be added with minimal impact on scheduling (see Fig. 15). Since fault accommodation is handled externally there is greater time available in which to schedule tasks making the approach more flexible to future additions. The adoption of external fault tolerance requires changes to the scheduler and assessment of the impact of accommodation algorithms on all tasks within the current minor cycle. The amount of time available in a current minor cycle is variable dependent on whether a fault or multiple faults exist in the system.

As can be seen the calculation of delays allows analysis of performance degradation. In a multiple sample rate system such as a gas turbine engine controller the effects will be different in different loops. This tool can be used to evaluate system performance and allow the control engineer to redesign the controller to ensure that specifications are met.

CONCLUDING REMARKS

In this paper a methodology for evaluating the impact on controller performance of faults within a distributed system which contains smart elements and fault tolerance has been considered. It is important to understand how delays introduced by databus communication and fault tolerant mechanisms may impact controller performance. The use of a FSM approach allows a general purpose simulation to be developed which can utilise characteristics for different databuses and fault tolerant strategies with relative ease. This can be used to understand the effects from time delays, topologies or even the technology of the databus itself.

ACKNOWLEDGEMENTS

The authors would like to acknowledge the support of Rolls-Royce plc and CONACYT, Mexico in connection with this work.

REFERENCES

Barret, W.A. and Couch, J.D., (1979), Compiler construction: Theory and Practice, Science Research Associates, USA.

Burns, A., Audsley, N. and Wellings, A., (1995), Real-Time Distributed Computing"; Proceedings of 5th IEEE Computer Society Workshop Future Trends of Distributed Computing Systems.

Garg, K., (1985), An Approach to Performance Specification of Communication Protocols Using Timed Petri Nets, IEEE Trans. Soft. Eng., Vol. SE-11, No 10, pp 1216-1225.

Gersting, J.L. (1982), Mathematical Structures for Computer Science, W.H. Freeman and Company, UK.

Joseph, M. and Pandya, P., (1986), Finding Response Times in a Real-Time System, BCS Computer Journal, Vol. 29, No 5, pp 390-395.

McLaughlin, R.T., (1997), CAN Overview, IEE Workshop, CANopen Implementation, Digest 97/ 389, pp1-27.

Pleinevaux, P. and Decotignie, J.D., (1988), Time Critical Communication Networks: Fieldbus, IEEE Network, Vol 2, No. 3, pp55-63.

Reynolds A.P., (1996,) Overview of Databuses Commonly Used in the Aircraft Industry, IMechE Part G. Journal of Aerospace Engineering, Vol. 210, pp 157-165.

Scholefield, D., Zedan H. and Difeng H., (1994), A Specification-Oriented Semantics for the Refinement of Real-Time, Theoretical Computer Science; Vol. 131, No 1, pp 219-241.

Thompson, H.A., (1991), Parallel Processing for Jet Engine Control, Springer-Verlag.

Thompson, H.A., (1996), Commercial Off-The-Shelf Databuses - Applicability to Aero-Engine Control, RRUTCShef/TN/96006.

Tindell, K.W., Burns, A. and Wellings, A. J., (1994), An Extendible Approach For Analysing Fixed Priority Hard Real-Time Tasks; Real-Time Systems, Vol. 6, No 2, pp 131-151.

DISTRIBUTED AERO-ENGINE CONTROL SYSTEMS ARCHITECTURE SELECTION USING MULTI-OBJECTIVE OPTIMISATION

H.A. Thompson and P.J. Fleming,

*Rolls-Royce University Technology Centre in Control and Systems Engineering,
Department of Automatic Control and Systems Engineering,
University of Sheffield, Mappin Street, Sheffield, S1 3JD, United Kingdom.*

Abstract: Over the past 10 years the cost of embedding intelligence into sensors and actuators directly has dramatically reduced. This has led to the recent explosion of smart sensors and actuators available from manufacturers. Initially, these have been developed for the process control industries but increasingly applications in aerospace are being found. Integration of intelligent components is being done in an ad hoc manner by incorporating smart elements in inherently centralised architectures. This paper discusses the application of multi-disciplinary, multi-objective optimisation to a military gas turbine engine control system architecture design where implementation benefits need to be traded off against implementation penalties. *Copyright © 1998 IFAC*

Keywords: Multi-Disciplinary Multi-Objective Optimisation, Distributed Systems

1. INTRODUCTION

Designing control systems architectures is extremely difficult involving a mixture of quantitative and qualitative decisions bringing together data and experience from many different design disciplines to improve safety through adding redundancy, while at the same time reducing weight, acquisition cost and operating support costs of the system. In this paper we consider how multidisciplinary, multi-objective optimisation can be used in architecture selection. In particular, the work has investigated the design of real-time distributed control systems for an aero-engine application. For this problem there are many parallel requirements, practical issues and trade-offs which need to be considered.

By incorporating intelligence in the sensor or actuator to create "smart" technology wiring weight can be reduced and routing problems simplified giving lower actuator installation and life-cycle costs. For actuators, if two-way communication with the main computer is employed local control, self-calibration, error detection and health monitoring can also be built-in. This would remove the tight coupling found in present systems with numerous dumb peripherals making alterations and maintenance easier. The design cycle is an iterative and time consuming process repeated many times until convergence to an optimal/acceptable solution is achieved. This is exacerbated by the number of candidate distributed approaches that are possible which must each be analysed to obtain their relative advantages and disadvantages. This research is producing multi-objective optimisation tools in conjunction with Rolls-Royce, to aid the designer in systems architecture evaluation.

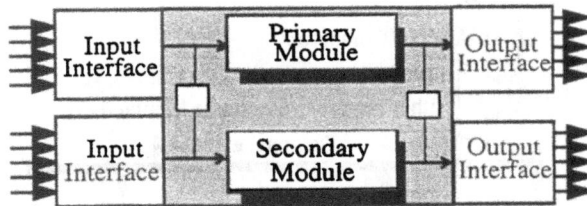

Fig. 1 Dual-Lane Controller Configuration

2 THE NEED FOR A DISTRIBUTED CONTROLLER

The dual-lane architecture (See Fig. 1) is a well proven concept with several million hours of operational experience. Changing the architecture thus requires careful justification in terms of benefit to be gained. Much of the desire to change the architecture is due to demands for improvements in control and monitoring which is now possible with modern digital control systems. The example considered in this paper is an engine controller architecture for a conceptual ASTOVL aircraft as shown in Fig. 2 which utilises complex engine geometry to enable control for vertical, hovering and forward flight. This concept requires significant increases in functionality.

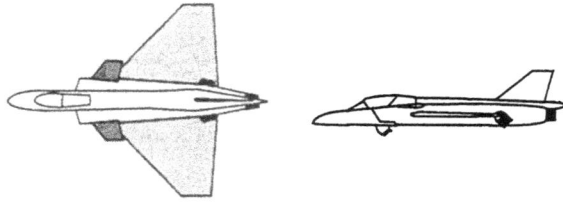

Fig. 2 Conceptual ASTOVL Aircraft Design

A limitation of the current centralised architecture is the need to alter the main processor modules to accommodate new sensor inputs and actuator outputs. As the functionality and I/O requirements have increased improvements in electronics over the years has resulted in progressively less and less "real-estate" within the unit being needed. As a consequence the connections to the unit now dictate the controller box size. In addition, there is a large amount of harness cabling to sensors and actuators in this centralised approach which increases the weight and cost of the system.

3 FUTURE ARCHITECTURES

Adopting a totally distributed system using smart technology (Thompson, 1994) is not at present possible because there are still technological barriers which need to be overcome. In some cases smart actuators and sensors which meet requirements are already available, but in other cases further development is needed to meet the harsh environmental conditions. As a result, any new architecture must be flexible enough to cope with retrofitting technology as it becomes available and commercially viable.

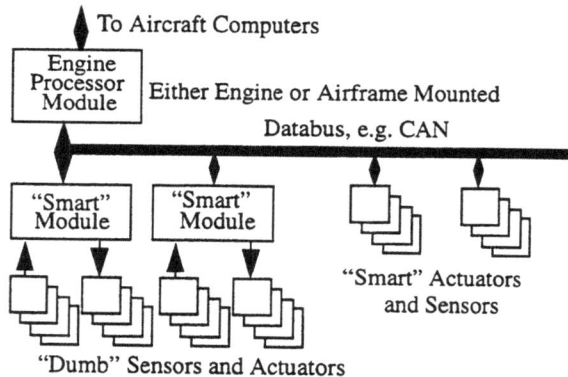

Fig. 3 Likely Future Architecture

Future architectures (See Fig. 3) will thus use a mixture of smart sensors and actuators connected directly to a databus and smart modules that interface to dumb sensors and actuators.

4 MULTI-OBJECTIVE OPTIMISATION OF ARCHITECTURES

Multi-disciplinary, multi-objective optimisation recognises that a number of design criteria from different engineering disciplines need to be satisfied simultaneously (Kroo *et al.*, 1994). It also recognises that there is no one ideal "optimal" solution but rather a set of Pareto-optimal solutions for which an improvement in one design objective will lead to degradation of one or more of the remaining objectives. These solutions are known as non-inferior or non-dominated solutions to the problem. The traditional approach adopted is to formulate the problem as a nonlinear programming problem. This, however, requires the engineer to set weights and goals. In addition, it is difficult to handle multimodality and discontinuities in the function space. Evolutionary algorithms are different in that they do not require derivative information or an initial estimate of the solution region. Instead, a stochastic search is performed over the entire solution space giving more likelihood of finding a global optimum than conventional methods. Evolutionary algorithms also allow noisy, discontinuous and time varying function evaluations with a mixture of continuous and discrete parameters. This is highly important when considering systems architecture optimisation.

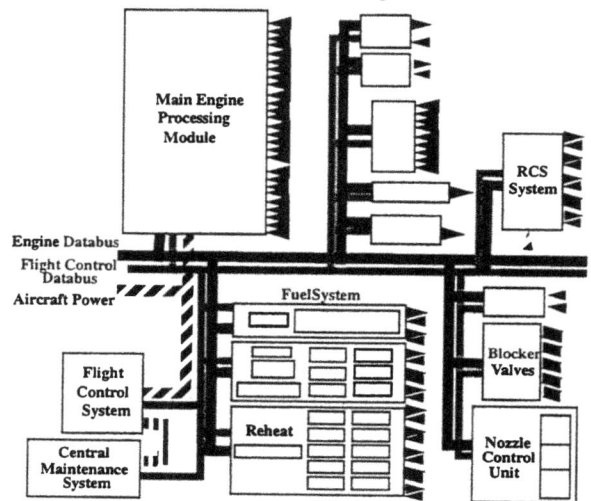

Fig. 4 Baseline Distributed Control System

An idea of the complexity of the controller is given in Fig. 4 which shows the main system functionality. To meet safety requirements many of the inputs and outputs are duplicated, triplicated or quadruplex. Additional replication may also be used to improve availability. As a result it is necessary to consider around 240 sensors and actuators which can be simplex, duplex, triplex, smart or dumb. Overall the optimisation process needs to consider:

1) The number of smart interface units
2) The bus interconnection topology
3) The mix of dumb/smart sensors/actuators.

Fig. 5 shows the approach adopted for systems architecture optimisation. At the heart of the optimisation process is a Multi-Objective Genetic Algorithm (MOGA) (Fonseca and Fleming, 1993). This generates and assesses competing architectures through a process mimicking natural selection.

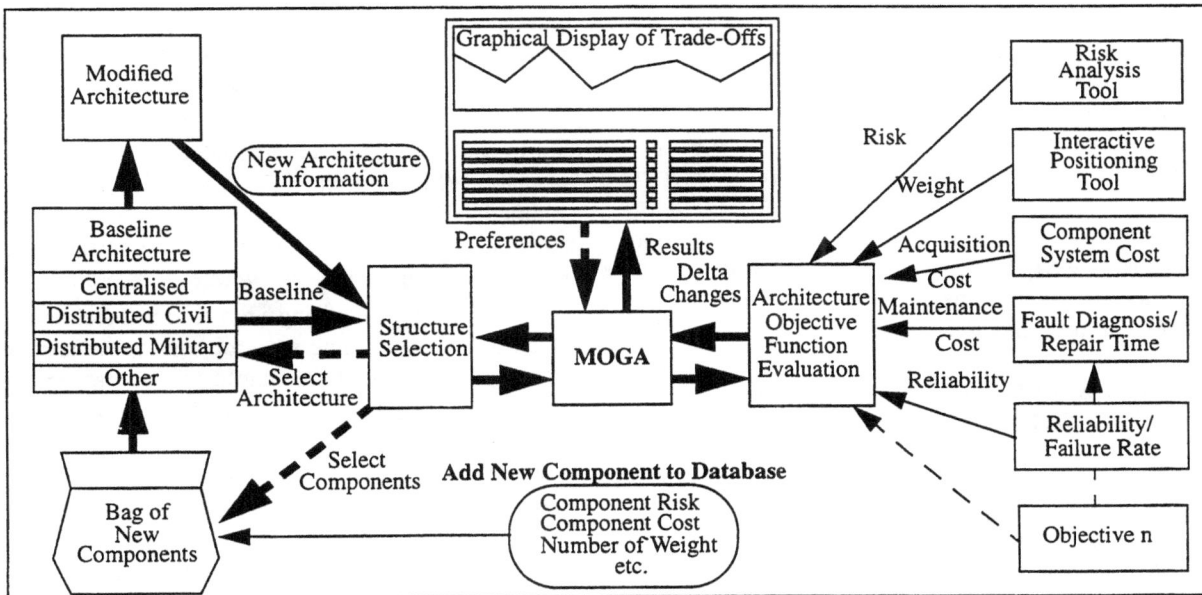

Fig. 5 Multi-Objective Optimisation of System Architectures

Competing architectures are evaluated with respect to a number of parameters, such as risk, weight, acquisition cost, maintenance cost and reliability. In order to provide justification that a new architecture is better than an existing architecture baseline architectures are defined so that changes made can be assessed in a relative manner.

5 ASTOVL SYSTEM OPTIMISATION

The distributed architecture of Fig. 4 was encoded as a set of binary strings such that actuators and sensors in the system could be selected from a "bag" of components (see Fig. 5) at random and incorporated into the system. This "bag" is updated with new components as and when they become available so that they can be included in the optimisation variables. For each unit in the system the input and output requirements were defined. Metrics were defined for dumb and smart elements to allow weight, cost, diagnostic capability, risk and maintenance to be evaluated. (Where maintenance takes into account potential for failure from complexity.)

Combined together the diagnostic capability and maintenance metrics can be used to define a metric for maintainability of the system. This is highly important for system operators who want to be assured of low operating support cost. Parameters for simplex, duplex and triplex sensors and actuators (both smart and dumb) were defined. From Fig. 4 the architecture was defined to initially have 13 "on-engine" smart units which can be connected via different bus topologies. In practice, the use of 13 smart units is unlikely given the weight limitations of current technology. It is therefore necessary to cluster functionality into less units to meet requirements. If clustered the sensor and actuator interconnections are automatically rewired to the new combined unit.

This allows the system to vary between a totally centralised system or a system with 13 distributed smart units.

The interconnection architecture between smart units in terms of the number of databuses required and the topology was first considered. A graphical interactive positioning tool was developed which allows smart units to be placed at specific points on the engine. A local optimisation is then performed to connect the units via a star network, a ring network, a horizontal databus running the length of the engine and a vertical databus running around the circumference of the engine. The interconnects required to connect the main smart units in terms of the weight of the wire, connectors, clips and brackets required to secure the databuses from vibration is automatically calculated for simplex, duplex and triplex databuses. (The fastening of the databuses was found by Thompson and Fleming (1995) to be a key parameter in previous work investigating system weight.) Fault tolerance of each topology with respect to failures that could be accommodated by the system was also considered. For the ASTOVL system ring architectures were found to be a significant advantage when considering weight and fault tolerance. In particular, for survival of two failures a duplex ring architecture or a simplex combination of star and ring networks provided the same fault tolerance for nearly the same weight. Interestingly, horizontal or vertical databuses performed badly with respect to weight, star architectures tended to perform badly with respect to both weight and fault tolerance. It should be noted that these results are for a specific military gas turbine engine. For a large civil turbofan engine the size of the engine and interconnection routing may result in a different topology being selected.

Within the optimisation fault tolerance requirements for replicated sensors are considered to be a hard constraint and so sensor selection was restricted to whether they should be smart or dumb. Adding smartness to sensors results in more cost and weight and so clear justification has to be made for a change. This is likely to be largely based on improvement in diagnosis possible. For actuators there is more latitude since adding smartness incurs a much lower percentage weight and cost penalty. Actuators can be simplex wound, dual-wound or triple-wound dependent on their safety criticality.

6 SIMPLEX, DUPLEX AND TRIPLEX ARCHITECTURE EVALUATION

Of particular interest in this study was whether a future distributed architecture should be duplex or triplex. To enable this three baseline architectures were used; a simplex architecture, a duplex architecture and a triplex architecture. Note that although simplex architectures do not meet safety requirements their evaluation allows comparative measures to be formed. This allows architectures to be compared to give "good" or "bad" classifications. In order to compare and classify architectures it is highly important to define objectives that take into account the many diverse design interactions from a variety of disciplines. In the following sections some key points of the objective function formulation are highlighted.

6.1 Cost and Weight Metrics

The cost and weight of future systems are extremely difficult parameters to estimate since accurate values for a smart unit, sensor or actuator are difficult to establish when the entry into service date may be 10 or 20 years in the future. As a result the problem was formulated such that percentage increases in cost and weight over existing dumb elements were considered. These can be varied and the effect on the system architecture selection can be evaluated. It should be noted that it is also possible to view the problem in another way and use the exploration of percentage increases in cost and weight to define specifications for future smart units, actuators and sensors relative to the viability of a system. As one might expect as the percentage overhead in cost is reduced there is a blurring in boundary between the simplex, duplex and triplex systems. This highlights that as the cost of adding smartness to system components is reduced it will become increasingly difficult to evaluate competing designs with respect to these objectives.

6.2 Risk Metric

The risk of the system takes into account the technologies used within the units, sensors or actuators and also where they are located. For instance, a sensor or actuator located on the fan case of the engine experiences relatively cool

temperatures. Location of smart components on the core of the engine requires use of high temperature electronics (Lewis, 1995). In some cases this is around 200^0C and can be accommodated by Silicon on Insulator (SOI) technology already becoming commercially available. In other cases extremely high temperatures can be experienced, for instance in nozzle vectoring actuators, where temperatures of 500^0C are possible requiring a move to SiC technology which is still undergoing research and hence is more risky. The status of the technology was categorised into the areas given below and intuitive scores assigned where negative values were given to negative factors and positive scores to positive factors. As a consequence the more positive the risk metric is, the less risky the solution.

- Certified
- Demonstrated on System
- Demonstrated in related Application (e.g. Automotive)
- New Technology
- Predicted Technology
- Difficulty to Certify
- Development Prog - Supplier
- Development Prog - External
- Productisation Cost
- Development Cost
- Alternative available

6.3 Maintenance Metric

For smart systems a clear trade-off exists between increases in fault diagnosis capability and increase in sources of potential failure from added complexity which would result in increased maintenance. It would not be useful for instance to add a component that will fail more often because of an increase in complexity. To take this into account a maintenance metric was defined related to the complexity of the system based upon experience of sources of failure and causes of No Fault Found. NFF is a problem where faults are flagged and identified by the diagnostic system in a particular unit but cannot be reproduced when the removed unit is bench tested.

6.4 Fault Diagnosis Metric

The fault diagnosis metric takes into account the ability of an element for:

- Local Compensation
- Self Diagnosis
- Remote Diagnosis
- Reconfiguration on Failure
- Time Limited Despatch

It should be noted that improvement in fault diagnostic capability is a major selling point of distributed smart components.

7 RELATIVE RANKING

Key to selection of good and bad architectures is the ability to rank competing designs. Consider Fig. 6

which shows results of comparison of fault diagnosis and maintenance for 150 simplex, duplex and triplex architectures selected at random. It can be seen that a duplex architecture that gives the same fault diagnosis but has a higher maintenance cost than a simplex architecture can be classed as "bad". Likewise, a similar argument can be made when considering triplex architectures against duplex architectures.

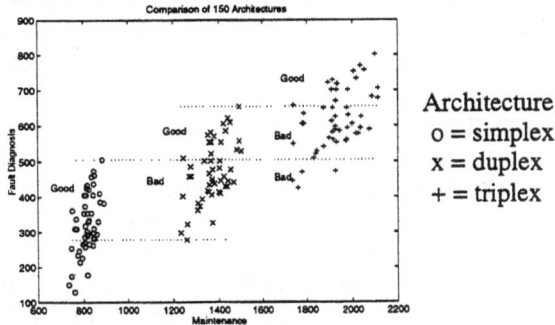

Fig. 6 Fault Diagnosis vs Maintenance

Conversely, duplex architectures that give better fault diagnosis than simplex systems can be considered to be "good". Likewise triplex architectures that give better fault diagnosis than duplex architectures are "good". It is interesting to note that some triplex architectures perform worse than some of the simplex architectures indicating that they are "very bad". For engineers the interesting next step is to look at the architectures that perform badly to see whether there is a fundamental characteristic that makes them bad. Likewise good characteristics of architectures can be highlighted.

8 UTILISATION OF GENETIC ALGORITHMS

In the example of Fig. 6 just two objectives are considered. Considering more than 3 objectives at a time is difficult to visualise in a meaningful way. Groups of three objectives can be plotted to give different viewpoints of the system but knowing how best to visualise and interpret the data is not a simple task. The MOGA technique (Fonseca and Fleming, 1993) was developed at Sheffield University for multi-objective optimisation. It has been applied to a number of applications (Chipperfield et al.,1997, Schroder et al., 1997) and has proved to be very successful when tackling complex problems. The MOGA tools are highly interactive and intuitive allowing the design engineer to adjust objective goals on-line and visualise the effect on the solutions to the problem. A family of trade-off graphs and preference articulation windows are shown in Figs. 7, 8 and 9 for simplex, duplex and triplex architectures respectively. This follows a parallel coordinates methodology for display. The x-axis shows the design objectives and the y-axis the performance of different architectures. The system architectures are optimised with respect to cost, risk, maintenance and fault diagnosis capability. Each line represents a preferred non-dominated

solution found for the design goals as indicated by the cross marks. Design goals can be varied interactively via the sliders and objectives can also be ignored or turned into constraints. In each case results show the solutions obtained for a population of 40 individuals after 100 generations of a genetic algorithm. The problem was organised such that objective parameters as described in the previous sections were minimised.

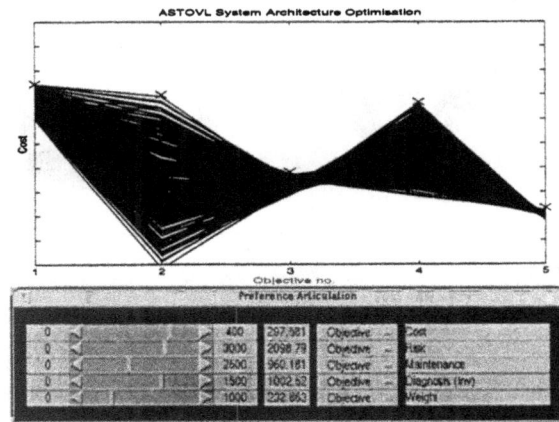

Fig. 7 Simplex Architecture Optimisation

Note that since fault diagnosis capability needs to be maximised an inverse metric of "lack of" fault diagnosis was used in the optimisation. The risk metric was also rescaled so that it could be minimised. Using the design interface the engineer can explore trade-offs in the system. Where lines connecting objectives cross, e.g. between objectives 3 and 4, maintenance and diagnosis capability, a trade-off exists. This is important design information and highlights relationships between parameters. The objectives can also be interactively moved via mouse input allowing, for instance, objective 1, cost, to be directly compared with objective 5, weight. This is a very powerful way of establishing the existence and severity of trade-offs which exist in the system.

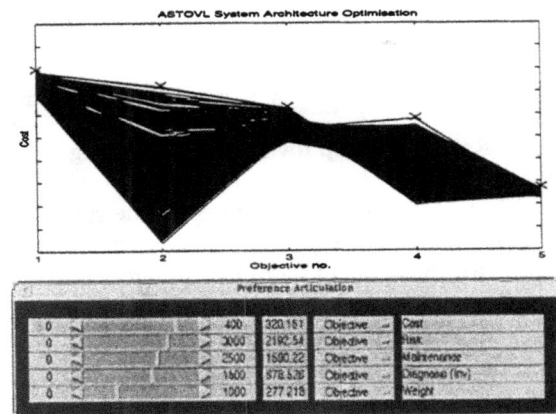

Fig. 8 Duplex Architecture Optimisation

It is also possible to tighten objectives until no solution to the problem exists. This can be used to establish achievable system requirements for specification of a system to outside suppliers. The technique can thus be used in a number of ways in the design process. A major advantage of the technique is that many more architectures can be evaluated in much shorter time scales. Traditional methods of teams of expert engineers evaluating architectures is a laborious process which restricts the number of potential architectures that can be considered. For this example the MOGA considered 4000 different architectures.

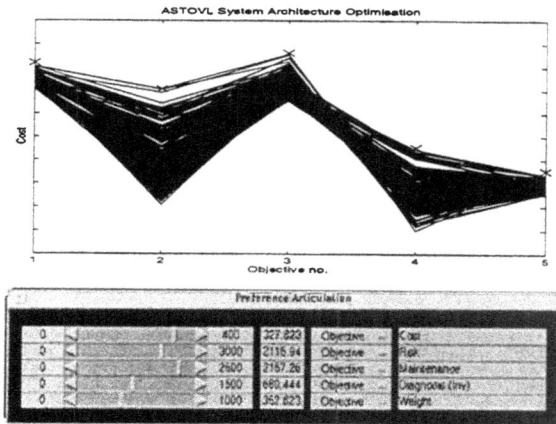

Fig. 9 Triplex Architecture Optimisation

As expected system cost, risk, maintenance, diagnostic capability and weight all increase as the architecture is changed from simplex, to duplex to triplex. It can be seen, however, that overlaps exist between different architectures (see Fig. 10). For instance, when considering cost the range of solutions for simplex, duplex and triplex systems overlap. This allows relative ranking to be used to give "good" and "bad" classifications with respect to this objective.

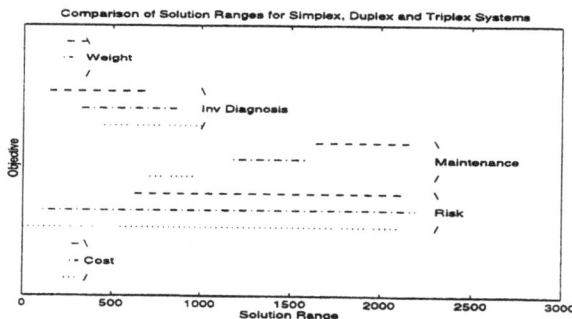

Fig. 10 Range Overlap for Relative Ranking

For risk it can be seen that there is a wide spread of values in the case of simplex and duplex architectures but less so when going to a triplex architecture. This highlights that the importance of risk increases as one moves to a triplex architecture. Maintenance is an interesting metric as there is no overlap between the three architecture types. Therefore relative ranking cannot be used in this case. For diagnostic capability clear overlaps exist showing the importance of this parameter. For weight the differences between the three architectures does not appear to be very great. Weight is highly important in this application, however, and small changes have a dramatic impact on military aircraft performance. Here, it is important to remember that the tool is only an assistant. Expert

engineer skill is still needed to drive the optimisation process and select solutions.

9 CONCLUDING REMARKS

Multi-disciplinary multi-objective optimisation has been applied to the design of a distributed gas turbine engine controller architecture for an ASTOVL aircraft. It has been shown that the MOGA technique combined with an interactive interface is an extremely powerful way of tackling this problem. The approach adopted reflects the interactive nature of the design process which is highly important for design engineers. The idea of relative ranking of architectures into "good" or "bad" categories has also been introduced. This is fundamental to architecture evaluation and can also be used to highlight good design characteristics to be encouraged and also bad design characteristics to be discouraged. Future work will address addition of further objectives within the optimisation process, sensitivity analysis and enhanced visualisation techniques to extract good and bad design practices.

10 ACKNOWLEDGEMENTS

The authors would like to acknowledge the support of Rolls-Royce plc and the assistance given by Rolls-Royce Military Aero-Engines Ltd.

11 REFERENCES

Chipperfield, A.J., Fleming, P.J. and Betteridge, H., (1997), Evolutionary Control Configuration Design, AIAA Joint Propulsion Conference, Seattle.

Fonseca C.M. and Fleming, P.J., (1993), Genetic Algorithms for Multiple Objective Optimization: Formulation, Discussion and Generalization, Proc. International Conference on Genetic Algorithms 5, pp416-423.

Kroo, I., Altus, S., Braum, R., Gage, P. and Sobieski, I., (1994) Multidisciplinary optimization methods for aircraft preliminary design, 5th AIAA/USAF/ NASA/ISSMO Symposium on Multidisciplinary Analysis and Optimization, AIAA 94-4325.

Lewis, T.J., (1995), Distributed Architectures for Advanced Engine Control Systems, AGARD/PEP 86th Symposium on Advanced Aero-Engine Concepts and Controls, Seattle.

Schroder, P, Chipperfield, A.J.,and Fleming, P.J., (1997), Multi-Objective Optimisation of Distributed Robust Active Magnetic Bearing Controllers", GALESIA'97.

Thompson, H.A. (1994), Parallel Processing Architectures for Aerospace Applications, IFAC Control Engineering Practice, Pergamon Press, Vol. 2, No. 3, pp. 509-520.

Thompson, H.A. and Fleming, P.J, (1995), Evaluation of Distributed Fault Tolerant Architectures for Gas Turbine Engine Control, Aerotech 95.

ON THE SUITABILITY OF OBJECT-ORIENTATION
FOR DISTRIBUTED REAL-TIME SYSTEMS DEVELOPMENT

Carlos Eduardo Pereira

Department of Electrical Engineering - DELET
Federal University of Rio Grande do Sul - UFRGS
Porto Alegre - RS - Brazil
Email: cpereira@iee.ufrgs.br

Abstract

This paper discusses issues related to the use of object-oriented concepts to the development of distributed real-time systems. It supports the argument that object-oriented concepts provide a powerful framework for developing large and complex distributed real-time applications, basically because of its abstraction mechanisms, concurrence and distribution support, reuse of models and code, etc. However, the paper also claims that most of existing object-oriented methods are lacking of very important concepts for hard real-time systems, specifically specification of timed behavior and temporal requirements as well as verification of timing properties. An approach for overcoming existing problems is presented.

Keywords: Real-Time Systems, Object-Oriented Paradigm, Industrial Automation

1 Introduction

The role played by real-time systems in our society is increasing with surprising speed, since our everyday life becomes more and more dependent on them. One of the main characteristic of a real-time system is a strong interaction with its environment. The dynamics of the physical system under control impose timing constraints that must be met and therefore dictate the temporal behavior to be achieved. Therefore, the assurance of the (functional and temporal) correctness of a real-time system, mainly for those embedded in safety-critical systems, is something that must be made possible. The behavior of a real-time system must be predictable, in order to guarantee that timing constraints will always be met.

However, not alone the assurance of temporal correctness system poses a problem during real-time system development. Large, dynamic real-time systems also require complex embedded software and hardware, that are difficult to understand, maintain and modify. The 'usual' difficulties present in building large systems are exacerbated in the real-time field, mainly due to the introduction of the additional complexity of time-dependent behavior (Bihari et al, 1992). Moreover, one can expect that many of the 'next-generation' real-time systems are going to be even larger and more complex than those currently being developed. They will contain highly dynamic and adaptive behavior, involve complex timing constraints and have a long life-time. Because of this, aspects like maintainability and extendibility have to be taken into account during the development of real-time applications.

2 Advantages of using OO for Real-Time Systems Development

Among several approaches that have been proposed over the last years, the use of a concurrent object-model can be considered as the most suitable for developing large real-time systems. This statement can be supported by considering the following aspects:

- Understandability and maintainability: Object-oriented models allow a direct mapping of the problem domain semantics onto the model being built (Embley et al., 1992 and Borgida et al., 1985). Due to the strong interaction of a real-time system with its environment, this direct mapping is extremely beneficial, since objects in the model can be used to represent and, so far, behave like the corresponding elements in the real-world (Darscht et al., 1995). In this sense, objects are considered as logical machines and not only as abstract data types.

- Modularity: Identifiable objects/classes provide a logical partitioning of the problem domain, thus unrelated objects can be totally ignorant of each other (Gwinn, 1992). Each of the object classes represents a self contained piece of the problem domain which can be examined separately from, or within the context of the system.

- Concurrence: Real-time systems have to interact with concurrent processes going on in the technical plant. The requirement imposed by concurrency on a modeling language is the ability to model a collection of interacting control threads, to model scenarios involving these control threads, and a means to help resolve concurrency issues. By using the concept that objects may have a behavior captured by an independent thread of control, objects become natural units for concurrent execution. This allows the real-world concurrence of applications to be naturally modeled. (Wegner, 1991 and Pereira et al., 1995).

- Distribution: In an object-oriented model, the real-time system is organized as cooperative collections of objects, which may interact by passing messages one to another in a client-server fashion. This pattern leads to a very effective way for implementing object-oriented models in distributed and parallel hardware and software architectures - consisting of a set of computing sites loosely coupled through a communications network. Such distributed architectures are becoming widely used due to the recent advances in fields such as microelectronics, communication, and software (Bihari and Gopinath, 1992).

- Better management of complexity: Object-oriented techniques provide powerful abstraction concepts to deal with complexity. The different kinds of hierarchies (generalization/specialization-, aggregation-, and 'services'-hierarchy) facilitate the presentation and management of varying levels of details.

- Reuse: Reusing involves laying aside all the details of a part of a model except those that match details of some other part of the model, then segregating the common details so that they are constrained to be consistent across multiple appearances and multiple uses. The object-oriented paradigm eases reuse specially due to its inheritance and instantiation capabilities. Inheritance allows the definition of a type hierarchy, where the behavior and structure of most general classes can be reused and specialized (by simply adding new services) within the context of subclasses. Instantiation allows another kind of reuse, since it allows a replication of attributes and behavior pattern defined within the scope of a class to all instances.

- Increased extendibility: The refinement or extension of a superclass into a subclass is very nicely supported by using inheritance mechanisms.

- Traceability: The fact that the objects identified during the analysis phase can migrate to implementation in a object-oriented programming language enhances requirement's traceability.

3 Drawbacks of existing object-oriented methods

Take properly into account the 'formal' vs. 'understandable' trade-off is a key point in developing complex real-time systems. Only by means of an approach that is both easy to understand, while also being formal enough to allow the checking of the description correctness, and consistency, an effective support to the requirements engineering phase can be achieved. Within this context, the use of object-oriented techniques are a promising approach to tackle problems related to the development of complex real-time systems. As already mentioned, objects fit nicely with concurrency since their logical autonomy makes them a natural unity for concurrent execution. That implies a new and fruitful way of thinking, thereby expanding greatly our modeling power and enabling the parallelism in the real world to be expressed in a more natural and easily understandable way. However, most of existing objet-oriented techniques, even those developed for real-time systems, do not support the description and evaluation of timing aspects properly. Some of the issues that are usually not considered by object-oriented methodologists are:

- description of timing requirements, such as cyclic activation of object methods, deadlines, absolute time constraints between events, and so on. One should be able to specify these aspects directly using a proper specification language. To define a Timer Class, from which timer instances can be created, is surely not the best way to specify timing requirements.

- inheritance of behavior and inheritance of timing requirements: it is a common sense that every object-oriented method should support the inheritance concept. Although there are some people concerned with problems related to inheritance of behavior and the side effects that a small change in a class behavior can cause in the behavior of its sub-classes. However, there are few methods that discuss inheritance of timing requirements. In (Pereira, 1995) an approach has been proposed, in which inherited timing requirements can only be made more restrictive in comparison to the super-cell. By doing that, we can guarantee that if some timing property holds for a given class, it will hold for all their sub-classes.

One of the key aspects that still prevent some real-time designers from adopting the object paradigm is regarding the determinism of object-oriented programs' temporal behavior. Here some considerations:

- dynamic binding and temporal determinism at runtime: it is true that languages supporting 'real' dynamic binding could not give any a priori guarantee that all timing constraints would be fulfilled, since it would always be possible that a new sub-class could be created, whose behavior would not respect timing constraints. However, since in many object-oriented languages dynamic binding implementation does not really means 'late binding' but also a possibility of implementing polimorfic functions (such as virtual functions in C++), it would be possible to check off-line if timing constraints would be met even when using virtual functions (in this case a worst case strategy should probably be adopted). In this case the strong typing mechanism usually present in object-oriented languages would even ease the checking process.

- temporal determinism and 'Quality of Services': deadlines and others timing requirements can be explicitly considered as key factors in the 'quality of services' offered by a given instance of a class. By doing this, an instance would be able to negotiate the execution of a service within a given time interval. A typical example would be an instance of a class Controller, where different implementations for the method control_algorithm() could be available (for instance, one implementing an on-off, other a PID, and other a state-based control strategy). The instance itself could be able to determine, at run-time, which method should be called.

- real-time communication protocols and operating system services: distributed objects rely upon an operation system and communication protocol in order to have time bound communication. Unfortunately, most of the existing systems do not have an integrated solution, where programming languages, communication protocol, and real-time operating system build an harmonic trio. The use of time parameters in the specification of a quality of a service may represent a solution to that problem.

- load balancing: since the object model supports very nicely concurrency and distribution, it allows the implementation of distributed real-time systems with dynamic load balancing. These characteristic is specially interesting for developing adaptive and dynamic real-time systems.

4 The ADOORATA Approach

ADOORATA stands for A Distributed Object-Oriented ReAl-Time Method for Industrial Automation. It has been developed within the scope of a cooperation program between Germany and Brazil. ADOORATA extends a previous project called OORTAC (Pereira et al., 1995), where an object-oriented framework for specification and design of distributed real-time systems was proposed. While OORTAC has basically focused on the earlier design phases, ADOORATA aims to cover the whole life cycle. For instance, it deals with the definition of a new real-time object-oriented language as well as of a predictable runtime environment.

An ADOORATA project starts with a careful problem domain analysis and requirements definition. For that purpose, usual modeling notations for describing class hierarchies, state machines, and object communication diagrams, as those provided by existing real-time object-oriented analysis methods can be employed (such as (Shlaer and Mellor, 1992), (Selic et al., 1994), or the real-time extensions to the unified modeling language, UML-RT). Special care is taken in order to achieve a specification that is hierarchical, layered, and compositional. The objects and classes behavior is first depicted using a extended version of conventional state-machines (in fact, extended automata are applied which may access the current object's state and the current value of object's attribute before performing a state transition). States' hierarchy is introduced in order to model behavior's inheritance (inheritance is considered as strong sub-typing, where subclasses may only extend the behavior of super-classes by refining their states). This state model is then enriched with time predicates in an event-based language (Pereira, 1995b). Timing requirements are described in the scope of objects, classes, and their relationships as assertions about the time of the occurrence of events (assertions between time intervals are allowed as well, they are indeed mapped to time point assertions). Both absolute, i.e. related to a global clock, or relative assertions are allowed. Additionally, the event-based language issued for timing requirements specification introduces two types of statements. There are statements to describe the actual behavior of the environment in which the system will be embedded and which the real-time system has to react to. These statements specify (temporal) facts, i.e. temporal relations present in the technical plant and which the system has to be aware of. There are additional statements to clearly specify the expected behavior to be achieved by the automation system under development. These are declarative, in the sense that they only describe what has to be achieved without imposing a previous solution to the problem. The obtained specification can then be simulated, using for instance the SIMOO framework (Copstein et al., 1996).

The obtained model tends to be easier to understand, since it encompasses concepts, as well as the structure and concurrence of the real world. It also enforces components reusability and extensibility.

During design, the identified classes are then mapped to a programming and runtime environment which supports the concept of active objets or actors (autonomous entities that can behave concurrently). Unfortunately, most OO-languages cannot be considered real-time programming languages, since they lack real-time features and mechanisms for expressing timing constraints. Therefore, three different alternatives for such a programming and runtime environment have been investigated within the scope of ADOORATA:

- AO-C++ (active object C++): it proposes extensions to C++ in order to obtain a real-time concurrent language, from which an object code incorporating IEEE-POSIX real-time operating systems facilities is generated (Pereira and Frigeri, 1996);
- Use of standards for communicating distributed objects, such as CORBA-RT;
- Definition of a new real-time object-oriented language, PEARL*. This language can be considered an object-oriented extension of the real-time programming language PEARL. It also includes some constructions for dealing with safety critical applications (Frigeri et al., 1997);

The proposed programming model should 'fill the gap' between the high-level real-time object-oriented model, based on concurrent objects, and a real-time distributed execution model (as provided by real-time operating system kernels).

5 Case Study

In order to better illustrate the proposed approach, some aspects of one of the case studies developed within the scope of the ADOORATA project, the automation of work-cell for distributing work-pieces in a flexible manufacturing system, will be described in this section. The work-cell consists of a *conveyor belt* on which *work-pieces* are transported in order to be classified by a *robotic manipulator* and stored into one of the four bins of *a indexing table*. Work-pieces are classified according to their geometric properties. Information about incoming work-pieces is acquired using a *CCD camera*.

The first step in the ADOORATA approach is the creation of an object-oriented diagram, where components of the technical plant are mapped to classes and objects. As already mentioned such diagram can, a priori, be created using any of the existing CASE tools that support object-oriented methods. The class behavior is then depicted using state machine models and timing requirements are specified using the event-based language. Following classes could be identified:

- **Robotic manipulator**: a robot arm with 6 degrees of freedom (5 axes plus a gripper). Its function is to pick-up a work-piece and to store it into a corresponding bin. In order to be able to catch the work-piece properly, the robot must know the exact geometry of it.
- **Conveyor belt**: it uses a rubber coated belt driven by a geared down motor. Operates at a constant speed, transporting the pieces.
- **CCD Camera**: A black-and-white CCD camera. Its function is to detect a piece and send its geometric information to the robot arm. It also makes possible the detection of work-pieces outside a predefined standard. Those work-pieces must be discarded.
- **Indexing table**: A solid aluminium platen driven by a geared stepper motor. It contains bins for storing the classified pieces.
- **Work-piece**
- **Parts dispenser**: Stores defective work-pieces

The robotic manipulator, the conveyor, and the indexing table, were described as active, concurrent objects. They are autonomous entities that behave concurrently with others, having its own thread of control. These classes are not only hardware and components wrappers but also incorporate control and coordination algorithms (the desired behavior that the automation system has to fulfill). For instance, the robotic manipulator class encompasses a control algorithm for properly positioning itself in order to load and unload work-pieces. It has a method called *load_piece*, which receives as parameter a work-piece instance. Based on the work-piece attributes, the robotic manipulator object is able to load and store it on the indexing table. It also communicates with both the conveyor and indexing table objects, in order to synchronize and coordinate their behaviors.

With such encapsulation of the control strategy within active classes/objects a distributed system architecture is obtained, which allows a more natural mapping of the structure and concurrence present in real world. This strategy makes also possible to incorporate several polimorfic control functions with different performance and quality of control properties. The decision about what function should be activated can be determined based on the (temporal) situation at the moment of its call. So, if an object has enough time left until the next deadline a more sophisticated algorithm can be activated. Instead, in case of a more critical situation, a faster algorithm can be used.

The system has been implemented using the AO-C++/QNX environment (Pereira et al., 1996), whose main features are:

1. the environment supports the definition of active, concurrent, and distributed objects;

2. since the number of added primitives to C++ is minimal, C++ programmers can become familiar with the syntax very quickly. The environment supports all existing C++ object-oriented features, such as multiple inheritance, polymorphism and overloading;

3. the C++ inter-object-communication is mapped to the inter-process-communication provided by QNX (actually any other RT-UNIX kernel supporting the POSIX standard could be adopted). That means, the semantics of the new constructs were defined as close as possible of those for sequential, single-machine applications. Therefore, a program written in AO-C++ can be translated without any major modification to different hardware architectures, so that the definition of the number of processing units as

well as the adopted communication protocol can be defined based on the specified timing requirements;

In our case study, the final system architecture consisted of a computer network with three IBM-PC compatible microcomputers running QNX. Computational intensive objects, such as the *CCD camera* object were allocated to dedicated hardware. Some objects, such as the robotic manipulator object, were split into two objects, running on different processors, one responsible for the manipulator position control and another for the coordination with other active objects.

6 Conclusions

An object-oriented approach has been presented. Besides the usual advantages of adopting the object-oriented paradigm in the development of real-time systems (listed in section 2), the approach aims to overcome some drawbacks of existing real-time object-oriented methods (as discussed in section 3). The results obtained so far have been quite encouraging. The approach has allowed a smooth transition among the models adopted in the different development phases. Starting from a concurrent object model of the technical plant, the desired temporal behavior and timing requirements are incorporate into active objects. The obtained model keeps the semantic and allows a natural mapping of the structure and concurrence present in the real world. The model is then converted to a distributed hardware and software architecture, so that the obtained executables should behave deterministically.

Ongoing work focuses on the definition of a high-level real-time programming language and an environment for programming safety-critical applications. Another work that has been carried out deals with the definition of a predictable run-time environment for distributed real-time object oriented systems.

7 Acknowledgments

This work has been partly supported by CNPq and FAPERGS (Brazilian research agencies). Thanks are also given to Leandro Buss Becker, Valter Bianchi Filho, and Rodrigo Caimi, who have implemented the described case-study.

8 References

Bihari, T. and Gopinath, P. (1992). *Object-Oriented Real-Time Systems: Concepts and Examples* IEEE Computer, Decemb pp. 25-32.

Borgida, A. Greenspan, S. and Mylopoulos,J. (1985). *Knowledge Representation as the Basis for Requirements Specifications.* Computer April 1985, pp. 82-91.

Copstein, B., Pereira, C. and Wagner, F. (1997) *SIMOO Environment for the Object-Oriented Discrete Simulation.* 9th European Simulation Symposium, Passau Germany, October 1997.

Darscht, P., Pereira, C. and Frigeri, A.H. (1995) *Building up Object-Oriented Industrial Automation systems : Experiences interfacing Active Objects with Technical Plants.* 1st IEEE - WFCS'95, Leysin, Swizerland, 1995.

Embley, D., Kurtz, B., Woodfield, S. (1992) *Object-Oriented Systems Analysis. A Model-Driven Approach.* Yourdon Press, Prentice Hall, Englewood Cliffs, 1992.

Gwinn, J. (1992) *Object-Oriented Programs in Real-time.* ACM SIGPLAN Notices, Vol. 27, No. 2, Feb 1992, pp. 47- 56.

Pereira, C., Frigeri, A., and Halang, W (1997) *Modelling Real-Time Applications through Interfaces and Active-Objects.* 22nd IFAC-IFIP Workshop on Real-Time Programming, Preprints, pp. 101-106. Lyon France, Sept. 1997.

Pereira, C., Frigeri, A., Darscht, P. and Halang, W (1995) *Object-Oriented Development of Real-Time Industrial Automation Systems.* IFAC Triennial World Congress, San Francisco USA, July 1996. pp. 321-326. ISBN:0-08042923-8

Pereira, C. (1995) *Proposal for Timing Requirements Specification and Consistency Checking during the Development of Real-Time Industrial Automation Systems* (in German). Institute for Industrial Automation and Software Engineering, University of Stuttgart, Germany, Feb. 1995. (165 pages).

Pereira, C. E.(1995b) *Temporal Reasoning on Object-Oriented Real-Time Specifications by using Constraint Propagation Techniques.* 20th IFAC/IFIP Workshop on Real Time Programming. USA, November, 1995.

Pereira, C. and Frigeri, A. (1996) *Applying Active-Objects to the Development of Real-Time Systems on a C++-QNX Environment,* Special Issues in Object-Oriented Programming - Workshop Reader of the 10th European Conference on Object-Oriented Programming, ECOOP'96. Linz, Austria, July, pp. 470-473.

Selic, B.; Gullekson, G. and Ward, P. (1994) *Real-Time Object-Oriented Modeling.* John Wiley & Sons, Inc., ISBN 0-471-59917-4.

Shlaer, S. and Mellor, S. (1992) *Object Life Cycles: Modeling the World in States.* Yourdon , NJ. 1992.

Stankovic,J. (1988). *Misconceptions about real-time computing: a serious problem for next-generation systems.* Computer, October 1988, pp. 10-19.

Wegner, P. (1990). *Concepts and Paradigms of Object-Oriented Programming.* OOPS-Messenger, Vol. 1, No 1, Aug 1990, pp.8-87.

TOWARDS REAL-TIME, NON-LINEAR QUADRATIC OPTIMAL REGULATION

Robert F Harrison and Stephen P Banks

Department of Automatic Control and Systems Engineering, The University of Sheffield, Sheffield, UK.

Abstract: A novel non-linear design method based on linear quadratic optimal control theory is presented that applies to a wide class of non-linear systems. The method is easy to apply and results in a near optimal solution that has the potential to be implemented in real-time in the sense that the solution is causal, in contrast to the conventional, quadratic regulator solution, which is anti-causal (backwards in time) and must be computed off-line. The key feature of the design method is the introduction of state-dependence in the weight matrices of the usual linear quadratic cost function, leading to a non-linear control policy, even for linear dynamics. To demonstrate the method, a simple vehicle suspension model with a cubic damping force is used, in conjunction with non-linear state penalties that better reflect the engineering objectives of active vehicle vibration suppression. A number of simulations is conducted and compared with a passively mounted vehicle. *Copyright © 1998 IFAC*

Keywords: Non-linear control systems; quadratic optimal regulators; active vehicle suspension; Riccati equations; stability.

1. INTRODUCTION

Linear quadratic (LQ) optimal control theory is a highly developed approach for the synthesis of linear optimal control laws that has been widely applied. In particular, the infinite-time-horizon solution has appeal for the regulation of processes that are well modelled by linear time-invariant dynamics because the solution comprises a set of static gains that are calculated once, off-line, and are implemented causally thereafter. The finite-time-horizon solution, while being more general, and admitting of time-varying dynamics or weighting parameters, is essentially an off-line procedure because the associated (differential) Riccati equation must be solved backwards in time. Incorporation of such a scheme in a closed-loop, real-time process is not therefore possible. The receding horizon approach attempts to overcome this difficulty by repeatedly solving the open-loop, finite-time-horizon problem for short periods into the future, and using this as the feedback gain over the time step. It does this at the

expense of optimality, which may or may not be important from the point of view of the practising engineer.

A further disadvantage of the LQ philosophy is that, being a linear feedback, the control signal is affected in the same way by small and large signals. In many applications it may be preferable to ignore small error signals (due, for instance, to measurement noise) as far as possible, while responding optimally to large errors. It may also be desirable to switch attention between control objectives depending on their current values. The receding horizon strategy approximates the former behaviour to a certain extent, but is unable to address the latter.

For non-linear dynamics the situation is exacerbated, with few explicit solutions for their optimal quadratic control as yet known except those based on series expansions. These are unrealisable, unless via truncation, leading to a loss of optimality and possibly stability. Normally optimal quadratic control for non-linear systems is conducted numerically and tends, inherently, to be non-causal.

In this paper we make use of a new result that generalises the LQ theory to non-linear systems to provide a non-linear design method that overcomes some of the difficulties mentioned above. This non-linear quadratic (NLQ) method applies to systems having a broad class of non-linear dynamics with state-dependent weighting matrices (the design degrees of freedom). In brief, it turns out that the infinite-time-horizon LQ regulator problem when solved afresh at every point on the state trajectory leads to a near-optimal control policy (Banks and Mhana, 1992). For admissible system dynamics, the weighting parameters can be made to be functions of the state variables. Thus, in addition to handling non-linear dynamics, the design stage allows for the introduction of state-dependence in the weighting matrices, leading to a more flexible control strategy.

Our method is causal, but has considerable computational overhead. However, by using a solution to the Riccati equation based upon the matrix sign function (Gardiner and Laub, 1986), it is possible to derive a parallel algorithm (Gardiner and Laub, 1991) that may be suitable for real-time implementation.

In order to demonstrate the approach we consider the active control of a vehicle suspension system in order to reject disturbances induced by surface asperity: a problem that has received much attention in the past although usually for linearised dynamics. Clearly, the assumption of linearity may often be valid but some designs are inherently non-linear such as the oleopneumatic shock struts of aircraft landing gear.

While the LQ approach is attractive in that it is possible to penalise different variables so as to trade-off between, say, ride comfort and handling, or comfort and suspension travel, the way these variables are treated is essentially fixed – no provision is made to allow the suspension to distinguish between a smooth surface and a rough one. Evidently, while comfort might be a prime objective under normal circumstances, in extreme conditions the suspension should be stiffened to avoid hitting its limits, hence incurring damage. This is true even if the dynamics are linear up to this point. Although, in principle, time-varying weighting parameters are allowed in the (finite-time) LQ approach, lack of prior knowledge of the surface profile, and the acausal calculation for the solution makes the introduction of these difficult. The required amplitude dependence can never, therefore, be achieved through the LQ approach.

We illustrate our approach on a simple non-linear model incorporating a non-linear damping element, and compare our results with the passive system and the optimal NLQ solution using fixed matrices.

The remainder of the paper is organised as follows. In §2, to motivate the work, the LQ regulator problem is first set out, the generalised results are stated and the robust solution of the infinite-time-horizon LQ problem is outlined. In §3 the suspension model is presented and the choice of design parameters is discussed. The results of a series of experiments are described in §4, followed by conclusions in §5.

2. THE DESIGN METHOD

2.1 Linear quadratic regulator

The LQ optimal regulation problem is expressed as follows: minimise the cost function

$$J = \int_0^\infty \left(\mathbf{x}' Q \mathbf{x} + \mathbf{u}' R \mathbf{u} \right) dt \qquad (1)$$

subject to the linear time invariant dynamics:
$$\dot{\mathbf{x}} = A\mathbf{x} + B\mathbf{u} \qquad (2)$$
where $\mathbf{x}$ is an n-vector of system states, $\mathbf{u}$ is an m-vector of control variables, A and B are matrices of appropriate dimension and the superscript, t, indicates transposition. The matrices Q and R are positive semi-definite and definite, respectively, and are used to penalise particular states and controls according to the engineering objective.

It is well known, e.g. (Friedland, 1987), that the control policy which solves the above problem is a linear combination of the system states, given by:
$$\mathbf{u} = K\mathbf{x} \qquad (3)$$
$$K = -R^{-1}B'P \qquad (4)$$
$$0 = PA + A'P - PBR^{-1}B'P + Q \qquad (5)$$
where K is given by (4) and P is the positive definite solution of the algebraic matrix Riccati equation (5). A unique, positive definite solution to the above exists if the pair (A,B) is stabilizable and (A,Γ) is detectable, with $Q = \Gamma'\Gamma$.

2.2 Non-linear quadratic regulator

The extension of the above to non-linear systems looks identical, except that, instead of performing a single optimisation and applying the resulting gain-matrix for all time, the optimisation has to be carried out at every time step. Consider a non-linear dynamical system that can be expressed in the form:
$$\dot{\mathbf{x}} = A(\mathbf{x})\mathbf{x} + B(\mathbf{x})\mathbf{u} \qquad (6)$$
where the Jacobians of $A(\mathbf{x})$ and $B(\mathbf{x})$ are subject to some bounded growth conditions (Lipschitz), then at each point, $\bar{\mathbf{x}}$, on the state trajectory, a linear system is defined with fixed $A = A(\bar{\mathbf{x}})$ and $B = B(\bar{\mathbf{x}})$. In (Banks and Mhana, 1992) it is shown that solving the infinite-time LQ optimal control problem, point-wise on the state trajectory, results in the near-optimal, stabilising quadratic control policy for systems described by equation (6). Thus, by choosing the $\mathbf{u}$ that minimises the usual quadratic cost function at

every time step, we have a near-optimal control policy for a very wide class of non-linear systems. Evidently, $A(\overline{\mathbf{x}})$, $B(\overline{\mathbf{x}})$ and Q are subject, point-wise, to the same conditions as for the linear case. It is clear that the proposed solution is identical to the one obtained from equations (3, 4 and 5) when the dynamics are linear.

Because the control synthesis takes place point-wise, the designer is now free to select Q and R in ways which are more directly applicable to the control *engineering* objectives. These can be made to be functions of the instantaneous state variables, i.e.

$$J = \int_0^{\tilde{\infty}} \left(\mathbf{x}' Q(\overline{\mathbf{x}}) \mathbf{x} + \mathbf{u}' R(\overline{\mathbf{x}}) \mathbf{u} \right) dt \qquad (7)$$

subject to the requirements for the solution of the Riccati equation and the invertibility of R. Ensuring that $A(\overline{\mathbf{x}})$, $B(\overline{\mathbf{x}})$, $R(\overline{\mathbf{x}})$ and $Q(\overline{\mathbf{x}})$ satisfy these requirements *a priori*, is difficult in general, however, for polynomial functions which are not identically zero, the required properties will be lost only on sets of zero measure and will not, therefore, persist.

2.3 Solving the matrix Riccati equation

In order to calculate the optimal solution, it is necessary to solve the matrix Riccati equation at each point in time. In practice this will be done in a computer and it will be necessary to solve the equation at each discrete time-step. The usual approach to the solution of this problem is via an eigen-decomposition of the Hamiltonian matrix for the system (Laub, 1979). For sizeable dynamics such an approach is computationally intensive and may not be able to deliver solutions at the required sample rate (i.e. in "real-time"). It can also be sensitive, depending as it does on the numerical solution of an eigen-problem. It should be noted that, even though the problem is to be solved in *discrete-time*, we do not solve the discrete-time Riccati equation: the dynamics are essentially continuous-time.

The matrix-sign-function (MSF) is an appealing alternative to eigen-decomposition owing to its simplicity (Roberts, 1980), requiring only the operations of matrix inversion and addition, and multiplication of a matrix by a scalar (Gardiner and Laub, 1986). This simplicity also suits the MSF algorithm to parallel computation (Gardiner and Laub, 1896, 1991) as follows:

$$Z_0 = H$$

$$Z_{i+1} = \frac{1}{2}\left(\frac{1}{c_i} Z_i + c_i Z_i^{-1} \right) \qquad (8)$$

where $c_i = \left| \det(Z_i) \right|^{1/2n}$ is a scaling used to speed convergence, and n is the dynamical order. H is the

Hamiltonian matrix of order $2n$, given by:

$$H = \begin{bmatrix} A' & Q \\ BR^{-1}B' & -A \end{bmatrix}$$

Assuming H has no eigen-values on the imaginary axis Z_i converges to $\mathrm{Sign}(H) = S$, say, (Gardiner and Laub, 1986). For the solution of the Riccati equation we require the quantity (Roberts, 1980)

$$S^+ = \mathrm{Sign}^+(S) = \frac{1}{2}\left(I_{2n} + S \right)$$

Now, by decomposing S^+ thus: $S^+ = \begin{bmatrix} V & W \end{bmatrix}$ we write, P, the solution of the Riccati equation (5) thus:

$$P = V'W(W'W)^{-1} \qquad (9)$$

The convergence of the algorithm has been investigated in (Roberts, 1980) and its global convergence is established in (Balzer, 1980). In (Gardiner and Laub, 1991) an algorithm for diagonal pivoting factorisation – a form of Gaussian elimination with partial pivoting – is used to develop a parallel algorithm involving no sequential computations. Complexity is of order $(2n)^3$, which dominates the communication burden. The speed-up achieved over the conventional method (Laub, 1979) is demonstrated on a hypercube architecture. More recently (Bunse-Gerstner Faßbender, 1997), a Jacobi-like method has been proposed with a simple function of the matrix size to predict the number of iterations needed for convergence. This of course does not guarantee that a solution can be found within any arbitrary time period.

We do not implement the parallel solution here. The discussion is included simply to underline that real-time operation may be possible.

3. SUSPENSION MODEL

The two-degree-of-freedom, quarter-car model of figure 1 has been widely studied in the literature. It represents an active element operating in parallel with passive elements – a linear spring, k_1, and a non-linear damper with characteristic $f_d = c_1 \xi + c_2 \xi^3$, where f_d represents the damping force and ξ, the relative velocity between the sprung and unsprung masses. The model parameters are based on the one published in (Lin and Kanellakopoulos, 1997).

The motions of the body and wheel (sprung, m_1, and unsprung, m_2, masses, respectively) are denoted by y_1 and y_2 respectively, while the deviation of the surface from some datum is denoted by d. The tyre is represented by a linear spring, k_2, with no damping, for simplicity. We assume that the control force, f, can be applied directly as a result of the control signal, with negligible actuator dynamics. Again this is chosen for simplicity, so as not to obscure the main point of the paper.

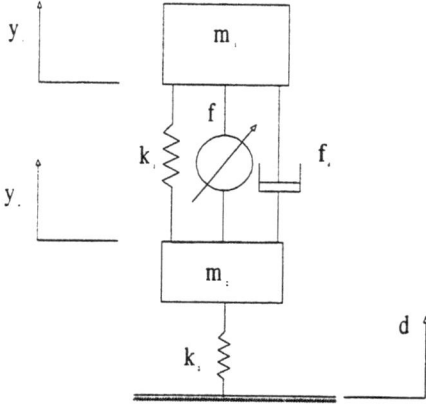

Fig. 1. Schematic of the two-degree-of-freedom, quarter-car model.

The equations of motion for the quarter car are given by:

$$\ddot{y}_1 = -\frac{k_1}{m_1}(y_1 - y_2) - \frac{c_1}{m_1}(\dot{y}_1 - \dot{y}_2)$$
$$-\frac{c_2}{m_1}(\dot{y}_1 - \dot{y}_2)^3 + \frac{1}{m_1}f$$
$$\ddot{y}_2 = \frac{k_1}{m_2}(y_1 - y_2) + \frac{c_1}{m_2}(\dot{y}_1 - \dot{y}_2)$$
$$+\frac{c_2}{m_1}(\dot{y}_1 - \dot{y}_2)^3 - \frac{k_2}{m_2}(y_2 - d) - \frac{1}{m_2}f$$

(10)

We choose state variables thus: $x_1 = y_1, x_2 = \dot{y}_1, x_3 = y_2, x_4 = \dot{y}_2$, and identify the control signal, u, with the force, f. Evidently equations (10) can be put into the required form (6), and the Jacobians of both A and B are clearly Lipschitz.

3.1 Design objectives

For the purposes of this paper let us suppose that our primary objective is to minimise passenger discomfort. We do this by attempting to reduce the accelerations to which the passenger is subject – vertical only, in this simple case. Thus a candidate for the cost function is $\ddot{y}_1 = C_a(\bar{\mathbf{x}})\mathbf{x} + D_a(\bar{\mathbf{x}})\mathbf{u}$, where $C_a(\bar{\mathbf{x}})$ is the second row of $A(\bar{\mathbf{x}})$ and $D_a(\bar{\mathbf{x}})$ is the second element of $B(\bar{\mathbf{x}})$. However, ride comfort can only take precedence when safety and integrity are not compromised. Thus it is necessary to penalise some measure which embodies these ideas, usually the "rattlespace deflection", $y_1 - y_2$ $= C_r \mathbf{x} = \begin{bmatrix} 1 & 0 & -1 & 0 \end{bmatrix} \mathbf{x}$. In the conventional LQ approach we construct a cost function thus:

$$J = \int_0^\infty \left(q_a \ddot{y}_1^2 + q_r (y_1 - y_2)^2 + ru^2 \right) dt$$
$$= \int_0^\infty \left(\begin{array}{c} \mathbf{x}'\left(q_a C_a' C_a + q_r C_r' C_r\right)\mathbf{x} \\ +2\mathbf{x}' q_a C_a' D_a u + \left(q_a D_a^2 + r\right)u^2 \end{array} \right) dt$$

(11)

Letting $N = q_a C_a' D_a$ and $R = q_a D_a^2 + r$ we accommodate the cross-term in the usual way $Q \leftarrow Q - NR^{-1}N'$, $A \leftarrow A - BR^{-1}N'$ with the original $Q = q_a C_a C_a' + q_r C_r C_r'$ (Friedland, 1987). q_a, q_r are used to control the trade-off between ride and handling. The new Q and A are now used in the standard equations (6) and (7).

In the non-linear design procedure we make q_r state-dependent, thus: $q_r(\bar{\mathbf{x}}) = 500\psi(y_1 - y_2, .02, .001)$ with

$$\psi(\xi, \theta, \delta) = \begin{cases} ((\xi - \theta)/\delta)^4, & \xi > \theta \\ 0, & |\xi| \le \theta \\ ((\xi + \theta)/\delta)^4, & \xi < -\theta \end{cases}$$

(12)

where $\theta \ge 0$ defines a dead-zone and $\delta > 0$, the distance within which ψ first reaches unity. The rationale for this functional form is as follows. The primary objective is to reduce body acceleration hence the constant $q_a = 10000$. The secondary objective, which can over-ride the first for safety reasons, is to reduce overly large excursions in the suspension strut. Thus, for a rattlespace of ±0.055m, a dead-zone of ±0.02m is allowed before control action is taken; the non-linearity increasing to unity within the next 0.001m of travel and dominating the cost function very rapidly as the limits are approached. We have been guided here by the function chosen in (Lin and Kanellakopoulos, 1997), however, any other suitable function is a candidate. In addition, acceleration or other variables could equally well be weighted in this way. For comparison, we also consider a conventionally weighted rattlespace deflection with $q_r = 1000$. $r = 0.0001$ throughout.

We use the passive system (equation (10) with $f(t) = 0$ for all t) as a reference and compare its behaviour with that of the actively controlled models for the profile (Lin and Kanellakopoulos, 1997)

$$d(t) = \begin{cases} a(1 - \cos 8\pi t), & 0 \le t \le 0.25 \\ 0, & \text{otherwise} \end{cases}$$

(13)

where a is one-half the height of the hump.

4. RESULTS

Each of figures 2 onwards shows the passive, conventionally weighted (CW) and state-dependently weighted (SDW) behaviour of the vehicle model. These are indicated by a dashed, a dotted and a solid curve, respectively. Figures 2–5 relate to a modest disturbance of maximum height 2 cm which does not

92

threaten an approach towards the rattlespace limits. Here, because $|\dot{y}_1 - \dot{y}_2| < 0.02$ for all time, the rattlespace deflection is never penalised, and because acceleration dominates strongly in the cost function for both CW and SDW, it is not possible to distinguish between these behaviours. Figures 6–9 show the behaviour for a much more severe disturbance, with a maximum height of 11cm that forces the rattlespace weighting into play.

Figures 2, 3 and 4 indicate the improvements attained by both CW and SDW, while figure 5 hints that these come at the expense of marginally increased unsprung mass displacement.

Figure 6 shows how, for a severe disturbance the transient acceleration is degraded while tending towards an improved steady-state. While the acceleration is penalised conventionally, there is an effect when the rattlespace SDW operates (~0.06s).

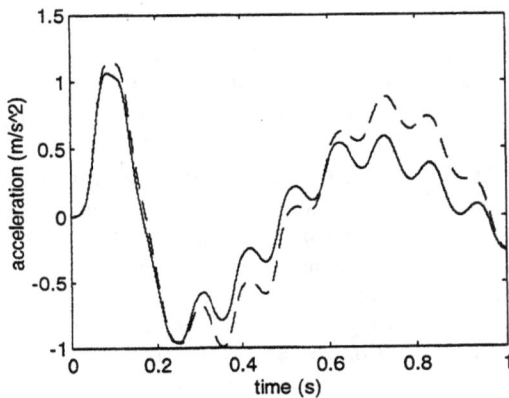

Fig. 2. Acceleration vs. time, a=0.01.

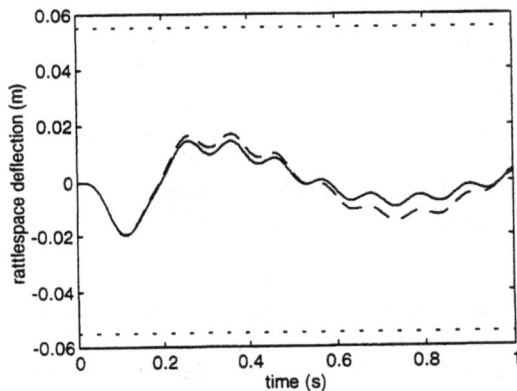

Fig. 3. Rattlespace deflection vs. time, a=0.01.

This sacrifices comfort for safety leading to a response which is worse than the passive system for the first ~0.5s. There is little difference between the passive and CW systems. Figure 7 highlights a more important difference in rattlespace behaviour. Again CW and SDW deliver improvements over the passive behaviour but now the value of SDW is evident.

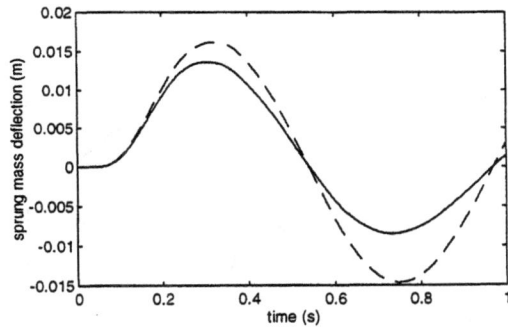

Fig. 4. Sprung mass deflection vs. time, a=0.01.

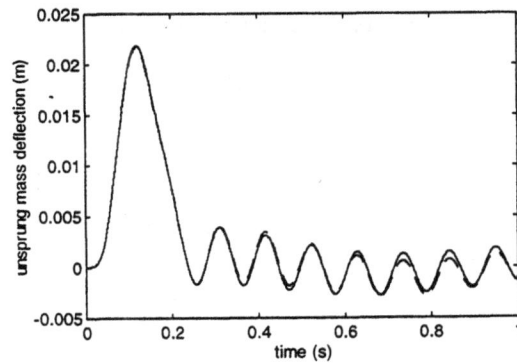

Fig. 5. Unsprung mass deflection vs. time, a=0.01.

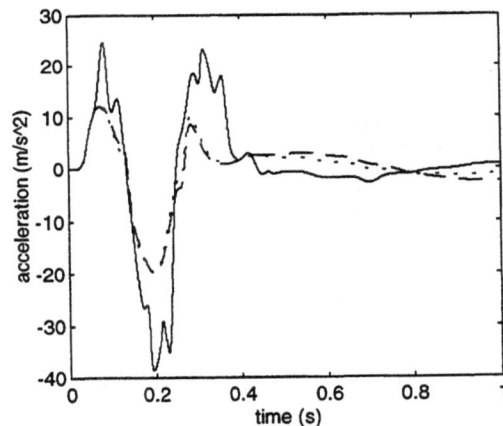

Fig. 6. Acceleration vs. time, a=0.055.

The CW system exceeds the limits of travel in the suspension strut – "bottoming" occurs – while SDW rapidly counteracts the approach (albeit for a deterioration in ride). Nonetheless, only by de-tuning the CW system can bottoming be avoided, and this could not be guaranteed.

Figures 8 and 9 indicate that improvements in ride and handling may come at the expense of increased excursions of the sprung and unsprung masses themselves. However, in a fuller design it would be straightforward to penalise these too.

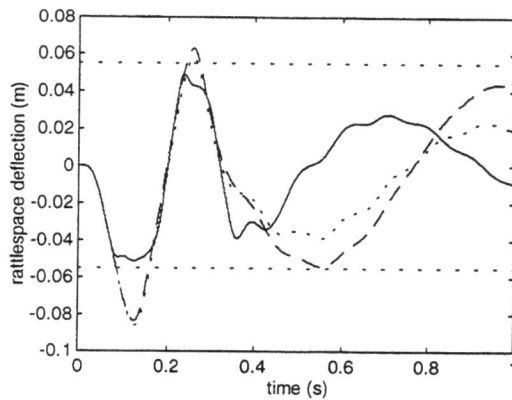

Fig. 7. Rattlespace deflection vs. time, $a=0.055$.

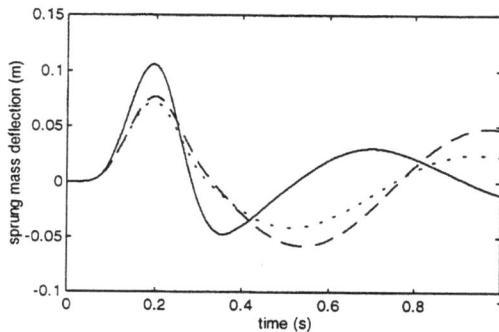

Fig. 8. Sprung mass deflection vs. time, $a=0.055$.

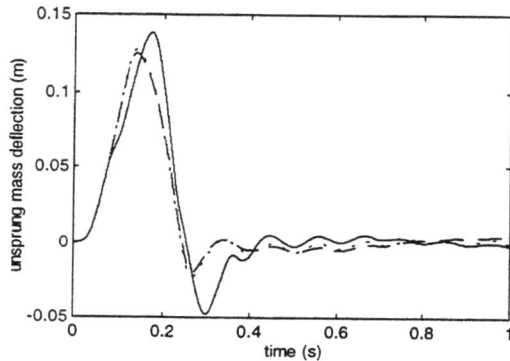

Fig. 9. Unsprung mass deflection vs. time, $a=0.055$.

Finally, we observe that, using the Matlab computing environment, with non-optimised code, the MSF method uses on the order of 5,000 flops per iteration and requires a maximum of nine iterations to converge at each time-step, with a minimum of six and a mean of 8.55. The total number of flops per time step is comparable with those required for the non-iterative eigen-decomposition-based solution (26,000 approximately).

5. CONCLUSIONS

A new method for the design and synthesis of near-optimal, non-linear control laws is proposed, based on a generalisation of LQ optimal control theory. The method is simple to apply and affords greater design flexibility (SDW) than the conventional approach (CW). The resulting controller is non-linear, even for linear dynamics, and can be implemented in real-time. To illustrate the method a simple two-degree-of-freedom quarter-car model with cubic damping has been studied using a non-linear penalty function. Preliminary results show that the method has applicability and could easily be tuned to provide desirable closed-loop behaviour.

The main issue to be addressed in the future is the real-time implementation of such a system. Clearly, without guarantees on the maximum number of iterations needed *per time step* for parallel solutions, the method is severely restricted. However, at the expense of a small loss of "optimality" the essential problem of real-time, non-linear optimal control, *viz.* the anti-causal nature of the solution, has been overcome and a way has been opened towards real-time optimal quadratic regulation.

REFERENCES

Balzer, L.A. (1980). Accelerated convergence of the matrix sign function method of solving Lyapunov, Riccati and other matrix equations. *Int. Jnl. of Control*, **32**, 1057–1078.

Banks, S.P. and K.J. Mhana (1992). Optimal control and stabilisation for non-linear systems. *IMA Jnl. of Math. Control and Information*, **9**, 179–196.

Bunse-Gerstner, A. and H. Faßbender (1997). A Jacobi-like method for solving algebraic Riccati equations on parallel computers. *IEEE Trans. on Automatic Control*, **42**, 1071–1084.

Friedland, B. (1987). Control system design: an introduction to state-space methods, McGraw-Hill Book Co., New York.

Gardiner, J. D. and A.J. Laub (1986). A generalisation of the matrix-sign-function for algebraic Riccati equations. *Int. Jnl. of Control*, **44**, 823–832.

Gardiner, J.D. and A.J. Laub (1991). Parallel algorithms for algebraic Riccati equations, *Int. Jnl. of Control*, **54**, 1317–1333.

Laub, A.J. (1979). A Schur method for solving algebraic Riccati equations. *IEEE Trans. on Automatic Control*, **24**, 913–921.

Lin, J-S. and I. Kanellakopoulos (1997). Non-linear design of active suspensions. *IEEE Control Systems*, **17**, 45–59.

Roberts, J.D. (1980). Linear model reduction and solution of the algebraic Riccati equation by use of the sign function. *Int. Jnl of Control*, **32**, 677–687.

CONTROL OF THE OUTPUT PROBABILITY DENSITY FUNCTIONS FOR A CLASS OF NONLINEAR STOCHASTIC SYSTEMS

Hong Wang

Department of Paper Science
UMIST
Manchester M60 1QD
U. K.

Abstract: Following the recent developments on the modelling and control of the output probability density functions for linear stochastic systems (Wang, 1997, 1998), this paper presents an extented solution to the control of the probability density function for the output of a class of nonlinear stochastic systems. This is based on the fact that there exist many control systems where the requirements are set to control the shape of the probability density function of the system output, rather than the actual values of the system output. At first, the representation of dynamic model is discussed. This is then followed by the construction of a nonlinear control algorithm. An application to a papermaking process has been made and desired results have been obtained. *Copyright © 1998 IFAC*

Keywords: Nonlinear systems, neural networks, probability density functions, stochastic systems control.

1.INTRODUCTION

As has been discussed in (Wang, 1997, 1998), over the past decades, research into the control for stochastic systems has been focussed on the control of the output of the system, rather than the probability density function of the system output. A number of well known algorithms have thus been developed and successfully used in many practical industrial systems. Typical examples are minimum variance control , self-tuning control and stochastic linear quadratic control (Astrom, 1970). In most existing approaches, it has been assumed that all the variables in the system obey a Gaussian-type distribution. However, this as-

sumption is restrictive for some applications. This is particularly true for many control processes in the wet end of papermaking machines (Wang, et, al, 1997; Parker, 1995), where the control of the probability density function of process variables is required.

It can thus be concluded in general that new techniques are needed for the modelling and control of the probability density functions of the output variables for stochastic systems.

Recently, the work has been focussed on linear dynamic systems where several modelling and control algorithms have been developed and are shown to work well for some systems (Wang. 1997, 1998). However, due to the wide existence of nonlinear systems in practice, it is also important to investigate the control algorithms for the control of the output probability density functions for general nonlinear stochastic systems. This forms the main purpose of this paper, where an extended solution to the control of a class of nonlinear systems will be described.

2.MODEL REPRESENTATION

Denote $z(k) \in [a, b]$ as a uniformly bounded random process variable defined on $k = 0, 1, 2, ...$ and assume that $z(k)$ represents the output of a stochastic system. For example $z(k)$ can be used to represent the paper grammage in the wire section of a paper machine. Denote $u_k \in R^1$ as the control input vector which controls the distribution of $z(k)$, then at each sample time k, $z(k)$ can be characterised by its probability density function $\gamma(y, u_k)$ which is defined by

$$P(a \leq z(k) < \xi, u_k) = \int_a^\xi \gamma(y, u_k) dy \quad (1)$$

where $P(z(k) < \xi, u_k)$ represents the probability of output $z(k)$ lying inside the interval $[a, \xi]$ when u_k is applied to the system. This means that the probability density function $\gamma(y, u_k)$ of $z(k)$ is controlled by u_k. For example, u_k can be regarded as a retention aid when the retention system (Wang, et, al 1997, 1998) in papermaking is considered.

Assume interval $[a, b]$ is known and the probability density function $\gamma(y, u_k)$ is continuous and bounded, then using the well known B-spline neural network, the following approximation is obtained

$$\gamma(y, u_k) = \sum_{i=1}^n w_i(u_k) B_i(y) + e_0 \quad (2)$$

where w_i are the weights which depends on u_k, $B_i(y)$ are pre-specified basis functions and e_0 represents the approximation error which satisfies $|e| \leq \delta$. δ is a known small positive number.

Figure 1 shows the principle of such an approximation (Brown and Harris, 1992; Girosi and Poggio, 1990; Wang, 1997). To simplify the presentation, in the rest of the paper term e_0 in equation (2) will be neglected.

Since most systems exhibit dynamic behaviour, there may exist a dynamic relationship between w_i and u (Wang, 1997, 1998). In this paper, a class of nonlinear dynamic relationship will be considered. For this purpose, let us denote

$$b_i = \int_a^b B_i(y) dy \quad (i = 1, 2, ..., n) \quad (3)$$

and

$$f(y, u_k) = \gamma(y, u_k) - L(y) \in R^{1 \times 1}$$

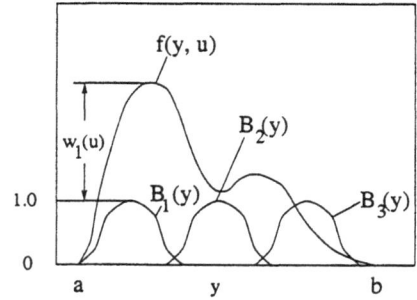

Fig. 1. B-splines expression for probability density function $f(y, u)$.

$$L(y) = b_n^{-1} B_n(y) \in R^{1 \times 1}$$

$$\begin{aligned}
C(y) = & \left(B_1(y) - \frac{B_n(y)}{\int_a^b B_n(y) dy} \int_a^b B_1(y) dy, \right. \\
& B_2(y) - \frac{B_n(y)}{\int_a^b B_n(y) dy} \int_a^b B_2(y) dy, \\
& ..., B_{n-1}(y) - \\
& \left. \frac{B_n(y)}{\int_a^b B_n(y) dy} \int_a^b B_{n-1}(y) dy \right) \\
& \in R^{1 \times (n-1)} \quad (4)
\end{aligned}$$

then the following nonlinear dynamic relationship (affine-model)

$$\begin{aligned}
x_{k+1} &= a(x_k) + b(x_k) u_k \\
V_k &= h(x_k) \\
f(y, u_k) &= C(y) V_k \quad (5)
\end{aligned}$$

can be proposed, where $x_k \in R^N$ is the state vector, $V_k = (w_1, w_2, ..., w_{n-1})^T$ is the output vector, N is the known order of the system and $\{a, b, h\}$ are known nonlinear functions.

Under the assumption that the system (5) has a *strong relative degree* (SRD), it can be further expressed in the following canonical form as

$$\begin{aligned}
V_{k+1} &= G(\phi_k) + H(\phi_k) u_k \\
\phi_k &= (V_k, V_{k-1}, \cdots, V_{k-n+1}) \quad (6)
\end{aligned}$$

where $\{G, H\}$ are also known nonlinear functions obtained from $\{a, b, h\}$.

3.CONTROL ALGORITHM DESIGN

The purpose of the control algorithm design is to choose control sequence $\{u_k\}$ such that the actual probability density function of the system output is made as close as possible to a pre-specified probability density function, $g(y)$, which is independent of $\{u_k\}$. As such, a nature choice of the performance function should be

$$J = \sum_{k=1}^M \left\{ \int_a^b (\gamma(y, u_{k+1}) - g(y))^2 dy + \lambda u_k^2 \right\} \quad (7)$$

where λ is a positive number and $\{u_k\}$ is selected such that J is minimised. Denoting

$$\Sigma = \int_a^b C^T(z)C(z)dz \in R^{(n-1)\times(n-1)} \quad (8)$$

and assuming that all the basis functions have been selected such that matrix Σ is nonsingular, then it can be shown from equation (5) that

$$V_k = \Sigma^{-1} \int_a^b C^T(y)f(y,u_k)dy \quad (9)$$

This means that the output vector V_k can be calculated from the weighted integration of the measured probability density function of the system output. As a result, in the rest of this paper, we will assume that V_k is measurable (or available to the controller). Since the pre-specified function $g(y)$ is continuous and defined on the compact set $[a,b]$, if we select the same basis functions $B_i(y), (i = 1, 2, ..., n)$, then the following expression can be obtained

$$g(y) = \sum_{i=1}^n g_i B_i(y); \quad \forall y \in [a,b]^n \quad (10)$$

where $g_i, (i = 1, 2, ..., n)$ are the fixed weights. As such, selecting the control sequence u_k to make $f(y, u_k)$ as close as possible to $g(y)$ can be regarded as equivalent to choosing u_k such that the weight $w_i(u_k)$ is as close as possible to g_i. Denote

$$V_g = (g_1, g_2, ..., g_{n-1})^T \in R^{n-1} \quad (11)$$

then it can be seen that

$$g(y) = C(y)V_g + L(y) \quad (12)$$

Denote

$$\tilde{V}_k = V_k - V_g \quad (13)$$

then when $M = 1$ the performance function (7) can be expressed as

$$J = \tilde{V}_{k+1}^T \Sigma \tilde{V}_{k+1} + \lambda u_k^2 \quad (14)$$

Using equation (6), it can be further obtained that

$$J = \|G(\phi_k) + H(\phi_k)u_k - V_g\|_\Sigma + \lambda u_k^2 \quad (15)$$

As such, by letting

$$\frac{\partial J}{\partial u_k} = 0 \quad (16)$$

it can be shown that the optimal control value for u_k is

$$u_k = -\frac{H^T(\phi_k)\Sigma(G(\phi_k) - V_g)}{\lambda + H^T(\phi_k)\Sigma H(\phi_k)} \quad (17)$$

Since $V_{k-i}, (i = 0, 1, 2, ..., n - 1)$ are obtained from equation (9), it can be concluded that the

optimal control thus obtained is related directly to the $C^T(y)$ weighted integrations of the probability density functions $f(y, u_{k-i})$, where again $i = 0, 1, 2, ..., n - 1$.

Indeed, this is a general form of the control signal which minimises performance function J. In the case that $\lambda = 0$, the minimisation of this performance function means that the control input u_k should be selected such that

$$V_{k+1} = V_g \quad (18)$$

By substituting equation (18) into the state space equation (6), it can be seen that

$$V_g = G(\phi_k) + H(\phi_k)u_k \quad (19)$$

As a result, by multiplying the both sides of equation (19) with $H^T(\phi_k)$, it can be further formulated that

$$H^T(\phi_k)V_g = H^T(\phi_k)G(\phi_k) + H^T(\phi_k)H(\phi_k)u_k \quad (20)$$

This leads to the following control strategy

$$u_k = -\frac{H^T(\phi_k)(G(\phi_k) - V_g)}{H^T(\phi_k)H(\phi_k)} \quad (21)$$

However, it is important to point out that control input (21) is only a necessary condition in order to satisfy equation (19). To summarise, the following real-time algorithm can be obtained.

1. At sample time k, collect a set of the output;

2. Form the estimate of the output probability density function;

3. Calculate V_k via the use of equation (9);

4. Evaluate the control input u_k by using either equation (17) or equation (21);

5. Apply the obtained control input to the system and go back to step 1).

The algorithm is further illustrated in Fig. 2.

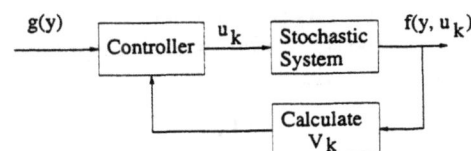

Fig. 2. The real-time control structure.

4. STABILITY ISSUES

With the control algorithm constructed as above, it is important to access the closed loop performance for the so-formed system. In this respect, stability of the closed loop system needs to be established. For this purpose, it is assumed that the constraints on the input energy is removed, leading to $\lambda = 0$. As a result, the closed loop system becomes

$$H^T(\phi_k)\Sigma(V_{k+1} - V_g) = 0 \qquad (22)$$

when the control input (17) is applied, or

$$H^T(\phi_k)(V_{k+1} - V_g) = 0 \qquad (23)$$

when the control input (21) is applied. This means that the requirement on the stability of the closed loop system is satisfied if we can conclude from either equation (22) or equation (23) that a uniformly bounded vector V_g can produce a uniformly bounded V_k.

This means that the closed loop system is stable if the solution of either equation (22) or equation (23) is uniformly bounded when the setpoint vector V_g is bounded. As such, the stability of the closed loop system depends on the structure of vector $H(\phi_k)$. Even though, it is still difficult to establish the global stability for the system. However, when the vector function $H(\phi_k)$ is differentiable, at least a condition which guarantees the local stability of the closed loop system can be formulated. For example, by using the standard linearization techniques, function $H^T(\phi_k)(V_{k+1}-V_g)$ can be linearised around a particular operating point to give a linear recursive equation. This linear equation can then be used to form a polynomial with respect to the unit back shift operator, q^{-1}. As such, the local stability can be established if the zeros of this polynomial are all inside the required region.

5. APPLICABILITY STUDY

In this paper, a typical application to the control of white water solid distribution for a papermaking process is made, where the input is the retention polymers used in the wet end of the paper machine.

Simulated models are used to generate the required data. For this purpose, let us consider the following stochastic system whose output probability density function is given by

$$\gamma(y, u_k) = V_k(39.02ye^{-13.6y} -$$
$$0.78e^{\frac{-(y-1.5)^2}{0.256}}) + L(y) \qquad (24)$$
$$L(y) = 0.78e^{\frac{-(y-1.5)^2}{0.256}} \qquad (25)$$

where $a = 0$, $b = 2$ and V_k is controlled via the following nonlinear dynamic system

$$V_{k+1} = \frac{1}{1 + 0.5V_k^2} +$$
$$\frac{1}{1 + |\sin V_k|}u_k \qquad (26)$$

From equation (24), it can be seen that

$$f(y, u_k) = V_k(39.02ye^{-13.6y} -$$
$$0.78e^{\frac{-(y-1.5)^2}{0.256}}) \qquad (27)$$

with

$$C(y) = 39.02ye^{-13.6y} -$$
$$0.78e^{\frac{-(y-1.5)^2}{0.256}} \qquad (28)$$

Assuming $V_0 = 0$, $V_g = 0.5$ and $\lambda = 0.05$, when the control input (17) is applied to the system, the responses of the control input and the vector V_k are shown in Figs. 3 and 4. Moreover, a 3-D presentation for the probability density function $\gamma(y, u_k)$ is shown in Fig .5, where all the units in y and u axises reflect the total sample numbers of the corresponding variables. It can be seen clearly that desired responses have been obtained. However, since V_k is calculated using equation (9), there is a noise component coupled with the responses.

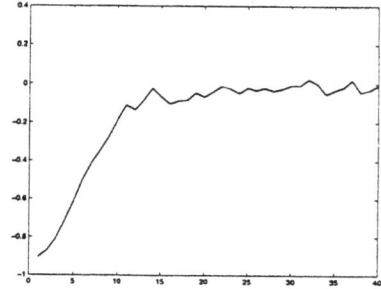

Fig. 3. The response of the control input u_k.

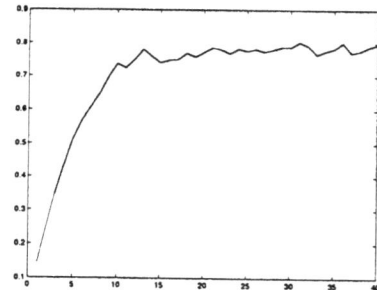

Fig. 4. The response of V_k.

6. CONCLUSIONS

In this paper a novel real-time control algorithm has been developed for the control of the output probability density function of a class of nonlinear stochastic systems. A B-spline neural network

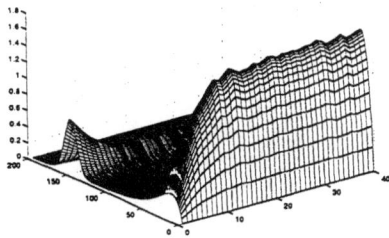
Fig. 5. The response of the controlled probability density function $f(y, u)$.

has been used to represent the approximation to the measured output probability density function. Under the assumption that the basis functions can be selected such that the matrix given in equation (8) is nonsingular, a nonlinear controller has been constructed which is a nonlinear function of the weighted integrations of the measured output probability density function (Wang, 1997, 1998). A simulated retention model of a paper machine is used to generate the required data and desired results have been obtained.

The problem has been originally formulated from the requirement on the solid distribution shape control of the white water system in paper making (Moor, 1997; Parker, 1995). Real applications are therefore necessary to verify the proposed algorithm. Indeed, this is the research work currently being carried out at the Control Systems Group in the Paper Science Department at UMIST.

7.ACKNOWLEDGEMENT

The work is partly funded by the on-going EPRSC research grant GR/K97721. This is gratefully acknowledged.

REFERENCES

Astrom, K. J., (1970). *Introduction to Stochastic Control Theory*, Academic Press.

Brown, M., and C.J. Harris (1994). *Neurofuzzy Adaptive Modelling and Control*, Prentice-Hall, Hemel-Hempstead.

Girosi, F and T. Poggio, (1990). "Networks and the best approximation property", *Biol. Cybern.*, Vol. **63**, pp. 169-176.

Moore, M. W., (1997). Practical applications of a nov- el web monitor for paper machines, *Proc. 1997 APPITA Conference*, Melbourne.

Noriega, R. and H. Wang (1998). A direct adap-

tive neural network controller and its applications, *IEEE Trans. on. Neural Networks*, Vol. **9**, No. 1.

Parker, J. R., (1995). On-line monitoring for performance and quality, *Proc. of TAPPI/CPPA Conference*, Montreal.

Wang, H., G. P. Liu, C. J. Harris and M. Brown, (1995). *Advanced Adaptive Control*, Pergamon Press, Oxford.

Wang, H., A. P. Wang, and S. Duncan, (1997). *Advanced Process Control for Paper and Board Making*, PIRA International.

Wang, H., (1997). On the modelling and control of the output probability density functions for stochastic systems, submitted to *Automatica*.

Wang, H., (1998). Control of the output probability density functions for multivariable stochastic systems: enhanced robustness and guaranteed stability, submitted to *IEEE Trans on Automat Contr*.

NEUROINTERFACES FOR HUMAN-MACHINE REAL TIME INTERACTION

Bernard Widrow[1], Edson P. Ferreira[2] and Marcelo M. Lamego[3]

Information Systems Lab., EE Dep., Stanford University

Abstract: This article aims at showing how neural networks can be employed in the generation of human-machine interfaces (neurointerfaces) for practical real time applications. In a great number of real world applications, due to technical and economic factors, full automation is not possible. In such cases, the human presence is essential and indeed, the system performance becomes highly dependent on human skills. Accordingly, an interface that modifies the problem, allowing unskilled human operators to perform the same task in a satisfactory way, becomes extremely useful. The adaptive nonlinear inverse modeling approach is employed as the basic methodology for specification and design of neurointerfaces. A successful application, a neurointerface that helps an operator to back up a scaled truck model connected to single-trailer and double-trailer configurations, is presented. *Copyright © 1998 IFAC*

Keywords: Neural Networks, Inverse Control, Adaptive Filters, Adaptive Control, Nonlinear Systems.

1. INTRODUCTION

Nowadays, the human-machine interaction has become a major concern. For many tasks, productivity, safety and liability conditions require a considerable degree of skills from human operators. In order to overcome the lack of skills, special human-machine neurointerfaces may be adopted. The basic idea is to change the operational space through a neural network, allowing the human operator to interact with the process through less-specialized commands. Accordingly, the operator devotes his attention to solve a less complex problem, directly at the task level. The objective is to improve the productivity, and safety levels of such tasks even in the case of unskilled operators.

This paper aims at showing how neural networks can be employed in the generation of human-machine interfaces for practical real time problems. Due to the complexity of such problems, neural networks are becoming a natural choice. Their abilities to reproduce highly nonlinear behaviors are described extensively in the literature.

[1] Professor.
[2] Visiting Professor sponsored by CAPES/UFES, Brazil.
[3] Ph.D. student sponsored by CNPq/UFES, Brazil.

In fact, the field of neural networks is experiencing a tremendous resurgence of activity. Successful industrial applications and favorable comparisons with conventional techniques are certainly the main causes of such interest. Fast growing low cost computational systems with considerable processing capabilities have provided an easy and reliable environment for the development of new training algorithms and more advanced topologies for neural network implementations.

This article is divided in 4 (four) sections. Section 2 presents the basic ideas concerning neurointerfaces. In Section 3, the adaptive inverse modeling approach, a framework utilized for neurointerface design, is described. A successful neurointerface application is presented in section 4. The neurointerface helps an operator to back up a scaled truck model connected to single-trailer and double-trailer configurations.

2. THE NEUROINTERFACE

A neurointerface may be thought of as an approximation of the system inverse model. Although such a statement may not be obvious, in fact, an operator develops with his experience a set of causal rules that map standard behaviors into control actions (cognitive model), and this is exactly what inverse modeling does. Thus, a neurointerface tries to reproduce the actions of an experienced operator by using the system inverse model.

While cases might exist in which the neurointerface provides just an approximation of those actions taken by an expert operator, the change of operational space made by the neurointerface, although it may not solve the problem completely, allows the human operator to interact with the process through less-specialized actions. For instance, in a complex industrial process, such as a steel plant described in Vescovi, *et al.* (1997), the operators can not observe directly the main variables of interest during operation (quality, production level, etc.). Instead, they control the process by reasoning about a set of related variables that they can directly observe (pressure, speed, humidity, etc.). The relations between these observable variables and the main variables of interest are not known precisely. A neurointerface applied to such a process may not be able to completely invert the process model due to the lack of information, and consequently may not solve the interface problem completely. Nevertheless, the productivity, safety and liability conditions may be far improved with its utilization.

There are cases, however, where the neurointerface can be fully specified. The main variables of interest are either directly available or may be expressed as some function of the observed variables. In addition, the mapping between standard behavior and control actions can also be achieved by using knowledge of the functional relations between the main variables of interest. This is the case, for instance, in backing a truck and trailer to a loading platform. Although, this constitutes a difficult task for all but the most skilled truck drivers, a neurointerface can be fully specified and the trailer truck operation exercise reduced to a much less complex problem. In this case, the neurointerface may be considered as a black box that takes commands from the driver (desired direction of the trailer back part) and provides the necessary actions (steer the wheels) in order to achieve such a goal. The truck speed and the angle between cab and trailer are sufficient information to obtain a precise inverse modeling of the system. We should note that the driver was not eliminated in this problem. Nguyen and Widrow (1990) proposed a neural network that provided the full automation in backing a trailer truck to a loading dock and indeed, eliminating the presence of the driver. In the present work, the human action is essential. In fact, the driver is concerned with providing the desired spatial trajectory, free of obstacles and normally the shortest one.

The truck-backing-up exercise is a kinematics inverse modeling problem. It means that the dynamic effects that may occur during the operation are not significant.
The neurointerface can also be applied to dynamic inverse modeling problems. A good example of this is the operation of a construction crane (Lamego and Rey, 1995). Since normally, a flexible cable does the coupling between the trolley car and the load, movements in the trolley car generate oscillations in the load. Thus, the crane operator is concerned in shifting the load from one point to another while achieving movement free of oscillations. Here, the neurointerface may be regarded as a black box that takes commands from the crane operator (desired trajectory of the load) and provides the necessary actions (actuation on each degree of freedom of the crane) in order to provide a smooth load movement.

It's also interesting to mention the case of robot arms because, in general, they do have inverse. For instance, Ferreira (1984) shows how to use the inverse nonlinear modeling to provide adaptive decoupling control for disturbance cancellation due to gravity, variable inertia and speed couplings (centrifuge and coriolis), in open chain robot arms. Using the Lagrange formalism the inverse model of an open chain robot arm is a very complex nonlinear vector equation. In general, there are many complaints about the complexity of this model. However, the inverse model has a nice hidden property: it is linear with respect to its parameters. Thus, the parameters can be easily identified and consequently, an inverse can fully specified. While this approach is general, a great analytical and computational effort is required to obtain a robot inverse model. In addition, each new robot requires a new model. A neurointerface applied to such a case may be able to completely invert the robot model, and consequently reduce the modeling effort. Here, the neurointerface may be considered as a black box that takes commands from the operator (desired spatial trajectory of the robot arm) and provides the necessary actions (actuation on each degree of freedom of the robot arm).

3. THE DESIGN OF NEUROINTERFACES USING ADAPTIVE NONLINEAR INVERSE MODELING

The nonlinear inverse modeling approach has been used for many years associated with linear feedback strategies. Basically, the objective is to cancel the plant nonlinear dynamical effects by using a nonlinear device that can reproduce an approximated inverse of the plant. The term "approximated" is employed to emphasize that, in general, a nonlinear system does not possess inverse. However, despite some pathological cases that might eventually exist, the methods of adaptive inverse modeling can often be applied to obtain acceptable inverse approximations of nonlinear systems.

The specification and design of a neurointerface use the fact that a nonlinear plant can be approximated by a neural network model, here represented by the function

$f : R^{(p+1)n+(q+1)m} \to R^n$, of the form

$$y_{k+1} = f(y_k, y_{k-1}, \cdots, y_{k-p}, u_k, u_{k-1}, \cdots, u_{k-q}, w_M)$$

$$y \in R^n, u \in R^m, w_M \in R^{\ell_M} \qquad (1)$$

Variables u and y are, respectively, the plant input and plant output vectors. w_M is the neural network weight vector.

The neurointerface can be regarded as a neural network approximation of the plant inverse model. It should be noted that the conventional and intuitive method, shown in figure 1, for adapting a linear filter to be the inverse of a linear plant does not apply to nonlinear cases. This is due to the fact that nonlinear systems do not commute. Roughly speaking, the block inversion shown in figure 1 does not work for nonlinear systems.

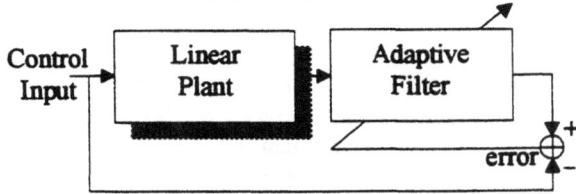

Figure 1. *Inverse modeling of a linear plant.*

Accordingly, the first step in a neurointerface design is to obtain a neural network model for the plant as defined in equation (1) and then, use it to obtain a neural approximation for the plant inverse (neurointerface). Figure 2 shows the nonlinear system identification procedure. The neural network uses the current and previous values of the plant input vector and also previous values of its output vector as its inputs. Its output represents an approximation of the plant output vector.

The neural model can be trained with a set of input-output data either acquired from the real plant or obtained from the plant mathematical model (if available).

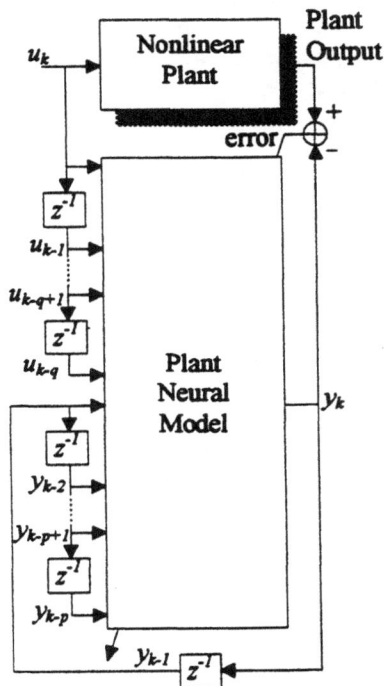

Figure 2. *Nonlinear system identification.*

The backpropagation-through-time, Werbos (1990), may be used to adapt the weights of the neural network

model. If the input of the neural network model does not include any connection to plant output (a feedforward neural network), the conventional backpropagation algorithm, developed by Werbos (1974) and rediscovered and popularized by Rumelhart, et al. (1986), may be employed.

The final step, shown in figure 3, is to train the neurointerface to compute an approximated inverse for the obtained plant neural model.

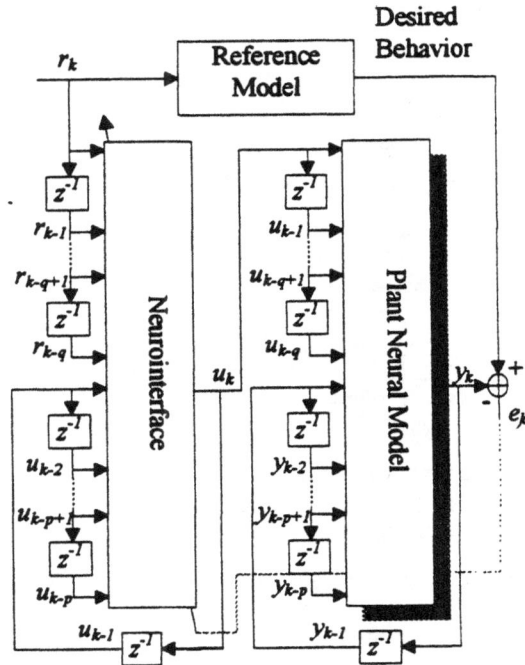

Figure 3. *Scheme used to adapt a neurointerface.*

Like the neural model case, the neurointerface computes a function $g : R^{(P+1)s+(Q+1)m} \rightarrow R^m$, of the form

$$u_k = g(u_{k-1}, u_{k-2}, \cdots, u_{k-Q}, r_k, r_{k-1}, \cdots, r_{k-P}, w_C)$$
$$u \in R^m, r \in R^s, w_C \in R^{\ell_c} \qquad (2)$$

Variables r and u are, respectively, the neurointerface input and output vectors and w_C is the neurointerface weight vector.

The vector r represents the ideal behavior of the plant neural model output vector y or a subset of it. Due to physical limitations, the ideal behavior, in most of cases, may not be achieved. Therefore, in order to provide a more realistic assumption concerning the desired behavior of the output vector y, a reference model can be adopted as shown in figure 3.

Because the plant is nonlinear, the neurointerface is trained in the configuration it will normally work with. The inversion shown in figure 1 is not allowed here. Consequently, to compute the mean square error gradient with respect to the neurointerface weights, information concerning the plant must be available. This is the reason why the plant is replaced by its neural realization during the neurointerface adaptation procedure.

The change in the neurointerface weights at each training step is in the negative direction to the gradient of the system error (e_k). To find the gradient, the chain-rule expansion for ordered derivatives is employed

$$\frac{\partial \|e_k\|^2}{\partial w_C} = -2 \frac{\partial^+ y_k}{\partial w_C} e_k \qquad (3)$$

with

$$\frac{\partial^+ y_k}{\partial w_C} = \sum_{i=0}^{q} \left(\frac{\partial y_k}{\partial u_{k-i}} \right) \left(\frac{\partial^+ u_{k-i}}{\partial w_C} \right) + \sum_{i=1}^{p} \left(\frac{\partial y_k}{\partial y_{k-i}} \right) \left(\frac{\partial^+ y_{k-i}}{\partial w_C} \right) \qquad (4)$$

and

$$\frac{\partial^+ u_k}{\partial w_C} = \left(\frac{\partial u_k}{\partial w_C} \right) + \sum_{i=1}^{Q} \left(\frac{\partial u_k}{\partial u_{k-i}} \right) \left(\frac{\partial^+ u_{k-i}}{\partial w_C} \right) \qquad (5)$$

Each of the terms in equations (4) and (5) is either a Jacobian matrix, which may be calculated using the *dual-subroutine* (Werbos, 1992) of the backpropagation algorithm, or is a previously calculated value of $\partial^+ u_k / \partial w_C$ or $\partial^+ y_k / \partial w_C$. To be more specific, the first term in equation (5) is the partial derivative of the neurointerface's output with respect to its weights. This term is one of the Jacobian matrices of the neurointerface and may be calculated with the dual subroutine of the backpropagation algorithm. The second part of equation (5) is a summation. The first term of the summation is the partial derivative of the neurointerface's current output with respect to a previous output. However, since the neurointerface is externally recurrent, this previous output is also a current input. Therefore the first term of the summation is really just a partial derivative of the output of the neurointerface with respect to one of its inputs. By definition, this is a sub-matrix of the Jacobian matrix for the network, and may be computed using the dual-subroutine of the backpropagation algorithm. The second term of the summation in equation (5) is the ordered partial derivative of a previous output with respect to the weights of the neurointerface. This term has already been computed in a previous evaluation of equation (5), and need not be re-computed. A similar analysis may be performed to determine all of the terms required to evaluate equation (4). After calculating these terms, the weights of the neurointerface may be adapted using the weight-update equation

$$\Delta w_{C_k} = 2\mu \frac{\partial^+ y_k}{\partial w_C} e_k \qquad (6)$$

The neurointerface is designed to operate in real time

without any adaptation algorithm. This implies that its adaptation procedure is performed offline. Figure 4 shows the basic configuration that a neurointerface is supposed to work with. Note that there is no external feedback in this topology. The neurointerface does not cancel disturbances that may occur in the plant. It just changes the operational space through a recurrent neural network, allowing the human operator to interact with the process through less-specialized commands.

Nonetheless, there are situations where the plant inversion supplied by the neurointerface is not enough to provide reliable operating conditions. Disturbance effects may occur during the plant operation and may lead it to risky operating regions. To overcome the disturbance effects and provide more reliable operating conditions to the operator, adaptive linear control schemes may be adopted.

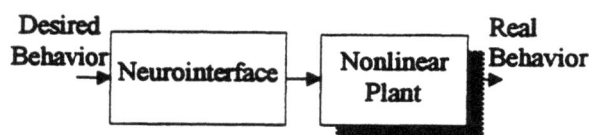

Figure 4. *Basic topology.*

Figure 5 provides a conventional adaptive linear control topology that can be used with a neurointerface.
In this case, the adaptive linear controller may have its parameters adapted using the same reference model adopted to the neurointerface training as a plant idealization. Of course, the reference model has to be linear and differentiable. The last restriction allows the error gradient computation with respect to the adaptive linear controller parameters and indeed, the adaptation of the linear controller.

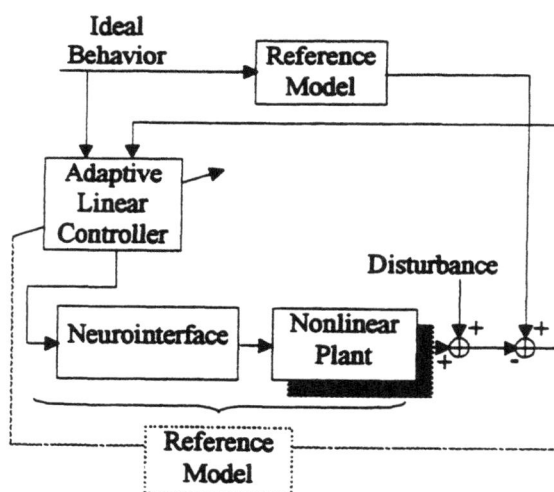

Figure 5. *Neurointerface working in closed loop.*

Another possible adaptive linear control scheme is shown in figure 6. It is the adaptive inverse control technique described in Widrow and Wallach (1996). In this scheme, an equivalent linear model for the neurointerface and the nonlinear plant combined is identified in real time. Then, using a digital copy of the linear model, the linear controller C and the linear disturbance canceler Q are calculated offline. The offline process can rum much faster than the real time, so that as the plant linear model is evaluated, Q and C are immediately obtained.

The same procedure used in adaptive linear systems can be employed here. The neurointerface is able to cancel most of the nonlinear effects the plant may have and indeed, it can be used in combination with adaptive linear control schemes for the control of nonlinear plants.

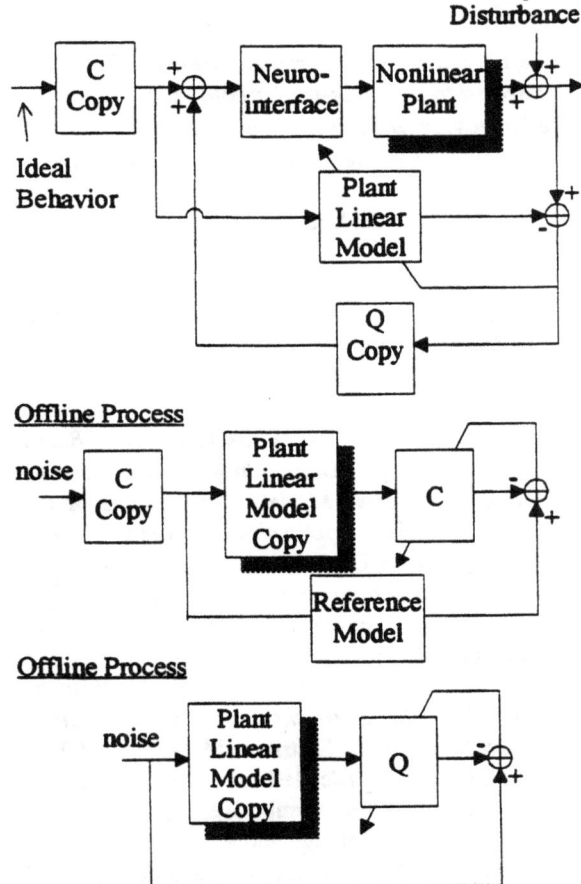

Figure 6. *Neurointerface combined with adaptive inverse linear control.*

4. EXPERIMENTAL RESULTS

Backing a truck and trailer to a loading platform constitutes a difficult task for all but the most skilled truck drivers. This section briefly presents the experimental results of a neurointerface that reduces the trailer truck operation exercise to a much less complex problem. Two configurations are considered: a scaled truck model connected to single-trailer configuration (see figure 7) and connected to a double-trailer configuration (see figure 8). The neurointerface may be considered as a black box that takes commands from the driver (desired direction of the trailer back part) and provides the necessary actions (steer the front wheels). For the single-

trailer configuration, the desired direction of the trailer back part is related to the angle between cab and trailer. For the double-trailer configuration, it is related to the angle between the first trailer and the second one. In both implementations, the neurointerface works in closed loop. The adaptive linear control topology presented in section 3, figure 6 is employed here. For the single trailer configuration, the neurointerface has, as its inputs, the truck speed, the desired value of the angle between cab and trailer (θ_2) and the previous value of the neurointerface's output (the front wheel steering angle, θ_1). Similarly, for the double-trailer configuration, the neurointerface has as its inputs, the truck speed, the angle θ_2, the desired value of the angle between the first trailer and the second one (θ_3) and the previous value of the neurointerface's output (the front wheel steering angle, θ_1). The neurointerface was designed following the steps described in section 3. Acquired data from the truck prototype were used to obtain the neural model for each configuration. The obtained neural models were used for the training of the neurointerfaces. In both cases, only the disturbance canceler Q, an adaptive linear combiner, is utilized. The involved dynamics are simple and, indeed, the controller C need not be implemented. The equivalent plant linear model is adapted in real time.

Figure 7. *Truck model connected to a single-trailer configuration.*

Figure 8. *Truck model connected to double-trailer configuration.*

The offline process for adaptation of Q is started at every 500 samples, sampling period of 30 ms. The data acquired in this interval (15 sec.) are used as the training set for Q. The experimental results are shown in figures 9 and 10. They correspond to sequences of data acquired from the truck model moving backwards in both configurations.

CONCLUSIONS

This article presents a new approach for generation of human-machine interfaces through neural networks (neurointerface) for practical real time applications. The adaptive nonlinear inverse modeling approach is employed as the basic methodology for specification and design of neurointerfaces. The neurointerface is able to cancel most of the nonlinear effects the plant may have and, indeed, it can be used in combination with adaptive linear control schemes for the control of nonlinear plants. A successful application, a neurointerface that helps an operator to back up a scaled truck model connected to single-trailer and double-trailer configurations, is also presented. This is an introductory work and a great effort will be needed to improve the neurointerface approach. However, the general aspects covered in this paper combined with the excellent quality of the experimental results lead to conclude that the fully utilization of neurointerfaces for real time applications seems to be very promising.

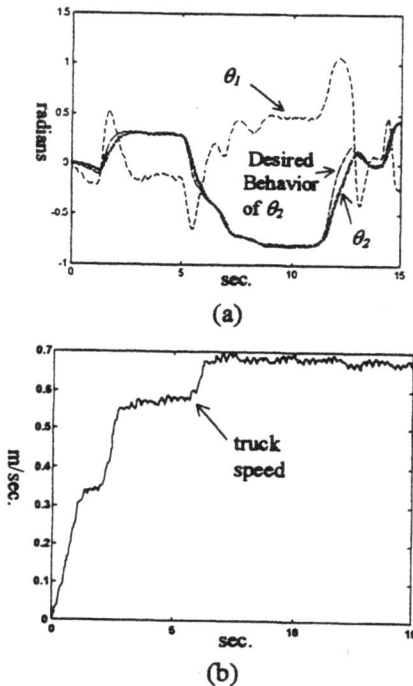

(a)

(b)

Figure 9. *Experimental results for the truck model connected to a single-trailer configuration: (a) desired behavior of θ_2, real behavior of θ_2 and neurointerface's output (θ_1); (b) truck speed.*

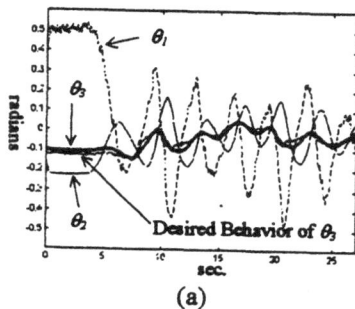

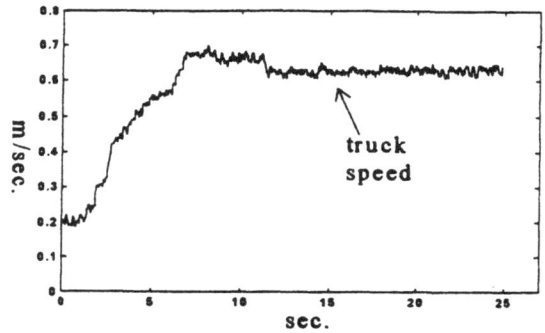

(a)

(b)

Figure 10. *Experimental results for the truck model connected to a double-trailer configuration: (a) desired behavior of θ_3, real behavior of θ_3 and θ_2 and neurointerface's output (θ_1); (b) truck speed.*

REFERENCES

Ferreira, E. P. (1984). *Contribution a L'Identification de Parametres et a la Commande Dynamique-Adaptative des Robots Manipulateurs.* Dr. thesis, LAAS du CNRS, Universite Paul Sabatier, Toulose, France, July.

Lamego, M. M. and J. P. Rey (1995). *The Interval Based Control Technique: Controlling Physical Systems Through Imprecise Models.* Proceedings of the IEEE Industrial Applications Society Annual Meeting, Florida, USA, October, pp. 1822–1827.

Nguyen, D. and B. Widrow (1990). *Neural Networks for Self-learning Control Systems.* IEEE Control Systems Magazine, April, pp. 18–23.

Rumelhart, D. E., G. E. Hinton and R. J. Williams (1986). *Learning internal representations by error propagation.* Parallel Distributed Processing. (D. E. Rumelhart and J. L McClelland editors), volume 1, chapter 8. MIT Press, Cambridge, MA.

Vescovi M. R., M. M. Lamego and A. Farquhar (1997). *Modeling and Simulation of a Complex Industrial Process.* IEEE Expert Magazine — Intelligent Systems & Their Applications, IEEE Computer Society, May/June, pp. 42-46.

Werbos, P. (1974). *Beyond Regression: New Tools for Prediction and Analysis in the Behavioral Sciences.* PhD thesis, Harvard University, Cambridge, MA, August.

Werbos, P. (1990). *Backpropagation through time: What it does and how to do it.* Proceedings of the IEEE, Vol. 78(10), October, pp. 1545–1680.

Werbos, P. (1992). *Neurocontrol and supervised learning: An overview and evaluation.* Handbook of Intelligent Control: Neural, Fuzzy and Adaptive Approaches. (D. White and D. Sofge editors), chapter 3. Van Nostrand Reinhold, New York.

Widrow, B. and E. Walach (1996). *Adaptive Inverse Control.* Prentice Hall PTR, Upper Saddle River, NJ.

SELF-TUNING NEURAL CONTROLLER

Alberto Aguado Behar. (1)
Alfonso Noriega Ponce. (2)
Antonio Ordaz Hernández. (2)
Vladimir Rauch Sitar. (2)

(1) Instituto de Cibernética, Matemática y Física, Cuba
(2) Universidad Autónoma de Querétaro, México

Abstract: In this paper, we presented a self-tuning control algorithm based on a three layers perceptron type neural network. The proposed algorithm is advantageous in the sense that practically a previous training of the net is not required and some changes in the set-point are generally enough to adjust the learning coefficient. Optionally, it is possible to introduce a self-tuning mechanism of the learning coefficient although by the moment it is not possible to give final conclusions about this possibility. The proposed algorithm has the special feature that the regulation error instead of the net output error is retropropagated for the weighting coefficients modifications. *Copyright © 1998 IFAC*

Keywords: Neural networks, feedforward, back-propagation, networks, self-tuning control.

1. INTRODUCTION.

The use of Artificial Neural Nets for systems identification and control, has been the subject of a vast amount of publications in recent years. It is possible to mention for instance, the paper of Narendra and Parthasaraty (1990) in which several possible structures of the neural controller that suppose an a priori knowledge of the plant dynamic structure are proposed. In Bhat and McAvoy (1990) and in many other papers, a scheme based in an inverse dynamic neural model is used. A similar approach is adopted by Aguado and del Pozo (1997) but here the inverse neural regulator is complemented by a self-tuning PID algorithm which guarantees that the stationary error goes to zero. In the above mentioned paper and in many others, it is required an exhaustive previous training of the neural net, which is a serious obstacle for the practical implementation of the proposed solution in the industry. At the same time, the controller structure in some cases, as in the mentioned paper of Narendra and Parthasaraty, is very complex, including several nets each one with two hidden layers of many neurons which must be trained using large data samples.

An important contribution to enhance the possibilities of neural nets in the solution of practical problems is the paper by Cui and Shin (1992). In that work a direct neural control scheme is proposed for a wide class of non-linear processes and it is shown that in many cases, the net can be trained directly retropropagating the regulation error instead of the net output error. However, in that paper it is not discussed the influence of some training parameters, particularly the learning coefficient, over the closed loop dynamics. It is also not remarked that practically with a direct neural control scheme the training stage can be substituted by a permanent and real time adaptation of the weighting coefficients of the neural net.

In the present paper, it is proposed a self-tuning neural regulator, inspired in the ideas of Cui and Shin, but with the particular feature that the previous training is substituted by a permanent adjustment of the weighting coefficients based on the control error. At the same time, it is shown the influence of the learning coefficient over the closed loop dynamics and some criteria are given about how to choose that parameter. Finally, some examples are given where the possibilities of the proposed method for difficult non-linear systems control is shown, specially when there exists a considerable pure time delay.

2. SELF TUNING NEURAL CONTROLER STRUCTURE

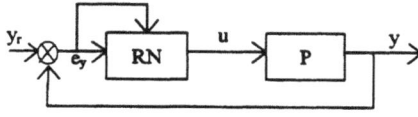

Figure 1. Scheme of the self-tuning neural regulator.

In figure 1, it is shown the proposed scheme of the self-tuning neural regulator. The neural net which assume the regulator function, is a 3 neuron layers perceptron (one hidden layer) which weighting coefficients are adjusted by a modified retropropagation algorithm. In this case, instead of the net output error :

$$e_u(t) = u_d(t) - u(t) \qquad (1)$$

it is used the process output error :

$$e_y(t) = y_r(t) - y(t) \qquad (2)$$

to adjust the weighting coefficients.

In figure 2, it is represented the neural net controller structure. The output layer has only one neuron because by the moment, we are limiting the analysis to one input-one output processes. In the input layer, the present and some previous values of the regulation error are introduced, it means that :

$$x(t) = [e_y(t) \quad e_y(t-1)........e_y(t-n)] \quad (3)$$

For the cases that has been simulated until now, the value n=2 was sufficient to obtain an adequate closed loop performance. It means that the number of input neurons can be 3. A similar number of hidden layer neurons was equally satisfactory.

3. WEIGHTING COEFFICIENTS ADAPTATION ALGORITHM.

In figure 2, the weighting coefficients w_{ji} and v_j for the hidden layer and output layer input connections respectively, are shown. In what follows, we will detail the adaptation algorithm for that coefficients, which ensures the minimization of a regulation error $e_y(t)$ function.

The output of the j hidden layer neuron may be calculated by means of :

$$h_j = \frac{1}{1+e^{-S_j}} \qquad j=1,2,3,.... \qquad (4)$$

where :

$$S_j = \sum_{i=1}^{3} w_{ji} x_i \qquad (5)$$

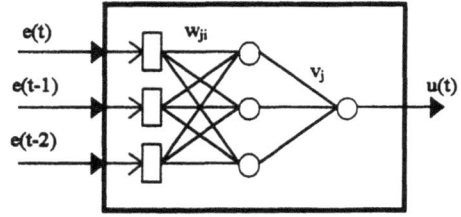

Figure 2. Neural net controller structure.

At the same time, the output layer neuron output value will be :

$$u(t) = \frac{1}{1+e^{-r}} \qquad (6)$$

where :

$$r = \sum_{j=1}^{3} v_j h_j \qquad (7)$$

As a criteria to be minimized, we defined the function :

$$E(t) = \frac{1}{2} \sum_{k=1}^{t} e_y(k)^2 \qquad (8)$$

where it is supposed that the time has been discretized by using an equally-spaced small time interval.

The minimization procedure consists, as it is known, in a movement in the negative gradient direction of the function E(t) with respect to the weighting coefficients v_j and w_{ji}. The E(t) gradient is a multi-dimensional vector whose components are the partial derivatives $\dfrac{\partial E(t)}{\partial v_j}$ y $\dfrac{\partial E(t)}{\partial w_{ji}}$, it means that :

$$\nabla E(t) = \begin{bmatrix} \dfrac{\partial E(t)}{\partial v_j} \\ \dfrac{\partial E(t)}{\partial w_{ji}} \end{bmatrix} \qquad (9)$$

Let us first obtain the partial derivatives with respect to the coefficients of the output neuron. Applying the chain rule, we get :

$$\frac{\partial E(t)}{\partial v_j} = \frac{\partial E(t)}{\partial e_y} * \frac{\partial e_y}{\partial e_u} * \frac{\partial e_u}{\partial u(t)} * \frac{\partial u(t)}{\partial r} * \frac{\partial r}{\partial v_j} \qquad (10)$$

$$= e_y \cdot \frac{\partial e_y}{\partial e_u} \cdot (-1) \cdot u(t)(1-u(t)) \cdot h_j \tag{11}$$

In (11) it is used the well known relation :

$$\frac{\partial u(t)}{\partial r} = \frac{\partial \left(\dfrac{1}{1+e^{-r}}\right)}{\partial r} = \frac{e^{-r}}{\left(1+e^{-r}\right)^2}$$

$$\frac{\partial u(t)}{\partial r} = \frac{e^{-r}}{1+e^{-r}} \frac{1}{1+e^{-r}} = u(t)(1-u(t)) \tag{12}$$

Let us define :

$$\delta^1 = e_y \, u(t)(1-u(t)) \tag{13}$$

Then :

$$\frac{\partial E(t)}{\partial v_j} = -\delta^1 h_j \frac{\partial e_y}{\partial e_u} \tag{14}$$

In the equation (14), it appears the partial derivative $\dfrac{\partial e_y}{\partial e_u}$, which can be interpreted as some kind of "equivalent gain" of the process. Further we will make some considerations about that term.

The partial derivative of function E(t) with respect to the weighting coefficients w_{ji}, can be obtained applying again the chain rule :

$$\frac{\partial E(t)}{\partial w_{ji}} = \frac{\partial E(t)}{\partial e_y} * \frac{\partial e_y}{\partial e_u} * \frac{\partial e_u}{\partial u(t)} * \frac{\partial u(t)}{\partial r} *$$
$$\frac{\partial r}{\partial h_j} * \frac{\partial h_j}{\partial S_j} * \frac{\partial S_j}{\partial w_{ji}} \tag{15}$$

$$= -e_y \frac{\partial e_y}{\partial e_u} u(t)(1-u(t)) v_j \, h_j \, (1-h_j) x_i$$

$$\frac{\partial E(t)}{\partial w_{ji}} = -\delta^1 v_j \, h_j \, (1-h_j) x_i \frac{\partial e_y}{\partial e_u} \tag{16}$$

Let us define :

$$\delta^2_j = \delta^1 v_j \, h_j \, (1-h_j) \tag{17}$$

and then:

$$\frac{\partial E(t)}{\partial w_{ji}} = -\delta^2_j x_i \frac{\partial e_y}{\partial e_u} \tag{18}$$

Using equations (14) and (18), the adjustments of weighting coefficients v_j y w_{ji} can be made by means of the expressions :

$$v_j(t+1) = v_j(t) + (\eta \frac{\partial e_y}{\partial e_u}) \delta^1 h_j \tag{19}$$

$$w_{ji}(t+1) = w_{ji}(t) + (\eta \frac{\partial e_y}{\partial e_u}) \delta^2_j x_i \tag{20}$$

where η is the so-called learning coefficient and $\dfrac{\partial e_y}{\partial e_u}$ is the " equivalent gain " of the plant. The main obstacle to apply the adjustment equations (19) and (20) is that in general the plant equivalent gain $\dfrac{\partial e_y}{\partial e_u}$ is unknown. However, in the above mentioned paper by Cui and Shin, it is demonstrated that it is only required to know the sign of that term to ensure the convergence of the weighting coefficients, because the magnitude can be incorporated in the learning coefficient η if the non-restrictive condition $\dfrac{\partial e_y}{\partial e_u} < \infty$ is accomplished.

Besides, the sign of $\dfrac{\partial e_y}{\partial e_u}$ can be easily estimated by means of a very simple auxiliary experiment, by instance, to apply an step function at the process input.

The assumption that the sign of the gain remains constant in a neighbourhood of the operation point of the process is not very strong and it is accomplished in most practical cases. Finally, in the worst of cases, real time estimation of the gain sign could be incorporated, without great difficulties, in the control algorithm.

Having in mind the above considerations, the equations (19) and (20) could be written as follows :

$$v_j(t+1) = v_j(t) + \eta \, sign(\frac{\partial e_y}{\partial e_u}) \delta^1 h_j \tag{21}$$

$$w_{ji}(t+1) = w_{ji}(t) + \eta \, sign(\frac{\partial e_y}{\partial e_u}) \delta^2_j x_i \tag{22}$$

The right value of learning coefficient η can be experimentally determined from the observation of closed loop system performance when some changes are made in the controlled variable set-point.

The equations (21) and (22) for the proposed neural controller structure, may be interpreted as the regulator adaptation equations instead of training equations as it is normally done. Indeed, the simulation study done until now using the scheme shown in fig. 1 and permanently adjusting the weighting coefficients v_j and w_{ji} by means of (21) and (22) allowed us to arrive to the next conclusions :

■ It is not in general required a previous training of the net and once the control loop is closed, the weighting coefficients self-adjust in a few control periods, carrying the regulation error $e_y(t)$ to zero.

■ The dynamical performance of the closed loop depends exclusively on the learning coefficient magnitude, corresponding to higher values of η, a faster response that can even present a considerable over-shoot. Diminishing the value of η, the system is damped up to the point in which the desired response is obtained. The system, however, keeps the stability for a wide range of η values.

■ It is very convenient to use a variable learning coefficient, using an expression as :

$$\eta' = \eta + \alpha \text{ abs}(e_y) \qquad (23)$$

In this way a small basic value of η can be used, for instance $\eta=0.1$, and the effective value η' is incremented depending of the regulation error magnitude. The right value of α can be tuned experimentally without troubles. The use of equation (23) gives the system the possibility to present a fast response when the errors are big and then to go slowly to the reference value. That behaviour is, indeed, very convenient in practice.

4. SOME SIMULATION RESULTS.

Although the above described algorithm has been tested in many examples, here we will only show the results obtained in two simulated cases corresponding to non-linear processes which can be described by means of the equation :

$$y(t) = \left(\frac{K e^{-T_d s}}{(T_1 s + 1)(T_2 s + 1)} \right)^2 u(t)^2 \qquad (24)$$

The simulation was carried out on a real time environment provided by the CPG System (Aguado,1992) so that the obtained results are very close to those that could be expected in a real process.

In fig. 3 it is represented the closed loop behavior corresponding to the next process and regulator parameters :

$$T_d=5 \text{ sec. } T_1=3 \text{ sec. } T_2= 5 \text{ sec.} \qquad (25)$$

$$\eta=0.6 \quad \alpha=0.4 \qquad (26)$$

As can be seen, a practically perfect response is obtained when positive and negative step changes in the reference output value are applied. Given that the time constants are relatively small, the values of η y α can be chosen relatively big without producing positive or negative over-shoots. This type of performance is observed in general, it means that for fast dynamics processes, it is possible and convenient a fast learning of the net. We have observed that for time constants in the order of milliseconds, values of η of 5 and more can be used.

In figure 4, it is represented the close loop performance for the next constants values:

$$T_d=60 \text{ sec. } T_1=10 \text{ sec. } T_2= 20 \text{ sec.} \qquad (27)$$

$$\eta=0.05 \quad \alpha=0.20 \qquad (28)$$

As can be observed, we have a large time-delay non linear system which can be hardly controlled by a conventional adaptive algorithm, for instance a self-tuning PID. However, with the adaptive neural regulator, it is obtained a response that can be considered as very good, given the process characteristics. Notice that in this case, the values of η y α are considerably smaller, as expected, given that the process dynamics is much slower.

CPG System. Ver 5.00
17/11/97   3: 1:32

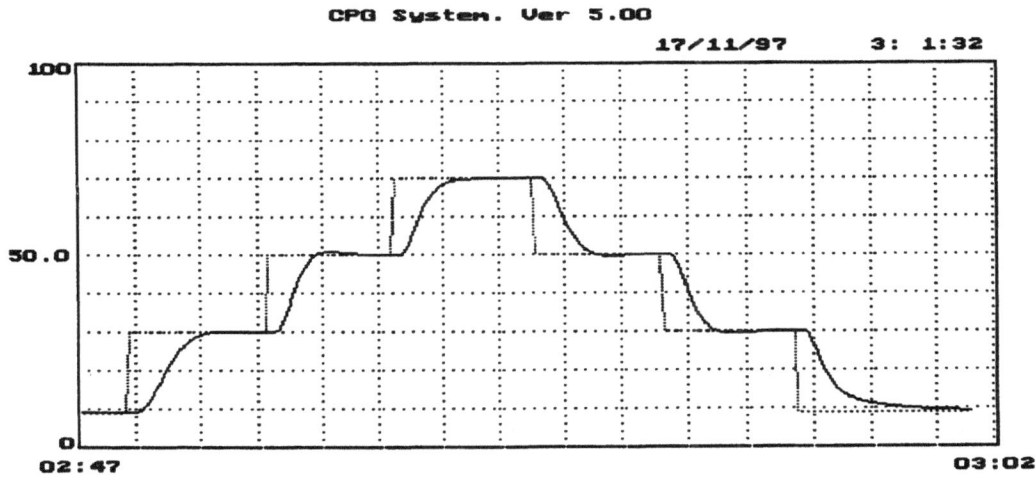

Figure 3. Closed loop behaviour for process in equations (25).

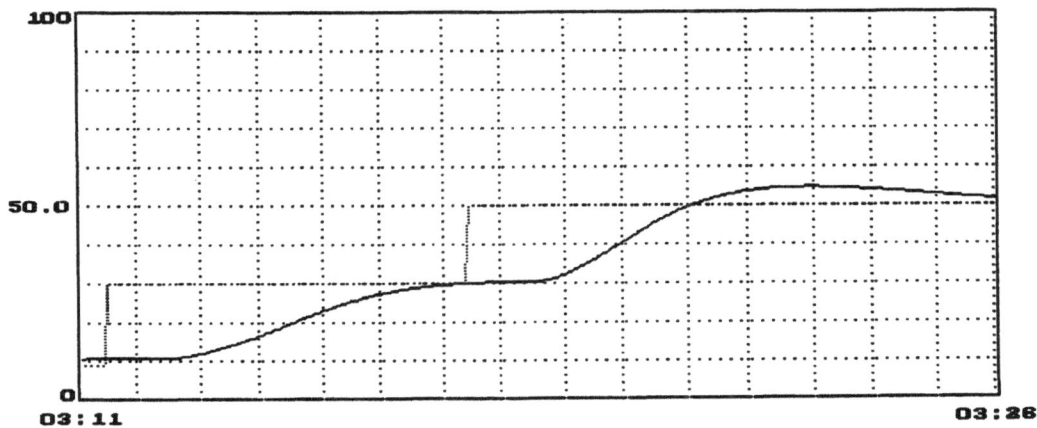

Figure 4. Closed loop behaviour for process in equations (27).

5. CONCLUSIONS.

The self-tuning control algorithm based on a neural net presented in this paper, promise to be a very interesting option for the control of processes with a difficult dynamics that could not be adequately controlled with PID regulators, even in their self-tuning versions, as it was shown in the above presented simulation cases. The class of processes in which the algorithm could be applied is very wide and it includes most of the cases that can appear in practice. In the near future, we plan to apply the algorithm to some real laboratory processes and to extend the obtained results to the case of multivariable and multiconnected systems.

REFERENCES.

Aguado, A. (1992): Controlador de Propósito General, Memorias del V Congreso Latinoamericano de Control Automático, La Habana, Cuba.

Aguado, A.; del Pozo, A. (1997): Esquema de Control Combinado usando Redes Neuronales, Memorias de CIMAF'97, La Habana, Cuba.

Bhat, N.; McAvoy, T.J. (1990): Use of Neural Nets for Dynamic Modelling and Control of Chemical Process Systems, Computers on Chem. Engng., Vol 14, No. 4/5, pp. 573-583.

Cui, X.; Shin, K.G. (1992): Direct Control and Coordination Using Neural Networks, IEEE Transactions on Systems, Man and Cybernetics, Vol. 23, No.3, pp. 686-697.

Narendra, K.S.; Parthasaraty, K. (1990): Identification and Control of Dynamical Systems using Neural Networks, IEEE Transactions on Neural Networks, Vol. 1, No. 1.

A ROBOTIC CLASSIFIER OF ROCKS: AN INTEGRATION OF ARTIFICIAL VISION AND ROBOTICS

René Vidal and Aldo Cipriano

Faculty of Engineering, Catholic University of Chile
PO Box 306, Santiago 22, Chile
Phone: 56-2-6864281; fax: 56-2-5522563
e-mail: rvidal@mat.puc.cl

Abstract: This paper presents the first stage in the development of a robotic classifier of the rocks that feed the grinding process in a mineral concentration plant. The system uses a sensor that determines the size of the rocks, and a robot that extracts them from a conveyor belt when they are too big to be ground. *Copyright © 1998 IFAC*

Keywords: Image processing, robotic manipulators, mineral processing.

1. INTRODUCTION

Robotics is rapidly being introduced into the industry, in all productive fields, and a big increase in the use of this technology is foreseen over the next years.

The mining industry is not very involved with this process, due to the difficulties imposed by the environment, the operation process itself, etc. Nevertheless, some aspects of mining activity can be considered to be, to some extent, robotized.

In Chile, mining is one of the most important activities and a lot of efforts have been done in order to introduce new technologies in mining processes. In a concentration plant, mineral granulometric distribution is one of the most important disturbances in the crushing and grinding processes, and it affects significantly both the amount of mineral to be processed and the power requirements. This situation has encouraged the utilization of the potentials of real time image processing in order to develop analyzers of thick granulometry.

One example of these analyzers is PETRA ADG-21 (Cartes *et al.*, 1995), which is an integration of a hardware system that, together with an application software, calculates the granulometric distribution of the mineral that feeds a semi-autogenous grinding mill (see Fig. 1). The mineral is transported on a conveyor belt and its size is usually in the range of 1/2" to 8".

PETRA's hardware system is based on an observation stage composed by a Closed Television Network (CTVN) and some elements that process the observed images. The CTVN consists of a black and white CCD camera, a black and white screen that reproduces the video signal coming from the camera, and a VCR. The image processing hardware requires a high performance computer running under Windows. The software was developed in Borland C++ for Windows.

An optimal control strategy for grinding plants based on these kind of analyzers, was proposed by Cipriano (Cipriano *et al.*, 1996). Simulation results show that the new strategy reduces the variability

Fig. 1. Rocks feeding the grinding process

of power draught and bearing pressure, increasing the mill throughput.

Another possible application of these systems is to determine the size of the rocks in order to extract the biggest ones before they enter the grinding process. This article presents a laboratory prototype of this application based on the Scorbot ER VII robot arm. The computation of the size of the rocks is accomplished through two different alternatives: a proximity sensor and an artificial vision system.

2. PROBLEM STATEMENT

The purpose of this paper is to present the first stage in the development of a laboratory prototype of a system that classifies the rocks that feed the grinding process in a mineral concentration plant. In this stage the following simplifications are considered:

- The rocks are transported on a conveyor belt one by one at a constant speed v.
- The robot is normally located in a fixed position P_1 over the conveyor. When the system detects a rock whose size S is bigger than a preset maximum size S_{max}, the robot grasps it when it is passing over P_1.
- The rocks are dumped on another fixed position P_2.
- The time spent in the grasping process is shorter than the time interval between two consecutive big rocks.

Consequently, the idea of this stage is to develop:

- A sensor that detects the size, position and speed of the rocks. The following alternatives are considered:
  · A proximity sensor that estimates the size of the rocks based on their height.
  · A 2D artificial vision system that estimates the size of the rocks based on their area (projection of the volume into the 2D image).
- A control system that makes the robot grasp the big rocks based on the information sent by the sensor.

The following sections describe the solution based on the proximity sensor, the solution based on the vision system, the Scorbot ER VII robot arm and the results obtained with the proximity sensor and the vision system.

3. THE PROXIMITY SENSOR

The proximity sensor is located in a fixed position P_s, which is located h [cm] over the conveyor and d [cm] from P_1 as it is shown in Fig. 2.

The proximity sensor is able to detect when the height of the rock is bigger than S_{max}, in this case represented by h. However, the sensor can determine neither the position nor the speed of the rock. Therefore, the speed is estimated based on the voltage applied to the conveyor motor by means of the linear relationship $v = aV + b$, where v is the speed, V is the conveyor motor voltage, and a, b are parameters to be estimated. Finally, the time T that the robot must wait before grasping the rock (in its middle point) after t_1 is:

$$T = \frac{d}{v} - \frac{t_1 - t_0}{2} - t_c , \qquad (1)$$

where t_0 and t_1 are the time instants in which the rock starts and finishes passing over P_s, respectively, and t_c is the closing time of the gripper, which needs to be estimated experimentally.

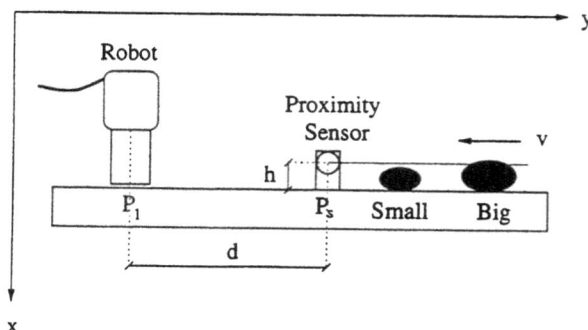

Fig. 2. Proximity sensor location

4. THE VISION SYSTEM

4.1 *Hardware*

The hardware consists of a color CCD camera, a color monitor where the video signal is displayed, a high performance computer with a SVGA monitor, a keyboard and a mouse to make the operation and/or interaction easier.

The information is usually delivered to the user by the computer through a SVGA monitor. Additionally, this information can be stored on hard disk, fed to a printer or transmitted through the RS-232C interface to another device (in this case the robot).

4.2 *Image Processing Algorithms*

A typical image captured by the acquisition system is shown in Fig. 3. This image can be divided in three different parts. The upper one contains the wall and the robot, the middle one contains the conveyor, the rocks and the robot's gripper, and the lower one does not contain relevant information. Therefore, it is not necessary to work with the complete image, because the rocks are located in the middle part.

Let $I_R(x,y) \in \mathbb{R}^{m \times n}$ be a reference image which only contains the conveyor and the robot's gripper (middle part). Because the gripper is black and the background is white, the horizontal position of the gripper can be obtained as:

$$y_{robot} = \frac{\sum\limits_{i=1}^{m} \sum\limits_{j=1}^{n} j \cdot I_{BR}(i,j)}{\sum\limits_{i=1}^{m} \sum\limits_{j=1}^{n} I_{BR}(i,j)} , \qquad (2)$$

where I_{BR} is the image shown in Fig. 4, which is obtained binarizing I_R.

Let $I_S(x,y) \in \mathbb{R}^{m \times n}$ be a sample image containing the conveyor, some rocks and the robot in its normal position P_1. In order to identify the rocks, the difference image $I_D = |I_S - I_R|$ is obtained and then binarized. Fig. 5 shows the results of applying these processes to the image shown in Fig. 3. It can be observed that the resulting image I_B contains only the rocks.

The size S and position (x,y) of the center of mass of the rocks, according to the axis convention of Fig. 2, can be determined as:

$$S = \sum_{i=1}^{m} \sum_{j=a_1}^{b_1} I_B(i,j) \qquad (3)$$

$$x = \frac{1}{S} \sum_{i=1}^{m} \sum_{j=a_1}^{b_1} i \cdot I_B(i,j) \qquad (4)$$

$$y = \frac{1}{S} \sum_{i=1}^{m} \sum_{j=a_1}^{b_1} j \cdot I_B(i,j) , \qquad (5)$$

where a_1 and b_1 are the cutting points between the vertical projection $f(y)$ of I_B (Fig. 6) and an horizontal straight line used as a detection threshold. The vertical projection function $f(y)$ is obtained as:

$$f(y) = \sum_{i=1}^{m} I_B(i,y) . \qquad (6)$$

Fig. 3. A typical image I_S

Fig. 4. Binarized reference image I_{BR}

Fig. 5. Binarized difference image I_B

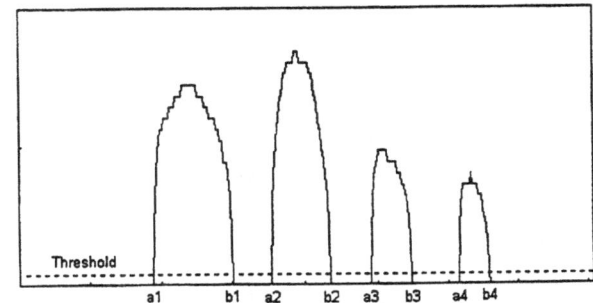

Fig. 6. Vertical projection of I_B

4.3 *Speed estimation*

In order to calculate the speed of the rocks, it is necessary to obtain M position estimations $y_1, y_2, \ldots y_M$. The number of estimations M must satisfy:

$$2 \leq M \leq \frac{l/|v| - t_c}{t_p} \qquad (7)$$

where l is the distance traveled by the rock, v is the conveyor speed, t_c is the closing time of the gripper and t_p is the processing time of one image.

The conveyor speed v can be calculated with different methods. A simple alternative is to assume a linear relationship between position samples y_i and time samples t_i, resulting:

$$v = \frac{M \sum_{i=1}^{M} t_i \cdot y_i - \sum_{i=1}^{M} t_i \cdot \sum_{i=1}^{M} y_i}{M \sum_{i=1}^{M} t_i^2 - \left(\sum_{i=1}^{M} t_i\right)^2}. \qquad (8)$$

Finally, the time that the robot must wait before closing the gripper is:

$$T = \frac{y_{robot} - y_M}{v} - t_p - t_c \qquad (9)$$

where y_{robot} is the robot position, y_M is the last position sample of the rock, t_p is the processing time of one image and t_c is the closing time of the gripper.

5. DESCRIPTION OF THE SCORBOT ER VII

As described by Eshed Robotec Ltd. (Eshed Robotec Ltd., 1982), the Scorbot ER VII is a vertical joined robot with five degrees of freedom. Each joint has a position sensor, an electric motor and a velocity reducer. The position sensor is an optical encoder which measures the angle and velocity of each joint and sends this information to the controller, which determines the torque to be applied to the joint motor. The arm's end effector is equipped with an electrical gripper which can grasp rocks of up to 2 [kg]. The robot weighs 30 [kg], has a repeatability of 0.2 [mm] and a work envelope of 850 [mm]. The robot is also equipped with some accessories such as a teach pendant, a conveyor belt, a rotary table and a slide base.

The Scorbot ER VII includes a PID control system which controls both position and velocity of each joint. The controller protects the system by specifying overcurrent limits, impact and thermic

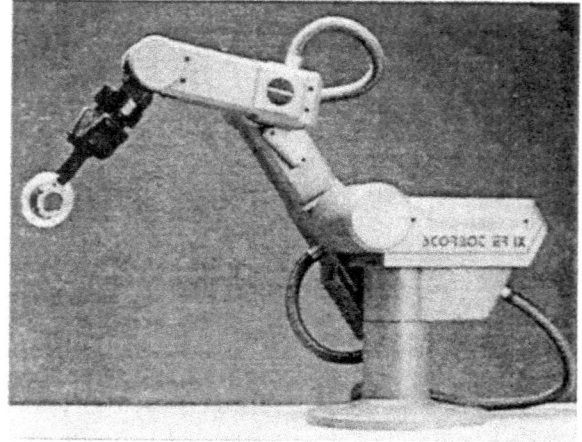

Fig. 7. Scorbot ER VII

protections, and motion envelope limits. All these parameters can be modified by the user.

On the other hand, the system includes different software packages for using and controlling the robot. ACL, Advanced Control Language, is an advanced multitasking robotic programming language that enables structured programming of complex tasks and applications through the use of variables, mathematical algorithms and functions.

6. EXPERIMENTAL RESULTS

6.1 *Results based on the proximity sensor*

In this case an inductive proximity sensor was used. The digital output of the sensor is connected to one of the digital inputs of the robot controller (labeled IN[1]). An ACL program was written in order to grasp the rocks. The program determines the conveyor speed as a linear function of the conveyor motor voltage, and then waits some seconds and grasps the rock as explained in section 3. The program listing is the following:

```
PROGRAM   TOMA1
*********************
SETP      P2=P1                Sets initial positions
SETP      P3=P4
SHIFTC    P2 BY Y   2500
SHIFTC    P2 BY Z   3000
SET       V = 2000             Sets the motor conveyor
SET       ANOUT[8] =-V         voltage to -2 [V] and
SET       V=71738 * V          calculates the conveyor
SET       V=V / 1000           speed.
SET       V=V - 61784
SET       D = 10
SET       D=D * 1000000
SET       D=D / V
LABEL     1
WAIT      IN[1] = 1            Determines when the
SET       T0 = TIME            rock starts and finishes
WAIT      IN[1] = 0            passing over P1
SET       T1 = TIME
SET       T1=T1 - T0           Calculates the waiting
```

```
SET      T1=T1 / 2        time considering a clo-
SET      T1=D - T1        sing time of 0.3 [sec]
SET      T1=T1 - 30
DELAY    T1               Waits before closing
GOSUB    TOMA2            the gripper
SHIFTC   P3 BY Z   270    Grasps, hauls and dump
GOTO     1                the rock
END


PROGRAM  TOMA2
*********************
CLOSE                     Grasps the rock
MOVECD   P3 P2            Hauls the rock to P2
OPEN                      Dumps the rocks
MOVECD   P1 P2            Returns to P1
END
```

It can be observed that ACL includes the command WAIT which stops the program flow until something occurs. In this case, it waits until IN[1] changes, which means that the rock is in front of the proximity sensor. Other interesting commands are SET ANOUT[8] = -V which sets to -V the motor conveyor voltage, SET T0 = TIME which stores in T0 the internal controller variable TIME (time in tenths of seconds since the computer was turned on), and GOSUB TOMA2 which executes a subroutine within the principal program.

Successful results were obtained with this laboratory prototype. The robot is able to grasp rocks of different size and at different conveyor speeds. However, the speed of the conveyor can change with the weight of the rocks without changing the conveyor motor voltage. This means that the system is not able to detect variations in the conveyor speed and in this case the system fails. Anyway, in a mineral concentration plant the speed of the conveyor is measured.

6.2 Results based on the vision system

The proposed algorithms were programmed in C++, Borland 4.02 compiler and Windows 3.11 operative system.

Together with the image processing and communication routines, a user interface was developed. The interface includes a processing window which displays the captured (I_S) and processed images (I_{BR} and I_B), and an information window which shows the results of the classification process: size, position, speed and waiting time. The user interface can be seen in Fig. 8.

Table 1 shows the experimental parameters of the experimental setup and vision system. It can be observed that the vision system spends 1.72 [sec] processing one image. Besides, the conveyor speed ranges from 6 to 15 [cm/sec], which implies that (equation 7) the number of position samples M

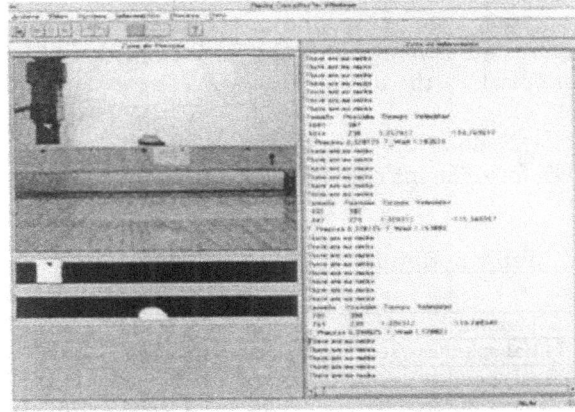

Fig. 8. User interface

ranges from 4 to 1, respectively. Therefore, theoretically speaking the system is not able to estimate speeds over 13 [cm/sec], because it can not obtain at least 2 position measurements.

Table 1. Experimental parameters

Parameter		Value	
Conveyor length	l	50.0	[cm]
Processing time	t_p	1.72	[sec]
Closing time	t_c	0.3	[sec]
Minimum conveyor speed	v	6.0	[cm/sec]
Maximum conveyor speed	v	15.0	[cm/sec]
Min. number of positions	M	1	[sample]
Max. number of positions	M	4	[samples]

In practice, the number of position estimations is variable, even at constant speed, because the real value of M also depends on when the first position y_1 is measured (equation 10). Therefore, in the real case the speed must be inferior to 10 [cm/sec] so that the system can obtain at least 2 position measurements.

$$\frac{y_{robot} - y_1 - vt_c}{vt_p} - 1 < M \leq \frac{y_{robot} - y_1 - vt_c}{vt_p}$$

(10)

If a higher speed is needed it will be necessary to reduce the processing time t_p. Because 64% of t_p is spent by the acquisition system, 27% by the pre-processing algorithms and only 9% by the algorithms of size, speed, etc., the reduction of t_p implies the use of a faster frame grabber.

Another possibility is to use the previous speed measurement. Thus, the estimation of the speed of the rock $v(t)$ can be obtained as:

$$v(t) = \begin{cases} v(t-1) & \text{if } M = 1 \\ \frac{y_M - y_{M-1}}{t_p} & \text{if } M \geq 2 \end{cases}$$

(11)

where $v(t-1)$ is the previous speed, y_M and y_{M-1} are the two latest position estimations and t_p is the processing time. This kind of speed computation requires an estimation of the initial speed, which can preset by the user.

Table 2 shows the results of the classification of rocks for different conveyor speeds.

Table 2. Results of the classifier of rocks

Experiment	1	3	5a	5b
Initial speed[cm/sec]	-7.16	-10.8	-15.1	-17.1
Final speed[cm/sec]	-6.60	-10.9	-15.1	-17.1
Number of rocks	16	16	16	16
Rocks with M = 4	7	0	0	0
Rocks with M = 3	9	0	0	0
Rocks with M = 2	0	12	0	0
Rocks with M = 1	0	4	16	16
Number of grasped rocks	16	16	16	0

As can be observed in experiments 1 and 3, the system calculates the speed of the rocks with at least 2 position samples. This situation allows the system to grasp 100% of the rocks whatever the selection of the initial speed may be. However, in experiments 5a and 5b, both with the same actual speed, the system obtains only one position sample. Therefore, the effectiveness of the system depends on the initial speed selected. If the initial speed is right (experiment 5a), the system grasps 100% of the rocks. If not (experiment 5b), the system does not grasp any rock at all.

Fig. 9 shows the effectiveness of the system for different initial speed errors and speeds. It can be observed that if the speed is inferior to 10 [cm/sec] the system is 100% effective. If the speed is superior to 10 [cm/sec], there exists a maximum initial speed error for which the system maintains his effectiveness. Over this error, the effectiveness decreases drastically.

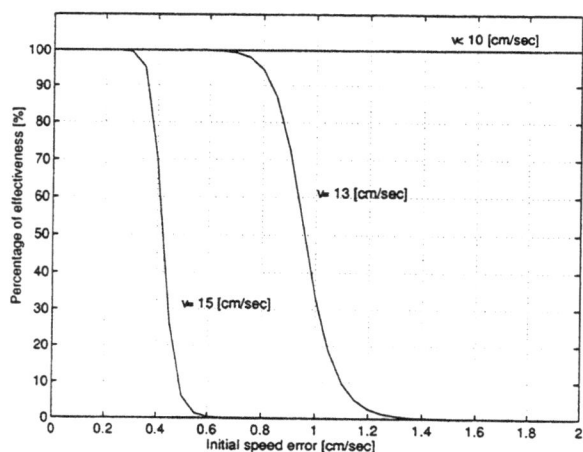

Fig. 9. Effectiveness for different speeds and speed errors

7. CONCLUSIONS AND FUTURE WORK

This paper presents the development of a laboratory prototype of a robotic classifier of moving objects. The system is applied to the classification of the rocks that feed the grinding process in a mineral concentration plant.

Two alternatives for measuring the size of the rocks are considered. The solution based on the proximity sensor is extremely simple, does not need any computer and works well in simple cases. The solution based on artificial vision has a better performance and offers a more precise measurement of position and size. However, it is more intensive in computing resources.

The results obtained with the vision system have encouraged us to continue with the project. As a second stage we plan to develop a 3D vision system in order to measure the size of the rocks with more precision. In a third stage we will intend to remove the assumption that rocks are moved one by one and we will try to identify big rocks within a pile that is transported along a conveyor belt.

8. ACKNOWLEDGMENTS

The authors gratefully acknowledge the financial support of FONDECYT, grant 1960394.

9. REFERENCES

Cartes, F., M. Telias, R. Barrientos and R. Améstica (1995). Coarse ore granulometric distribution analyzer PETRA ADG-21. In: *Proceedings of the COPPER 95 International Conference.* Vol. 2. pp. 233–245.

Cipriano, A., E. Pajares, R. Améstica and R. Barrientos (1996). Simulation analysis of control strategies for a semiautogenous mill by using on line granulometric distribution measurement. In: *Proceedings of 7th Latin American Congress on Automatic Control.* Vol. 1. pp. 359–365. In Spanish.

Eshed Robotec Ltd. (1982). *ACL Advanced Control Language v 1.43, Reference Guide.*

González, Rafael C. and Richard E. Woods (1993). *Digital Image Processing.* Addison-Wesley, Massachusetts.

Copyright © IFAC Algorithms and Architectures for
Real-Time Control, Cancun, Mexico, 1998

Evaluation of Real–Time Motion Controllers on a Direct–Drive Robot Arm*

Rafael Kelly[†] and Fernando Reyes[‡]

[†] División de Física Aplicada, CICESE, Apdo. Postal 2615, Adm. 1
Ensenada, B. C., 22800, MEXICO, *e-mail: rkelly@cicese.mx*

[‡] Escuela de Ciencias de la Electrónica, Universidad Autónoma de Puebla
Puebla, Pue., 72570, MEXICO

Abstract

Motion control of robot manipulators offers interesting practical and theoretical challenges to control researchers. This paper describes experimental comparison among four model–based control algorithms on a direct-drive robotic arm. All controllers include Coulomb and viscous friction compensation as well as cancellation of gravity torques; they are tested under common trajectory specification and performance index. Copyright © 1998 IFAC

Keywords: *Control of robot, Evaluation, Real-time, Computed torque control, PD control.*

1 Introduction

Robot manipulators offer interesting theoretical and practical challenges to control researchers due to the non-linear and multivariable nature of their dynamical behavior. However, despite a great amount of works in control algorithms for robot manipulators, most publications illustrate their potential performance benefits by simulations and only few have been accomplished with experimental results on actual manipulators. Below we provide an overview of previous efforts reported in the literature in this direction.

Control experiments using feedforward and compute torque control on a direct-drive robot were described in An *et al.* (1989) while Khosla and Kanade (1989) compared the computed-torque control and linear PD control schemes. Their results refer to experiments with a direct-drive robot. An experimental

analysis of model-based control techniques as well as the development of an evaluation environment have been presented in Leahy (1990). In Whitcomb *et al.* (1993), comparative experiments were presented with several adaptive controllers, and it was concluded that differences between them are not so significant to allow recommendation. Their results were obtained on two different robots. They propose the root-mean-squared tracking error as a measure to evaluate the control performance. Tarn *et al.* (1993) have presented experimental results on trajectory tracking performance of four servo schemes on a PUMA 560 arm. De Jager and Banens (1994) presented an experimental evaluation with several advanced robot controllers. These authors have demonstrated the superior performance of model-based robot controllers over conventional PD control.

Recently, Berghuis *et al.* (1995) presented an experimental study of three adaptation methods on a two rotary joint robot; they conclude that the tracking error driven gradient method is preferred for parameter adaptation in robotic applications. In Kim and Hori (1995) it was presented an experimental evaluation in tracking control with adaptive and robust control schemes on a direct-drive robot arm with two joints. Results obtained from implementation of a robust position/force controller on a two degrees of freedom direct–drive robot are described in Bridges *et al.* (1995).

In this paper, we discuss the experimental evaluation of four model–based control algorithms tested on a two degrees–of–freedom vertical direct–drive robot arm. Because the direct–drive nature of the arm, the robot nonlinear dynamics cannot be neglected for controller design purposes. This is particularly true in our case where high speed motions are requested. Model–

---

*Work partially supported by CONACYT, Mexico.

based controllers refer to those control algorithms requiring explicit knowledge of the closed–form robot dynamics for effective implementation. The control algorithms tested in this paper are: Computed–Torque control (Spong and Vidyasagar, 1989), PD+ control (Paden and Panja, 1988), PD control with computed feedforward (Kelly and Salgado, 1994) and linear PD control (Takegaki and Arimoto, 1981).

Throughout this paper, we use the notation $\lambda_m\{A\}$ and $\lambda_M\{A\}$ to indicate the smallest and largest eigenvalues, respectively, of a symmetric positive definite bounded matrix $A(x)$, for any $x \in \mathbb{R}^n$. The norm of vector x is defined as $\|x\| = \sqrt{x^T x}$.

2 Experimental system description

The experimental system consists of a direct-drive vertical robot arm with two degrees–of–freedom whose rigid links are joined with revolute joints (see Figure 1). It is equipped with position sensors, motor drivers, a Digital Signal Processor (DSP) motion control board, a host computer 486 PC and software environment which generates an user-friendly interface to provide the user with tools for development, implementation and study of control algorithms.

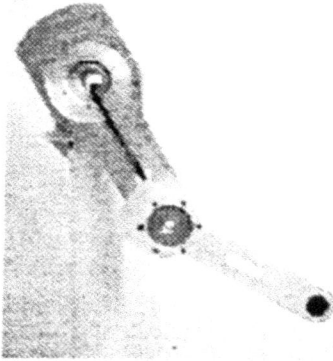

Figure 1: Robot

The servos are operated in torque mode, so the motors acts as torque source and they accept an analog voltage as a reference of torque signal. All controllers, including the test trajectories, have been written in the C language. The control algorithms are executed in the DSP control board at 2.5 msec. sampling rate although they are computed in a period smaller than 200 μsec. The standard backwards difference algorithm applied to the joint position measurements was used to generate the velocity signals.

3 Robot dynamics

The dynamics of a serial n-link rigid robot can be written as (Spong and Vidyasagar, 1989):

$$M(q)\ddot{q} + C(q,\dot{q})\dot{q} + g(q) + f(\dot{q}) = \tau \qquad (1)$$

where q is the $n \times 1$ vector of joint displacements, $\dot{q}$ is the $n \times 1$ vector of joint velocities, τ is the $n \times 1$ vector of applied torques, $M(q)$ is the $n \times n$ symmetric positive definite manipulator inertia matrix, $C(q,\dot{q})$ is the $n \times n$ matrix of centripetal and Coriolis torques, and $g(q)$ is the $n \times 1$ vector of gravitational torques, and the $n \times 1$ vector $f(\dot{q})$ stands for the friction torques.

The equation of motion (1) has associated several parameters which can be exploited to facilitate control system design. These parameters can be computed as follows (Kelly and Salgado, 1994):

$$k_M \geq n^2 \left(\max_{i,j,k,q} \left| \frac{\partial M_{ij}(q)}{\partial q_k} \right| \right), \qquad (2)$$

$$k_{c_1} \geq n^2 \left(\max_{i,j,k,q} \left| C_{k_{ij}}(q) \right| \right), \qquad (3)$$

$$k_{c2} \geq n^3 \left(\max_{i,j,k,l,q} \left| \frac{\partial C_{k_{ij}}(q)}{\partial q_l} \right| \right), \qquad (4)$$

$$k_g \geq n \left(\max_{i,j,q} \left| \frac{\partial g_i(q)}{\partial q_j} \right| \right), \qquad (5)$$

where $M_{ij}(q)$ is the ij-element of matrix $M(q)$, $C_{k_{ij}}(q)$ is the ijk Christoffel symbol and $g_i(q)$ is the i-element of vector $g(q)$.

Considering the values of the physical parameters of our robot arm, we obtain the following entries for the robot dynamics (Reyes and Kelly, 1997):

$$M(q) = \begin{bmatrix} 2.35 + 0.16\cos(q_2) & 0.10 + 0.08\cos(q_2) \\ 0.10 + 0.08\cos(q_2) & 0.10 \end{bmatrix} \quad (6)$$

$$C(q,\dot{q}) = \begin{bmatrix} -0.168 \sin(q_2) \, \dot{q}_2 & -0.084 \sin(q_2) \, \dot{q}_2 \\ 0.084 \sin(q_2) \, \dot{q}_1 & 0 \end{bmatrix} \quad (7)$$

$$g(q) = 9.81 \begin{bmatrix} 3.92 \sin(q_1) + 0.18 \sin(q_1 + q_2) \\ 0.18 \sin(q_1 + q_2) \end{bmatrix} \quad (8)$$

$$f(\dot{q}) = \begin{bmatrix} 2.288\dot{q}_1 + 7.17\mathrm{sgn}(\dot{q}_1) \\ 0.175\dot{q}_2 + 1.73\mathrm{sgn}(\dot{q}_2) \end{bmatrix} \quad (9)$$

where Coulomb and viscous friction have been considered to model the friction torque $f(\dot{q})$.

Using the expressions (2)–(5) together with (6)–(8), we get

$$k_M = 0.67 \text{ kg m}^2; \quad k_{c_1} = 0.33 \text{ kg m}^2 \quad (10)$$
$$k_{c_2} = 0.67 \text{ kg m}^2; \quad k_g = 40.28 \text{ kg m}^2/\text{sec}^2 (11)$$

4 Model–Based control strategies

Let us introduce the following notation. Let q_d, $\dot{q}_d$, $\ddot{q}_d$, be the desired joint position, velocity and acceleration trajectories which are chosen as bounded functions. Furthermore, $\|\ddot{q}_d\|_M$ and $\|\dot{q}_d\|_M$ denote upper bounds on the norms $\|\ddot{q}_d\|$ and $\|\dot{q}_d\|$ respectively. The joint position and velocity errors are denoted by $\tilde{q} = q_d - q$ and $\dot{\tilde{q}} = \dot{q}_d - \dot{q}$ respectively, K_p and K_v are $n \times n$ symmetric positive definite matrices (Proportional and Derivative gain matrices respectively).

4.1 PD control

The control algorithm against which all controllers are measured is the Proportional-Derivative (PD) linear controller. We label the torque control action by τ_{pd}. The control law is given by

$$\tau_{pd} = K_p \tilde{q} + K_v \dot{\tilde{q}} + g(q) + f(\dot{q}) . \quad (12)$$

The first formal stability analysis considering the full nonlinear robot dynamics was reported in Takegaki and Arimoto (1981). However, this controller does not provide exact tracking for time varying desired joint position, but it does guarantee a bounded position error $\|\tilde{q}\|$. Moreover, the steady-state magnitude of $\|\tilde{q}\|$ may be reduced by selecting high feedback gains (Whitcomb et al., 1993).

4.2 Computed–Torque control

Computed torque control is a special application of feedback linearization technique, which has been popular in modern systems theory. The computed-torque controller labeled τ_{ct} is given by the following equation (Spong and Vidyasagar, 1989):

$$\tau_{ct} = M(q)[\ddot{q}_d + K_v \dot{\tilde{q}} + K_p \tilde{q}] + C(q, \dot{q})\dot{q} + g(q) \\ + f(\dot{q}) . \quad (13)$$

Computed-torque controller achieves dynamic decoupling of all the joints using nonlinear feedback resulting. This can be easily seen by substituting the right-hand side of the dynamics (1) into (13); this

yields the closed–loop linear time invariant differential equation:

$$\ddot{\tilde{q}} + K_v \dot{\tilde{q}} + K_p \tilde{q} = 0 . \quad (14)$$

Since matrices K_p and K_v are symmetric positive definite, then the solutions ($\dot{\tilde{q}}(t)$ and $\tilde{q}(t)$) vanish exponentially. In practice, however, the presence of unmodeled high-frequency dynamics and the noise present in the velocity measurement conduct to deviations in trajectory tracking.

4.3 PD+ control

The PD+ controller, labeled τ_{pd+}, is given by (Koditscheck, 1984; Paden and Panja, 1988):

$$\tau_{pd+} = K_p \tilde{q} + K_v \dot{\tilde{q}} + M(q)\ddot{q}_d + C(q, \dot{q})\dot{q}_d + g(q) + f(\dot{q}) . \quad (15)$$

This controller was first introduced by Koditscheck (1984) and formally analyzed in Paden and Panja (1988) and Whitcomb et al. (1993). It has been proven that under exact robot and friction models, and symmetric positive definite matrices K_p and K_v, then the PD+ controller provides asymptotically exact tracking without exact linearization.

4.4 PD control with computed feedforward

This controller labeled τ_{pdcf}, consists of a linear PD controller and a feedforward of the nominal dynamics computed along the desired trajectory q_d. It is given by (An et al., 1988)

$$\tau_{pdcf} = K_p \tilde{q} + K_v \dot{\tilde{q}} + M(q_d)\ddot{q}_d + C(q_d, \dot{q}_d)\dot{q}_d \\ + g(q_d) + f(\dot{q}) . \quad (16)$$

The advantage of this controller is that once that desired trajectory (q_d, $\dot{q}_d$ and $\ddot{q}_d$) for a give task has been specified, then the feedforward terms relying on the robot dynamics $M(q_d)\ddot{q}_d + C(q_d, \dot{q}_d)\dot{q}_d + g(q_d)$ can be computed off–line reducing the computational burden.

The stability analysis of (16) has been reported in Kelly and Salgado (1994). In summary, under exact models for the robot dynamics and friction, if the symmetric positive definite matrices K_p and K_v are selected such that:

$$\lambda_M\{K_p\} > c_1 = k_M\|\ddot{q}_d\|_M + k_{c_2}\|\dot{q}\|_M^2 + k_g. \quad (17)$$
$$\lambda_m\{K_v\} > \frac{(c_1 + 2[k_{c_1}\|\dot{q}_d\|_M])^2}{4(\lambda_m\{K_p\} - c_1)} + \lambda_M\{M\}. \quad (18)$$

then the position error $\tilde{q}$ vanishes asymptotically in a local sense, i.e., $\lim_{t \to \infty} \tilde{q}(t) = 0$ at least for sufficiently small initial position and velocity errors.

5 Experimental evaluation

5.1 Trajectory specification and performance evaluation

The desired trajectory we have used has the a similar structure to those proposed in Dawson *et al.* (1994), de Queiroz *et al.* (1996), that is

$$q_d = \begin{bmatrix} 0.78[1 - e^{-2.0\,t^3}] + 0.17[1 - e^{-2.0\,t^3}]\sin(\omega_1 t) \\ 1.04[1 - e^{-1.8\,t^3}] + 2.18[1 - e^{-1.8\,t^3}]\sin(\omega_2 t) \end{bmatrix}$$

$$(19)$$

where ω_1 and ω_2 represent the frequency of desired trajectory for the shoulder and elbow joints respectively.

In our experimental tests, we use the expression (19) with a set of two frequencies for ω_1 and ω_2, that is, $\omega_1 = 15$ rad/sec and $\omega_2 = 3.5$ rad/sec represent the test trajectory I, while $\omega_1 = 7.5$ rad/sec and $\omega_2 = 1.75$ rad/sec denote test trajectory II.

We use $\mathcal{L}^2$ norm of tracking errors as criterion of tracking performance measurement as suggested in Whitcomb *et al.* (1993), Kim and Hori (1995), de Jager and Banens (1994), Jaritz and Spong (1996). The $\mathcal{L}^2$ norm measures the root-mean-square average of tracking error and it is given by

$$\mathcal{L}^2[\tilde{q}] = \sqrt{\frac{1}{T} \int_0^T \|\tilde{q}(t)\|^2 dt} \;. \qquad (20)$$

The better performance is for smaller $\mathcal{L}^2$ norm, it represents smaller tracking error. In the performance evaluation we have taken $T = 10$ sec. We compute the error norm average of five runs as suggested in Berghuis *et al.* (1995) with test trajectories I and II for each of the listed controllers.

5.2 Design of gain matrices

Using the numerical values of k_M, k_{c_1}, k_{c_2} and k_g given in (10) and (11), the maximum eigenvalue of inertia matrix (6) is $\lambda_M\{M\} = 2.533$ kg m². Also employing $\|\dot{q}_d\|_M = 8.0721$ rad/sec., and $\|\ddot{q}_d\|_M = 47.501$ rad/sec² for test trajectory I, we obtain the selection conditions for K_p and K_v gains in agreement with (17) and (18) as

$$\lambda_M\{K_p\} > 115.7 \text{ Nm/rad}, \qquad (21)$$
$$\lambda_m\{K_v\} > 9.23 \text{ Nm sec/rad}. \qquad (22)$$

During the preliminary experiments, we have first tested several K_p and K_v gains for the controller; they were chose in agreement with (21) and (22) and ensuring acceptable behavior in practice and without saturate the actuators. The controller gains K_p and K_v were chosen as

$$K_p = \text{diag}\{2000, 1000\} \text{ Nm/rad},$$
$$K_v = \text{diag}\{150, 15\} \text{ Nm sec/rad}.$$

The second group is for the computed-torque controller. The matrices K_p and K_v are selected to be diagonal whose entries are denoted by k_{p_i} and k_{v_i} for $i = 1, 2$ respectively. Our selection criterion was such that satisfy the critically damped condition: $k_{v_i} = 2\sqrt{k_{p_i}}$. The actual K_p and K_v gains of the computed-torque controller were selected as

$$K_p = \text{diag}\{1000, 5000\} \text{ 1/sec}^2$$
$$K_v = \text{diag}\{63.24, 141.42\} \text{ 1/sec}.$$

5.3 Experimental results

Because the lack of space, we will present the experimental results obtained using trajectory I. However, we will include later a table comparing all controllers according to both desired trajectories.

The experimental position errors and the applied torque using trajectory I for the four controllers are shown in Figures 2–5.

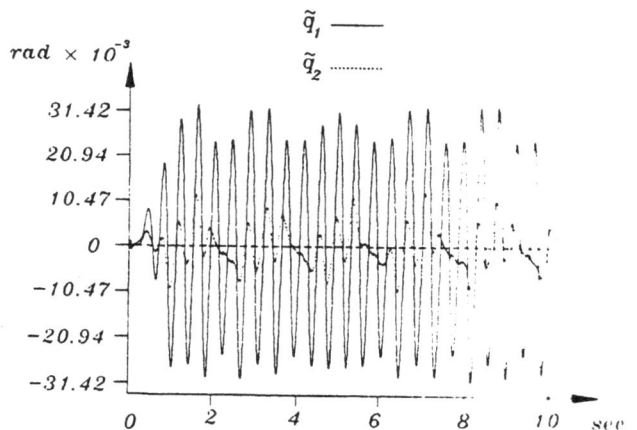

Figure 2: Tracking errors for the PD controller

The tracking error for the PD controller is depicted in Figure 2. Comparing this figure with those corresponding to the remaining model–based controllers (figures 3, 4, and 5), it can be seen that the maximum values of position errors $\tilde{q}_2$ and $\tilde{q}_1$ are approximately 4.7 times and 2.2 times greater.

On the other hand, model-based controllers with complete feedback and feedforward dynamics compensation reduce the peak tracking error when compared to PD controller alone. This is due to the compensation of all dynamic forces, including Coriolis and centrifugal, necessary to maximize performance.

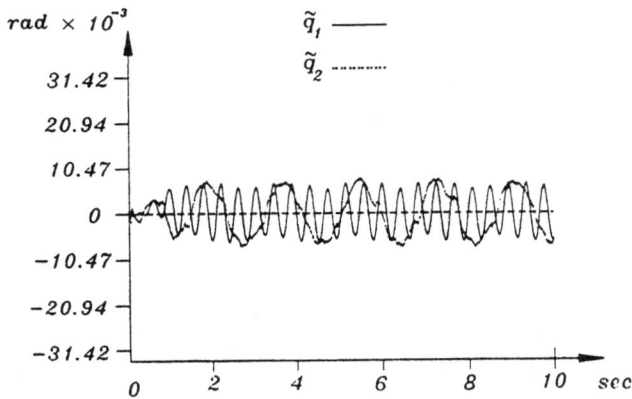

Figure 3: Tracking errors for computed-torque controller

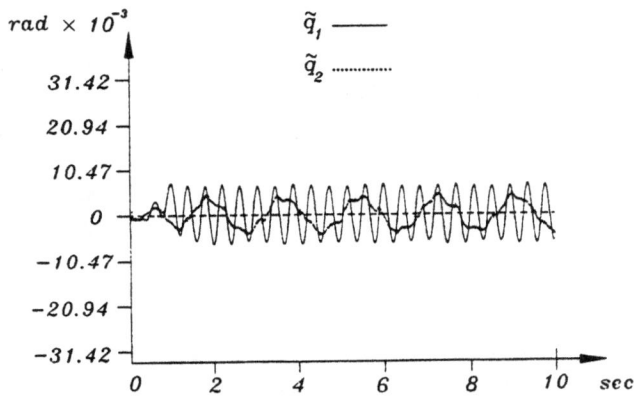

Figure 4: Tracking errors for PD+ controller

The position errors obtained with controllers: Computed-torque control, PD+ control and PD control with computed feedforward are depicted in figures 3, 4, and 5, respectively. These figures do not allow a clear conclusion about what controller has the best

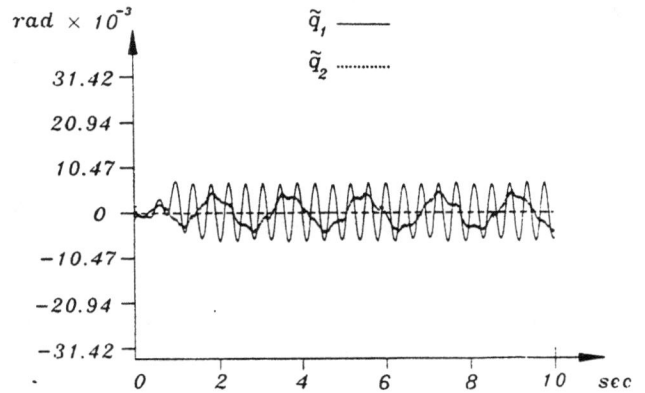

Figure 5: Tracking errors for PD control with computed feedforward

performance. However, the $\mathcal{L}^2$ norm criterion gives an useful performance index as shown in Figure 6. To average out stochastic influences, the data presentation in this figure represents the mean of root-mean-square position error vector norm of five runs; for clarity the data in this figure are compared with respect to $\mathcal{L}^2$ norm of PD controller. The results from one run to another was observed to be less 1% of their mean, which underscore the repeatability of the experiments.

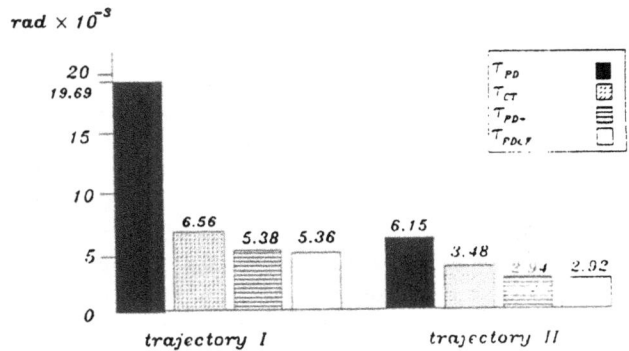

Figure 6: $\mathcal{L}^2$ norm for test trajectories I and II

Figure 6 includes the performance index for both test trajectories. In experiments with test trajectory II, we have observed that the absolute magnitude of tracking errors are smaller than these observed with the test trajectory I.

6 Concluding remarks

In this paper we have tested several control algorithms on a direct–drive robot arm. The performance of these controllers was compared to the linear PD control. As expected, from our experimental results we have observed that those controllers incorporating in some way the robot dynamics attained better tracking accuracy. This is particularly true when fast motions are requested to the arm, however, for slow desired trajectories, these controllers including the linear PD control may produce acceptable similar behavior.

References

1. An C. H., C. G. Atkeson and J. M. Hollerbach, (1988), *"Model-Based control of a robot manipulator"*, The MIT Press.
2. An C. H., C. G. Atkeson, J. D. Griffiths and J. M. Hollerbach, (1989), "Experimental evaluation of feedforward and computed torque control", *IEEE Trans. on Robotics and Automation*, Vol. 5, No. 3, pp. 368-373, June.
3. Berghuis H., H. Roebbers and H. Nijmeijer, (1995), "Experimental comparison of parameter estimation methods in adaptive robot control", *Automatica*, Vol. 31, No. 9, pp. 1275-1285.
4. Bridges M. M., J. Cai, D. M. Dawson, and M. T. Grabbe, (1995), "Experimental results for a robust position and force controller implemented on a direct drive robot", *Robotica*, Vol. 13, pp. 11-18, Cambridge University Press.
5. Dawson D. M., J. J. Carroll and M. Schneider, (1994), "Integrator backstepping control of a brush dc motor turning a robotic load", *IEEE Transactions on Control System Technology*, Vol. 2, No. 3, pp. 233-244, September.
6. de Jager A. and J. Banens, (1994), "Experimental evaluations of robot controllers", *Proceedings of the 33rd. Conference on Decision and Control*, pp. 363-368, Lake Buena Vista, Fl. U.S.A.
7. de Queiroz M. S., D. Dawson and T. Burg, (1996), "Reexamination of the DCAL controller for rigid link robots", *Robotica*, Vol. 14, pp. 41-49.
8. Jaritz A. and M. W. Spong, (1996), "An experimental comparison of robuts control algorithms on a direct drive manipulator", *IEEE Transactions on Control Systems Technology*, Vol. 4, No. 6, pp. 627-640.
9. Kelly R. and R. Salgado, (1994), "PD control with computed feedforward of robot manipulators: A design procedure", *IEEE Transactions on Robotics and Automation*, Vol. 10, No. 4, pp. 566-571, August.
10. Khosla P. K. and T. Kanade, (1989), "Real-Time implementation and evaluation of computed-torque scheme", *IEEE Trans. on Robotics and Automation*, Vol. 5., No. 2, pp. 245-252, April.
11. Kim K. and Y. Hori, (1995), "Experimental evaluation of adaptive and robust schemes for robot manipulator control ", *IEEE Transactions on Industrial Electronics*, Vol. 42, No. 6, pp. 653-662.
12. Koditscheck D., (1984), "Natural motion for robot arms" *Proceedings of the 1984 IEEE Conference on Decision and Control*, pp. 733-735, Las Vegas NV., December.
13. Leahy M. B. Jr., (1990), "Model-Based control of industrial manipulators: An experimental analysis", *Journal of Robotic System*, Vol. 7, No. 5, pp. 741-758, October.
14. Mills J. K., P. Baines, T. Chang, S. Chew, T. Jones, S. Lam and A. Rabadi, (1995), "Development of a robot control test platform", *IEEE Robotics and Automation Magazine*, Vol.2 , No. 4, pp. 21-28, December.
15. Paden B. and R. Panja, (1988), "Globally asymptotically stable PD+ controller for robot manipulators", *International Journal of Control*, Vol. 7, No. 6, pp. 1697-1712.
16. Reyes F. and R. Kelly, (1997), "Experimental evaluation of identification schemes on a direct-drive robot", *Robotica*, Vol. 15, Part 5, pp. 563-571, September–October.
17. Spong M. W. and M. Vidyasagar, (1989), *Robot Dynamics and Control*, John Wiley and Sons.
18. Takegaki M. and S. Arimoto, (1981), " A new feedback method for dynamic control of manipulator", *ASME Journal of Dynamic Systems, Measurement, and Control*, Vol. 102, pp. 119-125.
19. Tarn T. J., A. K. Bejczy, G. T. Marth, and A. K. Ramadarai, (1993), "Performance comparison of four manipulators servo schemes", *IEEE Control System*, Vol. 13, No. 1, pp. 22-29, February.
20. Whitcomb L. L., Rizzi A. A. and D.E. Koditschek, (1993), "Comparative experiments with a new adaptive controller for robot arms", *IEEE Transactions on Robotics and Automation*, Vol. 9, No. 1, 59-70, February.

EXPERIMENTAL AUTONOMOUS VEHICLE SYSTEMS
– REQUIREMENTS AND A PROTOTYPE ARCHITECTURE

Ole Ravn, Associate Professor,
Nils A. Andersen, Associate Professor

Department of Automation, Technical University of Denmark,
Building 326, DK-2800 Lyngby, Denmark, E-Mail: or@iau.dtu.dk

The paper describes the requirements for and a prototype configuration of a software architecture for control of an experimental autonomous vehicle. The test bed nature of the system is emphasised in the choice of architecture making re-configurability, data logging and extendability simple. The central element of the architecture is the 'global database' that serves several purposes, such as storing system parameters, making signals available for data logging and inter-process communication. Standard software components are used to a large extent, OS-9 as real-time operating system, a custom ANSI-C program extending the TCL system is used for plan execution and a combination of MATLAB and a custom made Java GUI as user interface on the remote operator console. The choice of these standard software components is explained and the individual components demonstrated. Examples of how specific tasks are implemented using the architecture are given. *Copyright © 1998 IFAC*

Keywords: Mobile robots, test bed systems, automation systems, software architectures.

1. INTRODUCTION

The construction of a number of experimental autonomous Systems for instance Autonomous Guided Vehicles (AGV) have been reported in the literature. [Halme and Koskinen, 1995], [ISER, 1997]. The main emphasis of these papers has been on the investigation of different algorithms and methods and not on the software architecture and the aspects related to the experimental nature of the AGV. Much software and many architectures seem to have been developed from scratch each time a new AGV has been build. This has made it more difficult to make experimental work than really needed, if emphasis from the beginning was on the experimental aspects of the software architecture. Our experience related to system architecture and programming of autonomous system software prototypes is presented in this paper. Examples of such vehicles are mobile land based robots, underwater vehicles (AUV) etc. As a means of verification and comparison of different algorithms a test bed Autonomous Guided Vehicle (AGV) has been build. For more information on construction and control of the test-bed AGV please refer to one of the following

references, see [Ravn and Andersen, 1993a], [Ravn and Andersen, 1993b], [Christensen and Lind, 1993], [Andersen et al., 1995], [Ravn et al., 1995a], [Henriksen et al., 1997].

The software system of the test bed should make experimenting simple from the algorithm developers point of view. A scalable complexity of implementing new algorithms should be possible. This means that a simple high-level specification of the task to be performed should be easily implementable and when the feasibility of the implementation has been shown a more efficient implementation could be done without much additional effort. The architecture should support a hierarchical approach to the specification of tasks and the specification in a high-level and simple way should work at both mission and task level. This means that it should be possible to specify a mission in high-level terms as well as new task level components, the latter could then later be implemented in a more efficient way. An example of such a high-level specification is the docking algorithm: First go to the point in front of the docking station at a specific distance, move towards the docking station at a fixed speed using the beacon navigation [Andersen et al., 1995] until the force exceeds a specific level, then stop.

Below are listed a few of the requirements of the software architecture given in the paper:

Safety. Changing parts of the algorithms should not make the basic parts (drive control etc.) fail. This is a problem when the system architecture is monolithic in the sense that one big program controls a operation of the vehicle. When new components or modifications of existing components are done there is a risk that programming errors will crash the whole program with the effect that control of the system may be lost. It is not possible to prevent programming mistakes especially on an experimental platform where several developers modify code. Using memory manager functions to protect against e.g. pointer errors is the simplest way to handle this. However this requires that the system is divided into several program modules that can be protected independently. Of cause the risk still exists that programming errors (e.g. pointer errors) will crash one module, but sensitive parts like the motion controller module should be in one module which is checked extensively.

Another important aspect of safety, besides the software related described above, is the issue of having backup controllers that are ready to intervene and bring the system to a safe state if the experimental controller fails. This is an very useful feature especially if highly complex systems are considered and are going to be operated and tested under realistic conditions.

Extendability. New algorithms, software modules and real sensors should be added in a natural way. For an experimental test-bed system to work well it is essential that the algorithm developers are able to do seamless integration of new algorithms into the system. This should be possible without having to understand the whole system in depth and simple rules for integration should exist. For instance if a new physical sensor is added to the system configuration it should be simple for the developer to write a driver module and access the information from the new sensor other places in the system. The control of the sensor should be decoupled from the use of it in a client-server fashion. New software modules should on the other hand be able to access all the data provided by other modules in the system in a simple and straightforward manner, not by reprogramming the modules that deliver the data.

Encapsulation. The algorithm developer should have a powerful script like language for describing new features based on already implemented modules. In developing autonomous vehicles the hierarchical architecture is one of the primary approaches. However in implementing commands in applications encapsulation and embedding commands into others is a very important feature. As outlined earlier when describing the docking procedure in high-level terms

a prototyping of new commands in high-level terms is important, as well as the possibility to re-implement commands in a more efficient way later. For a mission oriented architecture encapsulation is also important as it provides ability to specify the mission more clearly.

The outline of the rest of the paper is as follows, in section 2 the basic requirements of software for experimental test-bed systems are described in detail. Section 3 the basic components of the prototype architecture are outlined and in section 4 the central element the 'global database' is reviewed. In section 5 a few examples of the functionality that can be implemented in the architecture are given.

2. REQUIREMENTS FOR AUTONOMOUS SYSTEM SOFTWARE

The general requirements described in section 1 and the more specific requirements described below are conceived in the test-bed context but they are also generally applicable, maybe with some modifications, such as taking into account the implementational effort and the computational efficiency of the final system in production hardware. It seems desirable to be able to move from a test-bed implementation to the production system in a semiautomatic fashion, not having to re-implement everything again. No single architecture (or system) is foreseen, but different approaches to the design should be evaluated in relation to the requirements. The main issue is to put emphasis on system design for experimental autonomous system, not to come up with a single 'silver bullet'. On the other hand there is a need to make an operational *test-bed* system to enable experiments, to verify ideas and to be able to demonstrate the AGV.

2.1. Plug-ins

The basic user elements in the system software are plug-ins or components. This is seen as pre-defined places in the system where algorithm developers can incorporate their code with a clear interface to the rest of the system. What plug-in to use in a specific module is then determined at run time by the conditions of the systems and the user input. Plug-ins can be defined at different levels in the system architecture as shown in figure 1: at the plan generation level, the trajectory generator level or at the controller level.

The modules or plug-ins should honour a number of pre-defined functions. It should be noted that these plug-ins are typically called in a recursive way, in a sampled loop. The functions specified below is a traditional approach well-known from for instance device driver programming i the operating system context, but also in the controller prototyping area [Schneider et al., 1994]:

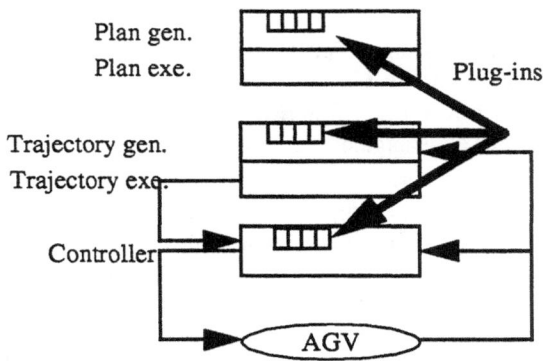

Figure 1. AGV system architecture using plug-ins

- Init: Executed when the plug-in is called the first time and initialises the internal states of the plug-in.

- Update: This function is called after the output has been calculated. For instance for updating the states of a controller and contains the less time critical updates to system variables.

- Output: This function called to calculate the output of for instance a controller between reading the input and writing the output.

- Get: Gets the plug-in state, i.e. the internal state of a controller.

- Set: Sets the plug-in state, i.e. the parameters of a controller.

- Term: Terminates the plug-in.

Furthermore the plug-in should be able to handle signal events of different types. The frame in which the plug-in is integrated will make different utility functions available to the plug-in, such as timing analysis, data-logging, event-logging, exception handling etc.

2.2. Design guidelines

Some design guideline and more specific requirements are listed and explained below. The considerations were made in connection with the design of the system software for the experimental AGV:

- Exception handling in a structured a complete way, multi-layer enabling graceful degradation. Exception handling is a very important feature for intelligent system in general where 'hours and hours of uninterrupted operation' is wanted to make the system robust. But especially in connection with experimental systems where the system is generally more exposed to errors in new experimental modules and high-level commands.

- All commands and actions should indicate their current and final status stating to what degree everything went OK. This is also essential for the intelligent operation in an unknown dynamic environment as the failing command should be taken into account in the further planning of the operation of the system.

- Modularity is essential. Both from a software safety point of view and from a software understandability point of view modularity is essential. This is a well-known fact from software engineering but is even more essential in this area.

- Clear understandable hierarchical structure with explicit layers to enable the user to easier understand the code. Of cause with a with build-in ability to enable reaction based behaviour. As with the before mentioned modularity it is essential that clear interface and software extensions guidelines are written down.

- In order to make the system more portable the specification of the system should be done in a generic framework i.e. inter-process communication, process interface etc. mapping these generic features on the actual system software in this case -OS-9. Furthermore these system specific functions should be collected in special well defined blocks of cause taken into account the computational efficiency

- Plug-in possibility for user code at pre-specified places. This point is treated above.

- Build-in timing analysis is an essential component for an experimental system. It enables the algorithm developer to evaluate the code and determine if timing problems are present. The timing analysis should be computationally efficient so that it is always present.

- State monitoring and event logging of every module or plug-in is a feature to help debugging of multi-tasking system. If an error occurs it should be possible to determine the previous state of the whole system in a simple and straightforward manner. Each module should have the possibility to log events that occur in its context.

- Configurable logging of signals, variable and states, that in turn can be used for visualization and uplink to remote host e.g. UNIX. Logging is essential in debugging, verification and validation of new modules and can in the running system give the basis for extension via uploading and communication to other modules. Logging variables can be divided into two categories, one where is speed in not essential i.e. the receiver and communication bandwidth can handle the data flow and one where the data has to be stored in a buffer for later uploading and where only limited data sizes are possible.

- Easy extension to include and use new hardware (sensors etc.). In experimental systems the hardware configuration should be easily extendable to allow the algorithm developer to concentrate on the algorithms and not be diverted by other issues such a integrating new sensors.

- Easy and flexible reconfigurablity of the system using the graphical user interface (GUI) or by configuring using the Global Database. New experimental algorithms always require tuning and it is essential that this task is make simple for the algorithm developer, so for instance re-compilation is not necessary for changing controller parameters.

- The system should support simulation of the vehicle and sensors using for instance (real Time Workshop (RTW) generated or other C code as 'hardware-in-the-loop' simulation The concept is shown in figure 2. The system software is easily tested, verified and validated against a simulation model of the actual system before the hardware is connected. In the figure the system is shown at three distinct levels, however in reality the situation is much more complex as several intermediate levels are possible. On the other side the controller could also be prototyped in at a higher level of abstraction in MATLAB or other languages before implementation in C and incorporation as a plug-in into the system.,

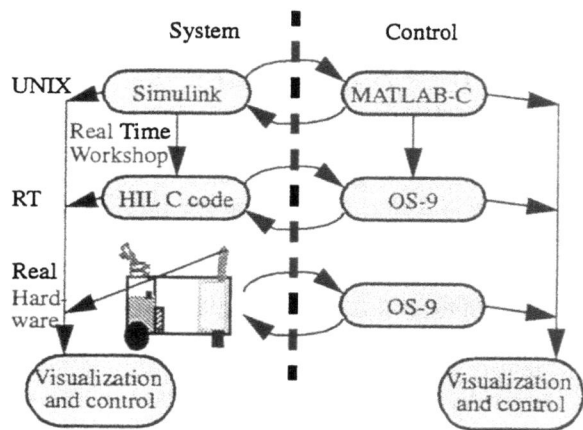

Figure 2. Hardware-in-the-loop simulation.

- Build-in system testing (hardware and software) features. In relation to localization of errors to specific hardware and software modules procedures that are capable of determining which modules are functional and which are not are essential.

The architecture that seem to match could something like the one shown in figure 1. It has a clear distinction between the controller layer, the reference generator layer and the command interpreter layer. The border interface between the layer has to be clearly defined. There is a clearly defined plug-in capability in all layers. For each layer the following applies.

3. BASIC COMPONENTS OF THE ARCHITECTURE

Most of the basic components of the software systems are ready-made or based on ready-made programs:

- Real-time operating system: OS-9, This forms the basis for the total system on the AGV giving

access to the usual low-level real-time and task synchronization functions. Hard real-time tasks like the motion controller module is made using these low-level function. The global database takes advantage of these functions as well.

- Plan Execution: Written as ANSI-C programs and incorporating TCL [Ousterhout, 1994]. As high-level plan execution module a standard software system (TCL) has been ported to the OS-9 system and tested. This gives access to statement control structures such as if-the-else, while and for-loops. The basic problem in this approach seem to be performance i.e. execution speed of the scripts (plans).

- Diagnosis: CLIPS, is a expert system shell developed at NASA to be used for diagnosis. [Giarratano, 1993] It can be run at different places in the system: as a process on the remote console machine or on the AGV on a separate processor or maybe as a low priority task on the main board. CLIPS will get its input signals from the 'global database'. See Section 5. The system should support the seamless placement of higher level tasks on different computers in this sense.

- User interface on remote station: MATLAB GUI and Java GUI designed using SpecJava. As graphical interface and basis for the plan generation system the standard package MATLAB is used. This provides platform independence as the interface can be executed on either PC's or UNIX systems. The MATLAB system provides easy access to data analysis, plotting etc. The Java GUI is designed using a GUI designer to make changes and extensions to the interface more straightforward.

Figure 3 shows the AGV and remote console and the distribution of components.

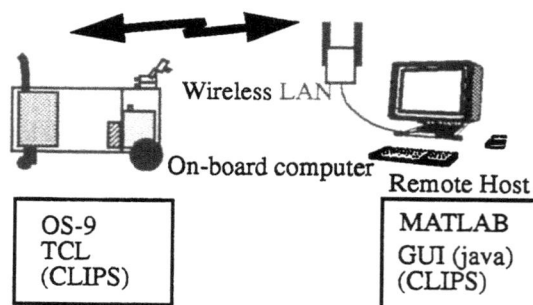

Figure 3. Basic software components of the architecture.

When designing an architecture for a complex real-time system such as an AGV different objectives has to be met. The execution of the hard real-time modules, i.e. Motion controller, Vision Controller and pan-tilt controller has to be correct with respect to time, the soft real-time modules are less critical and the aspect of easy development of new modules can be taken into account. Figure 4 shows the role of the global database as the central component of the system on the AGV. The hard real-time modules use the database for retrieving start-up parameter settings and export internal signal that other modules can utilize.

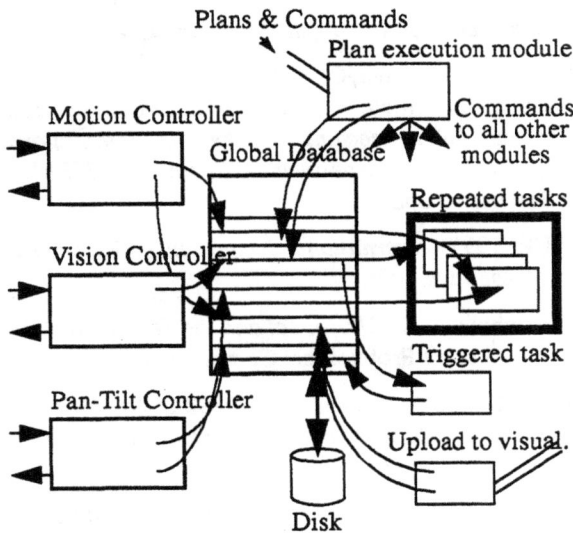

Figure 4. The global database as a central component in the of the architecture

It is important to note that special solutions for transferring data are made for critical high-performance links, to avoid the 'global database' acting as a bottleneck in safety and time critical communication between modules.

4. GLOBAL DATABASE

For the purpose of data collection, exchange and synchronisation of tasks a software module called, 'global database' has been written. It contains information for different purposes:

- Parameters for system modules.
- Signals made available for data logging
- Inter-process communication. i.e. a means of controlling the program flow.
- Linking the modules together.

The 'global database' is a library of functions with real-time database capabilities. The 'global database' has the possibility of controlling the program flow through inter-process communication and provides semaphore protection of common variables. Further more the 'global database' has the capability of sending signals when a variable value is changed i.e. if the variable value is rewritten. Protection of variables against unstable programs are provided through the use of a data module in combination with a memory management unit (MMU) and through a read/write access list for every variable. This encapsulation of data enhances system stability and modularity. Finally

the 'global database' contains a start-up synchronisation and a termination clean-up functionality.

The variables to be placed in the 'global database' are entered in an ASCII file using a plain text editor. The file must contain information about the names and initial value of every variable. A read/write access list to a variable is given either as specific names of the tasks with read/write access or with the string 'all' indicating that there are no restrictions on accessing the variable. Finally it is possible to enter the name of a 'watch' (if the variable is rewritten) semaphore that can be used together with different data logging facilities. [Ravn et al., 1995b]

A very important issue when working with a database in real-time systems is how time consuming the database is compared with the more unstructured but faster global variables. Time measurements with the 'global database' has been made on a Motorola 68040 single-board computer in order to determine the time loss by using the 'global database'. The given execution times are in microseconds, but only comparison between the given times is informative. It can be seen in the table that database calls are between 24 and 38 times more time-consuming than direct accesses to global variables. The function _gv_os9write is slower than _gv_os9read, due to the check for an existing watch semaphore.

	Global database		Direct access	
	Read (µs)	Write (µs)	Read (µs)	Write (µs)
No use of semaphore	13	19	0.8	0.8
Use database semaph.	29	30		

This time loss by using the 'global database' is considered to be small compared to the major enhancement that the 'global database' is offering to the real-time functionality, data encapsulation and program stability.

5. APPLICATION EXAMPLE

Several application programs for the AGV using the 'global database' have been developed. One example is a slow sampling battery monitoring module that is maintaining information about the condition of the AGV's battery in the database. The module is only connected to the basic AGV program through the 'global database', and is capable of writing the measured status in the corresponding entry in the database. If an error occurs in the module the program can not damage any other variables in the AGV database due to the write access list in the 'global database'.

Another application is a data uploading program. This program links to the 'global database' and adds the possibility of sampling different AGV database variables with a given sampling frequency. The variable names to monitor are given to the program at start-up and can be changed while the program is running.

This makes it simple to switch between the variables to monitor due to the fact that the program does not have to be re-compiled. The uploading program has been used to upload the pose (position and orientation) of the AGV to a visualisation program. The program is written using OpenGL and visualises the AGV in the laboratory environment. The visualisation can be connected to either the real AGV or a simulation of the AGV running in MATLAB/Simulink. The output from the visualisation is shown in figure 5.The simple development of data transferring part of this application illustrates the ease which the 'global database' give to developing such applications.

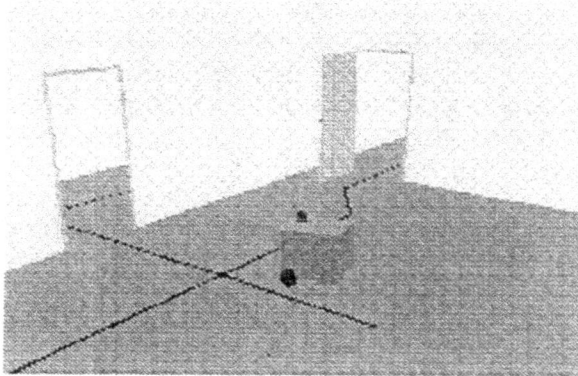

Figure 5. Visualization made with OpenGL of the AGV in the laboratory.

Another application of the up-loading program is the connection to the diagnosis program CLIPS running on the remote UNIX platform. Furthermore a simple module scheduler has also been made. This scheduler has access to the 'global database' on the AGV and it is then possible develop different system monitoring programs. See figure 4.

6. CONCLUSION

The paper has presented the requirements and a prototype software architecture suitable for use with experimental autonomous vehicles. The system consists of mainly ready made components that are in turn linked into a complete system. The main component is the so-called 'global database' that acts at a link between the different modules of the system. The global database also acts as an aid of data-logging, which is very important under experimental conditions.

Several examples that shows the versatility of the architecture is given. A battery power monitoring system, data up-loading module and a scheduler for non-critical tasks.

7. REFERENCES

Andersen, N., Henriksen, L., and Ravn, O. (1995). Navigation and control of an experimental intelligent vehicle. In *Proceedings of the 2nd IAV*, Espoo, Finland.

Christensen, A. and Lind, M. (1993). A modelling framework for integrated design of agv systems. In *Proceedings of the First IAV*, Southampton, England.

Giarratano, J. C. (1993). *CLIPS User's Guide*. NASA.

Halme, A. and Koskinen, K., editors (1995). *2nd IFAC Conference on Intelligent Autonomous Vehicles*.

Henriksen, L., Ravn, O., and Andersen, N. (1997). Autonomous vehicle interaction with in-door environments. In *Proceeding of the 5th International Symposium on Experimental Robotics — ISER'97*, Barcelona, Spain.

ISER (1997). *International Symposium on Experimental Robotics*.

Ousterhout, J. K. (1994). *Tcl and the Tk Toolkit*. Addison-Wesley Professional Computing Series.

Ravn, O. and Andersen, N. A. (1993a). An intelligent vehicle – indoor navigation. In *Proceedings of AMS 93*, München, Germany.

Ravn, O. and Andersen, N. A. (1993b). A test bed for experiments with intelligent vehicles. In *Proceedings of IAV93*, Southampton, England.

Ravn, O., Henriksen, L., and Andersen, N. (1995a). Visual positioning and docking of non-holonomic vehicles. In *Proceeding of ISER'95*, Stanford, California, USA.

Ravn, O., Pjetursson, A., and Andersen, N. A. (1995b). A system architecture for experimental autonomous vehicles. In *Proceedings of AMS 95*, Karlsruhe, Germany.

Schneider, S. A., Chen, V. W., and Pardo-Castellote, G. (1994). ControlShell: a real-time software framework. In *Proceedings of the International Conference on Robotics and Automation*. IEEE.

CONTROL SOFTWARE SYSTEM OF A TURBOGAS POWER UNIT

Carlos D. García

Raúl Garduño

Department of Electronics
CENIDET
P.O. Box 5-164
Cuernavaca, Mor. 62000 Mexico
Cgarcia@cenidet.edu.mx

Department of Electrical Engineering
Pennsylvania State University
104 EE East Building
University Park, PA 16802, USA
rxg147@psu.edu

Abstract: A third generation control software system for turbogas power units, in a series of modernization projects, is presented. Main objective was to reduce the complexity of its predecessors, to facilitate the integration of new control strategies to enhance the unit's performance. The automation, control and protection functions performed by previous generations served as requirement specifications. Adaptation of a generalized functional structure for industrial process control in defining the control system software architecture became the project key point. Real-time structured design, and programming lead to the desired higher quality system. Successful integration of a fuzzy speed controller fulfilled the objective. *Copyright© 1998 IFAC*

Keywords: Control software engineering; structured analysis and design; gas turbines; fuzzy control

1. INTRODUCTION

As a worldwide trend during the last years, one of the main actions taken to meet the demand of electric energy has been the modernization of control systems in aging power plants (Smith, 1993). This was the case for the distributed control systems at the Dos Bocas (DB-DCS) (Chavez and Delgadillo, 1990) and Gomez Palacio (GP-DCS) (Chavez and Diaz, 1993) combined cycle power plants.

In the original control systems (PACE 260 by Westinghouse), the automation, control and protection functions were implemented using three different technologies. Information and regulatory control functions were running in a digital system (DCC). Backup, protection and complementary control functions were realized in a subsystem built with analog electronic components (ACC). Logic sequences were implemented in a system based on electromechanical relays (Uram, 1977).

As a first generation update, the functions of the DCC and the relays system were implemented in digital modules in the DB-DCS. The ACC was preserved as well as the original BTG operation board. As a second generation update, the GP-DCS reviewed the implementation of its predecessor, absorbed the ACC in the digital system and replaced the BTG operation board by a set of SUN workstations. This yield a true all-digital control system (Garduno and Sanchez, 1995).

In principle, the DB-DCS and then the GP-DCS, were considered to be an extension of their respective predecessor. As a requirement, they must use the software already developed in the digital part, DCC, of the original control system. Thus, inherited the software architecture, data structures and programming style. As a consequence, the resultant software system grew in complexity. This in turn, generated an excessively personalized system, uneasy to validate and maintain, and most important, a system for which risk-free operation was unnecessarily more laborious to achieve. In this conditions, the implementation of new control strategies, that could exploit the digital system potential to achieve more efficient operation, became a potentially hazardous and painstaking task.

It is well known that software design and production has not catch up with hardware advances. This has been so because during years software production did not receive the importance it deserves. A reason for this is the wrong believe that software is easy to design, generate, and test. In fact, software can be more complex than the hardware in which it runs. Other important issue is the believe that software can

131

be updated at low cost and in short time. In this regard, all changes in a software system must be tested thoroughly to guarantee reliability. Thus, even if software is more flexible, its development and maintenance costs may supersede the cost of equipment by a large amount. One effective way to produce reliable software at reasonable cost is through the application of proven software engineering techniques.

The system reported in this paper, named Control Software System for TurboGas units (SPC-TG Spanish initials), constitutes a third generation development, now focused in improving the software quality through the application of proven real-time software engineering techniques.(Garcia, 1997). In part 2, the whole set of automation, control, and protection functions of a turbogas unit are summarized as requirement specifications for the SPC-TG. Part 3 describes a generalized functional structure for industrial processes control and how it was modified to define the system architecture. Then, part 4 describes how this structure was applied to organize all the functions of the SPC-TG. Part 5 contains some implementation remarks. Part 6 summarizes the validation tests at an overall system level. Part 7 depicts the integration of a fuzzy logic speed controller into the SPC-TG, and shows results of a startup test. Finally, part 8 makes some comments and concludes this work.

2. SYSTEM REQUIREMENTS

Turbogas units are considered among the most automated systems for power generation. Its profitable operation depends greatly on the control system. Compared to other kind of units, turbogas units operate at very high speed and temperature, being considered high risk systems. These facts pose stringent requirements upon their control systems. Control requirements were classified in three groups: Automation, Control, and Protection.

Automation requirements. Because of being part of a combined cycle, the turbogas unit must exhibit three operating levels: coordinated, automatic, and manual. In coordinated control level the turbo gas unit is operated in conjunction to the other units. The whole plant (two gas turbines, two heat recovery steam generators, and one steam turbine) can be commanded from one single workstation. From there, the operator can set the target power for the three generators in the plant, as well as the rate of change to achieve it. The turbogas unit generates its corresponding part of power. Also from there, a turbogas unit can be started, synchronized, loaded, and stopped in conjunction with the other generating units and with minimum operator intervention. In automatic control level, the turbogas unit is operated as an independent unit, without regard to the other units. In this level the operator can start and accelerate the turbine up to synchronous speed, synchronize the unit, load the unit at any value in its range, operate at free load under temperature supervision or regulate tightly at a given MW value, and shutdown the unit. In manual control level the turbogas unit depends the most on the operator, who can command it through buttons, selectors, and screen displays. In this level the operator is allowed to start the turbine, manually synchronize, change the MW generation by affecting directly the demand to the fuel control valve, and stop the turbine.

Control requirements. Generally speaking they consist in regulating the fuel and air flows according to the unit's operating stage, while keeping surge, acceleration and temperature under safe and efficient conditions. Fuel flow is regulated with a dual speed/load closed-loop control circuit, meanwhile air flow intake to the compressor is regulated with an open-loop circuit to position the inlet guide vanes (IGVs). Speed control is active during startup in two stages. First, it is used to track a given acceleration pattern from about 800 (ignition) to 3600 rpm (synchronous speed). Then the problem becomes a regulation one, speed must be regulated about 3600 rpm to match the system frequency to synchronize the unit to the power grid. MW load control is triggered during synchronization. There is two ways by which MW generation is controlled. First, the unit can be loaded at any MW value in its operating range, load rate change is set by the operator. Second, the unit can be loaded at a minimum load value or at base and peak load values under temperature supervision. IGVs position control is active in all operating stages in either manual or automatic mode. While in automatic the position reference is calculated according to characterization curves which take into account the operating state of the turbine and the temperature of the air entering the compressor.

Protection requirements. Interlocking must account for preserving the physical integrity of the unit. To do so it must implement many security verifications to minimize disaster risks. It is based on redundant measurement of both continuous and logic signals for the execution of several limit, alarm, hold, reject, runback, and trip logic procedures. Main points to be considered are vane and exhaust gas temperature limiting, maximum power limiting, acceleration limiting for surge protection, over-speed protection, ignition failure, alarming, and trip triggering for exceeding operating limits at the turbine itself or at the other units.

3. SYSTEM ARCHITECTURE

Once the requirements were specified, an architecture for the SPC-TG had to be defined. With the objective of complexity reduction in mind, a rather general functional structure for industrial batch-process control was adopted. This model has been applied successfully for batch process automation (Rosenof and Ghosh, 1987) and for building highly flexible batch process control software (Mundo, 1995). In this model all functions performed by a control system are grouped into seven basic functional groups (figure 1).

Logic/analog input/output signal handling functions: Constitute the interface between the control functions and the process. They are responsible for entering and sending contact signals, as well as continuous signals, from and to the process instrumentation at regular intervals or under demand. *Interlocks and protection functions:* Monitor critical variables to avoid the process

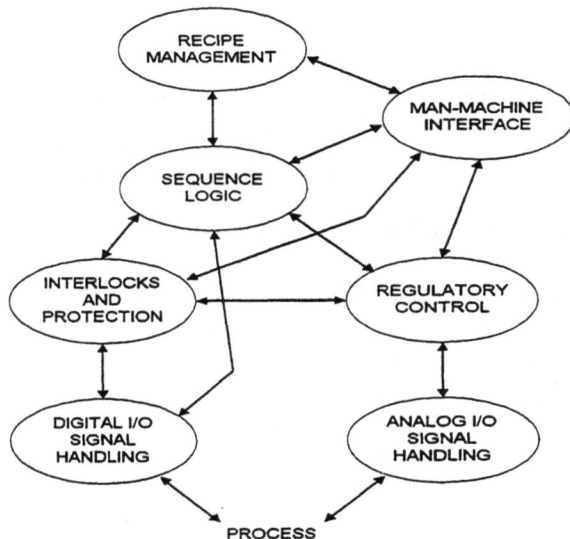

Fig. 1. Batch process control architecture

entering unsafe operating regions. They depend heavily on the physical characteristics of the plant and on the safety requirements of the process. *Sequence logic functions:* Specify how and in what order the plant operations should be executed. They allow the transition between the various operating states. Enable and disable the continuous control functions as required. *Regulatory control functions:* Evaluate control algorithms for driving the continuous varying signals in the process. *Recipe management functions:* Provide the characteristics of repeatability and flexibility required in automated batch processing to account for consistency and variety of the end product. *Operator interface functions:* Allow for interaction between operator and controller. They must be able to provide the operator with relevant information about the plant behavior. Must also provide a means to modify control parameters and strategies.

At least implicitly, all functions above are currently performed in many industrial control systems. Nevertheless, from the batch process control point of view, a turbogas unit produces only one product: electric power. This fact lead to the substitution of the recipe management functional group by what was named supervisory functions, which are mainly in charge of generating set-points for the control algorithms at the regulatory control functions group. Figure 2 depicts the general context diagram of the SPC-TG at its first level. The development embraced all functional groups, except the signal handling functions, due to their high dependence on the implementation hardware.

4. SYSTEM MODULES

A thorough functional analysis of the previous generations of control software systems, supported by the experience gained during their development and commissioning, allowed the smooth and logical accommodation of all the functions performed into the different functional groups, hereafter called tasks, as follows.

Supervision. This task is in charge of generating references (set-points) for the regulatory control loops (fig 3). Main functions at this level are: a) IGVs position reference generation; b) Speed reference generation; c) Load reference generation; d) Inlet vanes temperature reference generation; and e) Dynamic calculation of fuel valve opening limits.

Operator interface. Interaction between the operator and the controller was established via soft lamps and buttons for many relevant logic signals. Their state is determined within the controller, not in the man-machine interface, to reflect the true state in use by the controller. They are different from alarms or warnings generated by the protection logic. Soft lamps and buttons were classified in six different groups: a) Lamps/buttons of fuel valve auto/manual station; b) Lamps/buttons of IGVs auto/manual station; c) Lamps/buttons of MW/rate data entry; d) Lamps/buttons for startup and shutdown; e) Lamps/buttons for MW go (advance) and hold (detention), and f) Lamps/buttons for operation in load control mode (MW control, min/base/peak load).

Logic sequences. This task groups all functions concerning changes in the operating state of the unit. there are four basic sequences: a) Automatic startup sequence; b) Automatic synchronization sequence; c) Automatic shutdown sequence; and d) Gas/diesel fuel transfer sequence.

Regulatory control. This task (fig 4) includes all continuous control algorithms as well as the routines of the auto/manual stations. Strictly speaking there is only one closed loop, that of speed/load control.

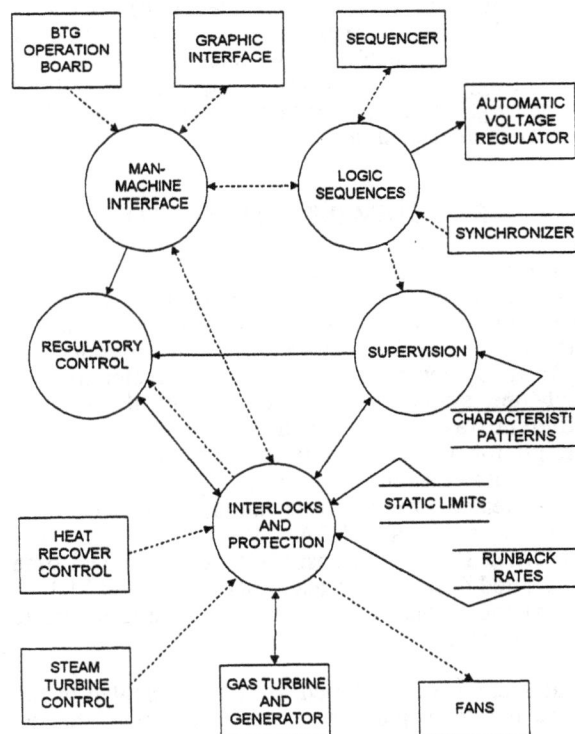

Fig. 2. SPC-TG context diagram

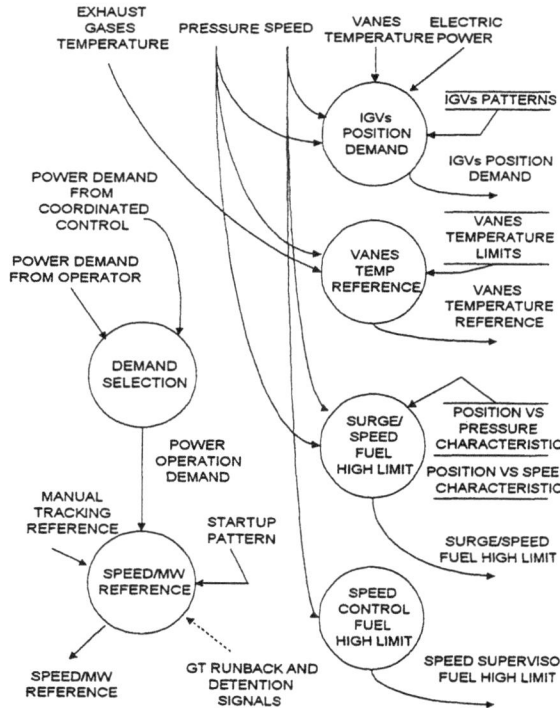

Fig. 3. Supervision module

Position of IGVs is controlled in open loop, but still has its auto/manual station. Temperature is controlled indirectly through the speed/load loop.

Protection. This task (fig 5) is responsible for preserving the physical integrity of the unit. Protection functions were gathered in eight groups.
a) Measurement reliability verification; b) Averaging of reliable measurements; c)Rejections to manual mode by non-reliable measurements; d) Hold state triggering; e) Runbacks triggering; f) Trips triggering; g) Alarms triggering; h) Control output demands interlocking.

5. SYSTEM IMPLEMENTATION

Software development for high risk digital control systems places a series of requirements that demand the application of specific methods, certainly different of those for management or scientific software systems (Leveson, 1990). Development of real-time systems has been approached in many different ways. Among them there is object oriented design for real-time systems (Neidert, 1993), and agent-based design (Boasson, 1993). Also there have been created programming languages for real-time systems such as ADA, JOVIAL, Pearl, and OCCAM. For the purposes of this project real time structured design (Gomma, 1984; Ward and Mellor, 1985) and standard ANSI C programming language were selected.

This system was designed to be implemented in the same hardware platform as its predecessor, that is, must maintain the same input-output frontier to guarantee its compatibility with all the other distributed control subsystems and to the process. It also preserved the data structure so that connectivity

to the GP-DCS man-machine interface workstations was also maintained.

During the development stage the execution of the tasks was performed by an upper level coordinator. Once in the target platform this coordinator is substituted by a real-time Task Scheduler. The hierarchical diagram in figure 6 shows, up to a second level, the functions of the system

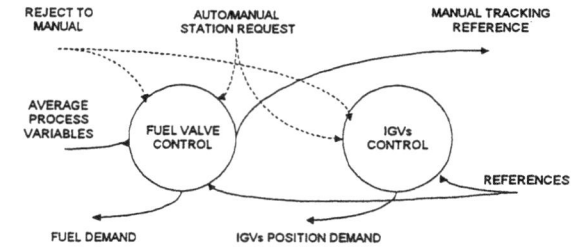

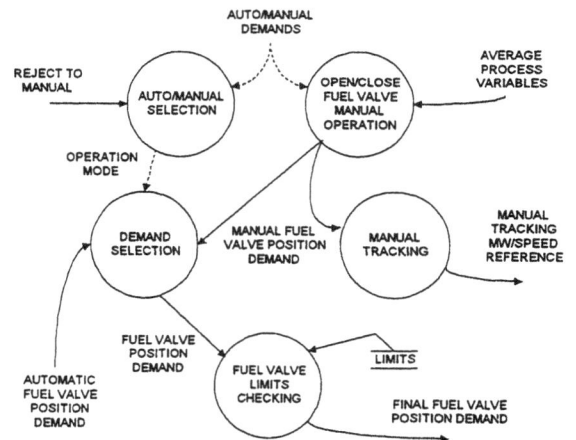

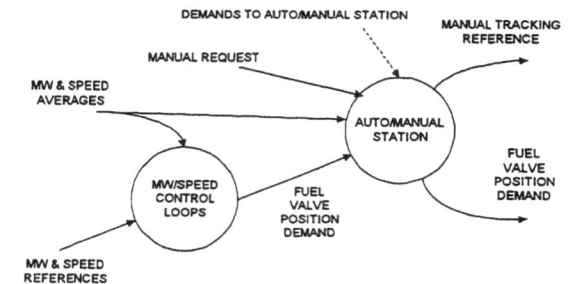

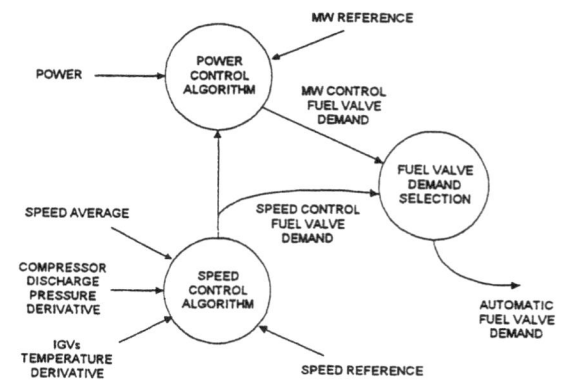

Fig. 4. Regulatory control module

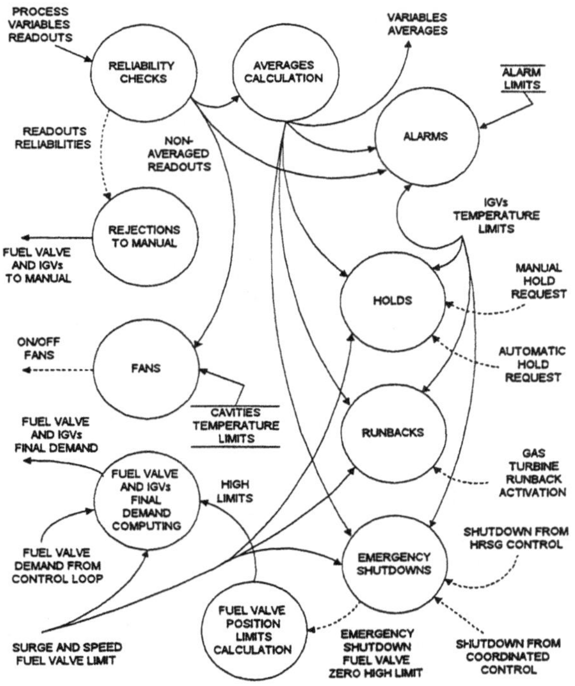

Fig. 5. Protection module

6. SYSTEM VALIDATION

Being the SPC-TG a high-risk system, error-free operation must be guaranteed. This happened to be the most demanding and time-consuming development stage. Performed in three successive phases, it allowed the correction of specification, analysis, design, and code errors. Once at the system level, validation was done through exhaustive simulations with a full scope mathematical model of the turbogas unit, which had been previously validated with plant data. The acceptance tests guidebook (Garduno, 1994) considers a minimum set of 47 operating tests, which covers all the operation maneuvers, and protection issues required by the unit. All tests must be satisfied without fault. Table 1 summarizes the tests and the number of tests in each category.

Results of tests for the new system yielded the same performance as the GP-DCS, which is currently in operation at the power plant. This shows that the new system satisfied all requirements with the added advantage of being a much less complex system.

Table 1. Control system acceptance tests

Test category	Number of tests
System initialization	1
Automatic startup	3
Synchronization and minimum load	3
Synchronization and minimum load	4
Fuel valve operating mode transfers	3
IGVs operating mode transfers	4
MW load control	6
Temperature load control	4
Runbacks	8
Go(advance)/Hold(detention)	2
Automatic shutdown	6
Trips, Fuel transfers	4

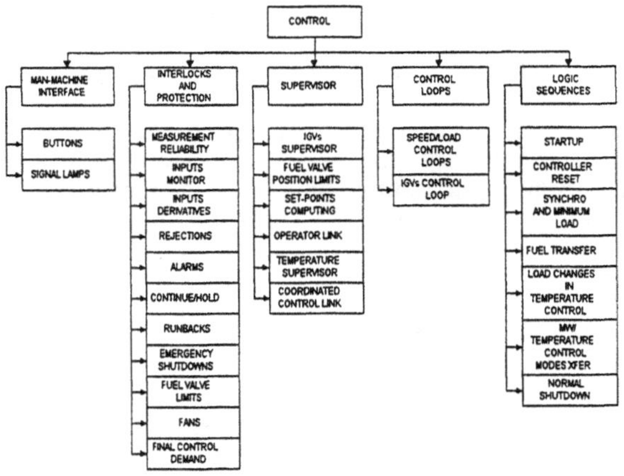

Fig. 6. SPC-TG hierarchical diagram

7. FUZZY SPEED CONTROL

Reduction in complexity facilitated updating the control strategies to a great extent. This was demonstrated by substituting the original PI speed control algorithm with a kind of fuzzy PID algorithm (Garcia and Garduno, 1997). The inputs of the fuzzy controller are the turbine speed tracking error and its two first derivatives, and it generates the demand to the fuel control valve.

Software implementation of the fuzzy algorithm demanded a brief analysis of the source code to make sure that the 'for' and 'while' loops do not introduced significant computational delay at run-time. Figure 7 presents the pseudocode for the integration of the fuzzy speed controller into the regulatory control task

Memory constraints in the target hardware platform motivated the use of a strategy for reduction in the number of rules without decreasing the controller performance. With three inputs, each one divided into seven membership functions, there would be 343 rules . The use of look-up tables would require a three dimensional rule matrix. The strategy selected for implementation reduced the rule matrix to two dimensions with savings of 73% in memory space (Abdelanour, et al, 1991).

```
SPEED/MW CONTROL LOOP
  AUTOMATIC-MANUAL STATION
  CONTROL ALGORITHMS
    IF (GENERATION MODE)
      POWER GENERATION CONTROL ALGORITHM
    ELSE
      SPEED CONTROL ALGORITHM
        ERROR COMPUTING, 1st and 2nd DERIVATIVES
        FUZZY CONTROL
          INPUT NORMALIZATION
          FUZZIFICATION
          RULE EVALUATION
          DEFUZZIFICATION
          OUTPUT SCALING
        END FUZZY CONTROL
      END SPEED CONTROL ALGORITHM
    END IF
  END CONTROL ALGORITHMS
END SPEED/MW CONTROL LOOP
```

Fig. 7. Integration of fuzzy speed controller

The use of the fuzzy speed controller improved tracking of the acceleration pattern during startup (fig. 8). ITAE and IAE performance criteria showed an 80% reduction compared to the original PI controller. Furthermore, the fuzzy controller allowed a faster recovery from runbacks when approaching the high temperature limit during startup. This implied a reduction of 7.9% percent in startup time period and 7.7% in fuel consumption.

8. CONCLUSIONS

In this generation of control system modernization for turbogas units attention was given to the software development process. Major emphasis was placed on the application of formal software engineering techniques to enhance the quality of the resultant control software system. Even if no attempt was made to quantify the betterment using software metrics, higher quality of the final system was demonstrated in a qualitative form by the following achievements. First, a simpler structure, which contributed to locate and correct discrepancies which were very difficult to solve in its predecessor, and to reduce the learning period by newcomers. Second, perfect compliance with all plant acceptance tests. Third, highly reduced development time of another digital controller for a different turbogas unit based on the SPC-TG (Ramirez, 1996). Finally, the implementation of a state-of-the-art fuzzy logic speed controller showed that the benefits of a better control strategy are more easily recognized when it is embedded in a better quality software system.

Even if the betterment process relied in the application of software engineering concepts, a sound knowledge of process engineering and control engineering was also required to produce useful improvements. This multidisciplinary approach cannot be overemphasized. The structure employed in this system resembles the current state of the practice in industrial process control software realization. A necessary next step is its revision and further extension to support the implementation of intelligent control functions.

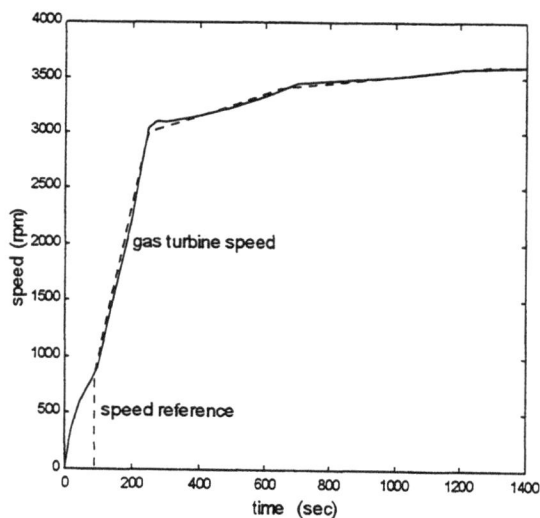

Fig. 8. Startup with fuzzy speed controller

9. ACKNOWLEDGMENTS

The authors wish to thank the facilities given by the Electrical Research Institute and CoSNET for the realization of this project

10. REFERENCES

Abdelanour, G.M., C.H. Chang and F.H. Huang (1991). Design of a fuzzy controller using input and output mapping factors. *IEEE Trans. on Systems, Man and Cybernetics*, **SMC-21**, 5, 953-960.

Boasson (1993). Control systems software. *IEEE Trans. on Automatic Control*, **AC-39**, 7, 1094-1106.

Chavez, R.T. and M.A.V. Delgadillo (1990). Modernization of a computer control system in a combined cycle power plant. *Proceedings ISA 90*. 165-170.

Chavez, R.T. and R.P. Diaz (1993). Upgrading of a computer control system in a combined cycle power plant. *Proceedings Power-Gen Americas 93*. **12-13**, 56-62.

Garcia, C.D.B. (1997). *Digital control for turbogas units*. *M.Sc. Thesis*. CENIDET. Mexico.

Garcia, C.D.B. and R.R. Garduño (1997). Gas turbine fuzzy speed control. *Proc. Second Joint Mexico-USA, on Neural Networks and Neurocontrol, Sianka 'an 97*. Mexico

Garduño, R.R. (1994). *Gas turbines all-digital control acceptance test guidebook. Internal Report*. Electric Research Institute. Mexico.

Garduño, R.R. and M.P. Sanchez (1995). Control system modernization: turbo gas unit case study. *Proc. Control of Power Plants and Power Systems, SIPOWER 95*. **II**. 245-250. Cancun, Mexico.

Gomma, A. (1984). A software design method for real-time systems. *Communications of the ACM*. **27**, 9, 938-949.

Leveson, N.G. (1990). The challenge of building process control software. *IEEE Software*. November, 55-66.

Mundo, J.A.M. (1995). *Water demineralization batch control in a thermoelectric plant. M.Sc. Thesis*. CENIDET. Mexico.

Neidert, R.P. (1993). A new paradigm for industrial control systems design. *ISA Transactions*. 225-233.

Ramirez, M.G. (1996). *Modernization of the speed, power, and temperature control of a turbogas unit. BSEE Thesis*. IPN. Mexico, D.F.

Rosenof, H.P. and A. Ghosh, (1987). *Batch process automation, theory and practice*. Van Nostrand Reinhold. New York.

Smith, D.J. (1993). Instrumentation and controls are going state-of-the-art. *Power Engineering*, November

Uram, R. (1977). Computer control in a combined cycle power plant. Part II, the digital gas turbine system. *Proc. IEEE PES Winter Meeting*, **A** 77, 078-9, 1-8.

Ward, P.T. and S.J. Mellor (1985). *Structured development for real-time systems. Vol. 1,2, and 3*. Yourdon Press.

A MATLAB TOOLBOX FOR SIMULATING TRANSPUTER AND DIGITAL SIGNAL PROCESSORS APPLICATIONS

F. Sustelo[1], A. E. Ruano[1,2]

1. Unidade de Ciências Exactas e Humanas, Universidade do Algarve
Campus de Gambelas, 8000 Faro, Portugal
2. Institute of System & Robotics, Portugal
email: fsustelo, aruano@mozart.si.ualg.pt

Abstract: The performance demands of modern control and signal processing systems is increasing beyond the capacity of conventional sequential processors, requiring parallel processing solutions to satisfy the real-time requirements.
In this paper a new version of Matlab toolbox for simulating homogeneous systems built with Inmos transputers or digital signal processors is presented. This toolbox extended the capabilities of a previous approach. Its development aims to help the designer to compare, effortlessly, the performance of alternative parallel solutions, and also to monitor the program execution, within each processing node.
This simulator is under further development to extend its applicability to parallel heterogeneous systems. *Copyright © 1998 IFAC*

Keywords: Parallel and distributed algorithms, digital signal processors, multiprocessor systems, homogeneous systems.

1. INTRODUCTION

The performance of modern control systems is increasing beyond the capacity of conventional sequential computers. Often, modern applications require parallel processing solutions to satisfy the real-time requirements. Additionally, the computational diversity expressed by the algorithms ideally requires different processor types, such as digital signal processors, vector processors, CISC and RISC processors, therefore comprising an heterogeneous architecture. Although a huge amount of effort has been applied to mapping and scheduling in homogeneous architectures, there isn't yet a proper methodology available. The increased complexity of heterogeneous systems makes this problem even more challenging [1]. Determination of the network topology according to the particular application is another challenging problem. Design tools are emerging for solving this type of problems

[2].
In the control and signal processing areas, when developing parallel applications, the engineer usually starts its design using Matlab [3] and its toolboxes to test and develop its sequential algorithm. Only then, when the sequential algorithm is fine-tuned for the application at hand, the parallelization problem is addressed. The programmer has usually to go through the following steps:

1. to consider the target architecture, whether an homogeneous or an heterogeneous system;
2. to partition the algorithm;
3. to determine the best network topology;
4. to code the algorithm on a suitable parallel language on the target architecture,
5. and finally to evaluate the performance of the parallel algorithm, by measuring its computational execution time.

In practice, and in almost every case, the

programmer goes through several loops within these five steps, the whole process taking a considerable amount of time.

On the other hand, in a significant number of cases, algorithmic partitioning, which seem to deliver a promising performance, achieve poor results, whether in terms of parallel efficiency or execution time, or both. Different reasons, specific for the problem at hand, can be pointed out for the poor performance obtained. Despite the different reasons, a point remains common to all: it is difficult, in a parallel program, to identify the reasons of inefficiency, which is clearly related with the difficulty of monitoring the parallel execution of the program.

This proposed toolbox offers the control engineer the possibility of investigating the performance o alternative algorithmic portioning schemes, without leaving the Matlab environment. The version presented here extends the applicability of an existing approach [4], by enabling the simulation of C40s Texas DSP networks [5], in addition to Inmos transputers [6].

2. LIMITATIONS OF THE SIMULATOR

This simulator was developed to enable the simulation of control applications in different parallel platforms, either homogeneous or heterogeneous (although only the homogeneous case is currently available). The user defines, thanks to a user-friendly interface in the Matlab environment, the topology and process mapping that will be employed. He can afterwards simulate the parallel execution and analyse it, therefore enabling comparisons between different approaches to be made and possible bottlenecks investigated. The aim here is not to achieve simulated execution times exactly equal to the ones obtained in the actual parallel platform, but to compare, effortlessly, different parallel solutions. In view of this objective, the simulator has some limitations:

1. As processes are written as Matlab functions, they must execute from the beginning of the code to a return statement or to the end of the function. This means that pre-empting a process (forcing it to release the CPU) is not possible to implement using Matlab. This limitation has two consequences:

1.1. Digital signal processors can support eight levels of priority and the transputer only two levels. For the purpose of this work we consider that each process has whether high or low priority. High priority processes are expected to execute for a short time. If one or more high priority processes are able to proceed their execution, then one is selected and runs until it has to wait for a communication, a timer input, or it completes processing. If no process at high priority is able to proceed, but one or more processes at low priority are able to proceed, then one is executed. If the possibility of simulating high and low priority processes is wanted, and as processes can not be pre-empted, we must consider that the CPU execution time of high priority processes is negligible compared with the CPU execution time of the low priority processes. By this it is meant that the longest period of time a high priority process can execute between consecutive communications is negligible compared with the smallest period of time a low priority process executes between communications, for processes within the same processor. This might not be a severe limitation, as high priority processes are usually employed mainly to route data, and not for computation purposes.

1.2 The impossibility of pre-empting a process also implies that the simulator does not employ the same CPU low priority scheduling strategy as the real processor. For instance, the transputer does not preempt processes of high-priority, but uses time-slicing for processes of low-priority [5]. As this is not possible in our simulator, we assume that, as in the case of high priority processes, a low priority process continues its execution until it needs a communication, or it finishes its execution.

2. The actual processor execution time is very difficult to simulate since it mainly depends, not only on the processor type and its clock frequency, but also on the actual CPU instructions generated by the compiler, on their existence in internal or external memory, etc. In terms of Matlab, the computing execution time may only be measured using the function *flops*, which roughly measures the number of generic floating point operations needed to execute a specific Matlab code. We therefore have to assume in our simulator that the execution time is only function of the number of floating point operations executed by the algorithm, where the execution time of one generic floating point operation is obtained by an average of time executions of different test codes, actually measured on the processor we are considering.

3. The main differences between inter-processor communication rely on the type of links available. The transputers communicate through serial bi-directional I/O ports, the communication is synchronised and unbuffered. The messages are transmitted as sequences of bytes, each of which is acknowledged before the next one is transmitted. The speed of transmission, in Kbytes/sec, depends on the setting of the link speed used, which can be 5, 10 or 20 Mbits/sec, and on the existence of data in internal or external memory [5]. The DSPs [4] communicate through parallel one-directional I/O ports, and its communication is asynchronous. Each link of DSP processor operates at 20 Mbytes/sec. As far as the Matlab simulator is concerned,

communications are divided into internal or logical communications, and external or physical connections. In the simulator, the product of the number of bytes sent, and a parameter denoting the time taken to communicate one byte give the time taken by an external communication. This parameter is obtained, for the different link speeds, by an average of communication times actually measured on the system. It is assumed that internal communications take no time.

3. USE OF THE MATLAB SIMULATOR

Processes are implemented as Matlab functions, written and identified by the user, and incorporating special functions available in the toolbox, briefly described below.

Before starting the simulation, the user must specify the configuration, i.e. the topology that will be used in the simulation, the allocation of processes to the available processors and the process priority. Executing the Matlab program, *PreSim*, does this. Provisions have been made to account for SIMD algorithms, by the use of replicas of code, whose number is specified by the user. Channels between processes are specified in terms of origin and destination processes, and the corresponding identifiers are automatically generated. As a result of *PreSim*, a program file and a Matlab data file are automatically generated.

In an improvement to the previous version, where the user subsequently had to edit the program file and the function files, these are now automatically updated. Provisions have been made to help the user to initialise the local variables of each process.

The simulation is then executed using a Matlab program available in the toolbox, *Sim*. This program, using the configuration files previously created, executes the simulated parallel program. After the simulation ends, information about the simulation can be plotted as Matlab graphs, upon user demand. The available information is concerned with:

• Parallel execution time, processor efficiency, sequential execution time, speedup and parallel efficiency. These two last informations are only available if there is a Matlab function which implements the sequential code;
• Graphs of idle and computing time for user-specified processors, along user selected periods of execution;
• Graphs of events for user-specified processes, along user-selected periods of execution. The events considered are related with channel communications, namely:
  • the process is waiting to receive data;
  • the process is actually receiving data;
  • the process is computing;
    the process is waiting to send data;

• the process is actually sending data.

3.1 External Simulator Functions

In the toolbox the following functions, specific for process communication, are available to the user. In these functions, the variable *flops* denotes the number of floating point operations performed by the calling process since the previous communication. For a more detailed description operation of these functions, please see [4]:

• wait (chan, flops, nv) - simulates the OCCAM ? primitive or the chan_in C function;
• send (var, chan, flops) - simulates the OCCAM ! primitive or the chan_out C function;
• par_send (chan_1, var_1, chan_2, var_2, ..., chan_6, var_6, flops) - simulates sending data (*var1*, *var2*, ...) through two or more channels (*chan_1, chan_2, ...*) in parallel;
• alt_wait(chan_1, nv1, chan_2, nv2, ..., chan_5, nv5,flops) - simulates the OCCAM ALT construct, or the alt_wait C function;
• [dest, source]=processes_id (chan_i) - returns the destination and source process identifiers associated with channel *chan_i*;
• HAS_REC_FROM, which stores in the element corresponding to the receiving process identifier the identifier of the sending process.

To save and restore the process workspace the toolbox uses the following functions, which are also available to the user:

• save_workspace (proc_number, i1, i2, ..., i21)
• [o1, o2, ..., o21] = restore_workspace (proc_number).

And finally the instruction:

• finish (proc_number, flops), is used to inform the simulator that the process has finished its execution.

As an example, let us suppose that we are simulating two processes, identified by *P1* and *P2*, which communicate through channels C1_to_2 and C2_to_1:

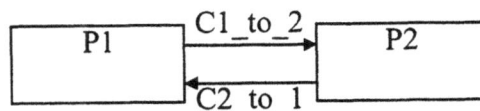

Fig. 1. Example topology

We shall assume that P1 and P2 would have an

OCCAM-like code:

```
PROC P1(C1_to_2,C2_to_1)
  ---variables u1,v1;
  SEQ
    ---first instructions
    C1_to_2 ! u1
    ---middle instructions
    C2_to_1 ? v1
    ---last instructions

PROC P2(C1_to_2,C2_to_1)
  ---variables u2,v2;
  SEQ
    ---first instructions
    C1_to_2 ? u2
    ---middle instructions
    C2_to_1 ! v2
    ---last instructions
```

The initialisation file, automatically generated, would contain (depending on the order the processes are introduced in *PreSim*) the following lines:

```
% Initializations for process P1(1)
%
% chan 1 has destination Proc_2(1)
chan1=1;
% chan 2 comes from Proc_2(1)
chan2=2;
%
after=0;
% Here other initializations
%
u1=...
v1=...
save_workspace(1,chan1,chan2,after,u1,v1);
insert_event(EVENT_START_EXE,1,0);
P1(1);
%
% Initializations for process P2(1)
%
% chan 1 has destination Proc_1(1)
chan1=2;
% chan 2 comes from Proc_2(1)
chan2=1;
%
after=0;
% Here other initializations
%...
u2=...
v2=...
save_workspace(2,chan1,chan2,after,...);
insert_event(EVENT_START_EXE,2,0);
P2(2);
```

The Matlab codes for P1 and P2 would be something like the following, where the instructions in italic represent what the user has typed, the rest of the instructions being automatically inserted as a result of PreSim:

function P1(proc_numb)
&&& u1, v1

```
[chan1,chan2,after,u1,v1]=restore_workspace(proc_numb)
ff=flops;
if after==0
  ---first instructions
  after=1;
  save_workspace(proc_numb ,chan1,chan2,after,u1, v1)
  send(chan1, u1,flops-ff)
  return
end
if after==1
  --- middle instructions
  after=2;
  save_workspace(proc_numb ,chan1,chan2,after,u1, v1)
  wait(chan2,u2, flops-ff,5)
  return
end
--- last instructions
finish(proc_numb,flops-ff)
```

function P2(proc_numb)
&&& u2, v2

```
[chan1,chan2,after,u2,v2]=restore_workspace(proc_numb)
ff=flops;
if after==0
  ---first instructions
  after=1;
  save_workspace(proc_numb ,chan1,chan2,after,u2, v2)
  wait(chan2,u2, flops-ff,4)
  return
end
if after==1
  --- middle instructions
  after=2;
  save_workspace(proc_numb , chan1,chan2,after,u2, v2)
  send(chan2, v2,flops-ff)
  return
end
--- last instructions
finish(proc_numb,flops-ff)
```

4. AN EXAMPLE

To illustrate the use of the simulator, two different parallelizations of the triangularization phase of an Householder QR algorithm, described in detail in [7], will be employed.

Considering an **A** matrix, with dimensions m*n, $m \geq n$, this factorization decomposes it into an orthonormal matriz **Q**, and an upper triangular matrix R_1, such that:

$$A = QR = \begin{bmatrix} Q_1 & Q_2 \end{bmatrix} \begin{bmatrix} R_1 \\ 0 \end{bmatrix} \qquad (1)$$

Considering just the triangularization phase (computation of the matrix R), Alg. 1 describes the sequence of the operations performed:

Alg.1 – Tiangularization phase(seq.)

1.for k=1 to n
    1.1 Compute the Householder vector h
    1.2 $\mathbf{R}_{k\cdots m,i} = -\sigma e_1$
    1.3 $r = \mathbf{h}^T \mathbf{R}_{k\cdots m,k+1\cdots n}$
    1.4 $\mathbf{R}_{k\cdots m,k+1\cdots n} = \mathbf{R}_{k\cdots m,k+1\cdots n} - \mathbf{hr}$
end

This algorithm is parallelized across a network of five transputers, in a ring topology. Each transputer has an high-priority router process, responsible for communications, and a low-priority worker process, responsible for the calculations.

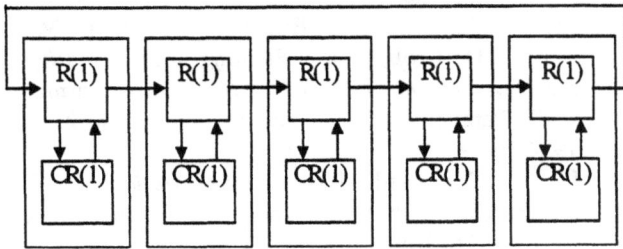

Fig. 2. Network topology

Considering the case of a matrix with dimensions m=100 and n=50, a first approach [7], using a conventional worker, achieves, in our simulator, a sequential execution time of 0.98 sec., a parallel execution time of 0.25 sec. and a speedup of 3.9. The actual values, measured in a transputer network, are 999 ms, 264 ms and 3.7. Fig. 3 shows some examples of the type of graphs available in the simulator. Fig. a) illustrates the efficiency of each processor, defined as the ratio between working time and total time per processor. Fig. b) shows, for processor 1, the working (1) and idle (-1) periods between 0 and 60 ms. To understand the reason of the existence of idle periods, a graph of events is represented in Fig. c), for the router and worker processes associated with processor 1, in the same time interval. The conventions are: waiting to receive (2), actually receiving (1), working (0), actually sending (-1), waiting to send (-2). When two graphs are shown in the same figure, as it is the case, the top graph is shifted by 3, and the bottom graph by -3. The reasons for the idle times can now be understood, since, in the first iteration, worker 1 computes his Householder vector, and continues the updating of its columns yet to be triangularized. In the next 4 iterations, he must wait to receive the Householder vectors, which are computed by the next workers. Fig. d) shows the equivalent events for processor 5.

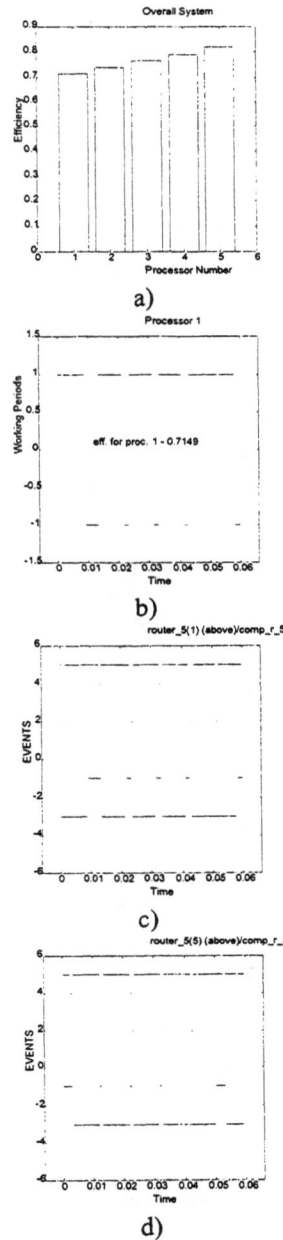

Fig. 3. Matlab toolbox graphs for the first approach of the example given

Employing now a more sophisticated worker [7], which computes its corresponding Householder vector one-step-ahead, another simulation was run. Now the parallel execution time obtained was 213 ms, resulting in a speedup of 4.6. The actual values, measured in the transputer network, were 235 ms and 4.3. The same kind of graphs of Fig. 3 are updated for this second approach in Fig. 4. We can observe that, due to the one-step-ahead computation of the Householder vector, the period of time each worker has to wait is dramatically reduced, or even annulled.

a)

eff. for proc. 1 - 0.8324

b)

c)

d)

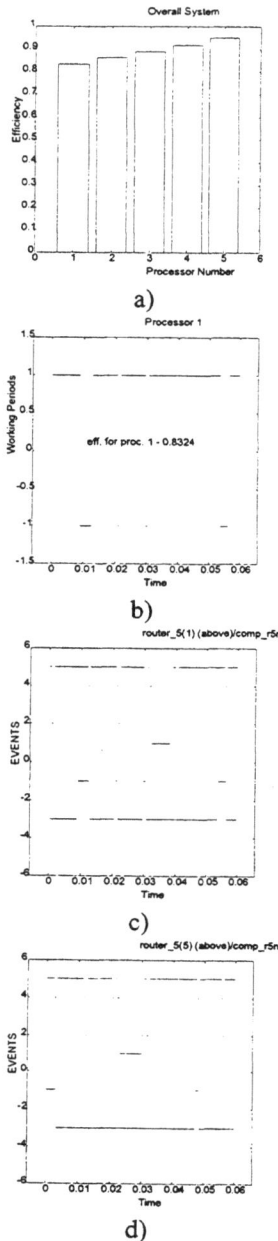

Fig. 4. Matlab toolbox graphs for the second approach of the example given

5. CONCLUSIONS

In this paper, a new version of Matlab toolbox simulating a multiprocessor network execution was described. This toolbox was given the extended capability of simulating an homogeneous system build with several processing elements, either Inmos transputers or Texas C40 DSPs, enabling the comparison of alternative parallel implementations and a close monitoring of simulated parallel execution, through the use of event graphs. This enables a previous study of possible parallelization schemes of a real-time algorithm on to a transputer or DSP network, before it is actually implemented. The simulator is currently under further

development, to extend its applicability to other DSPs, like Sharc processors, and to parallel heterogeneous systems.

REFERENCES

[1] Tokhi, M., Hossain, M., Baxter, M. and Fleming, P., *Performance Evaluation of Homogeneous and Heterogeneous Architectures for Real-Time Processing and Control*, 3rd IFAC Workshop on Algorithms and Architectures for Real-Time Control, Ostend, Belgium, June 1995

[2] Bass, J. and Browne, A., Hajji, M., Marriot, D., Croll, P., and Fleming, P., *Automating the Development of Distributed Control Software*, IEEE Parallel and Distributed Technology, **2**, 4, 1994, pp. 9-19

[3] The Math Works Inc.: *Matlab user's Guide*, The Math Works Inc., South Natick, MA 01760, USA, 1991

[4] Ruano, A. E., *A Matlab Toolbox for Simulating Transputer Applications*, Proc. 2nd International Meeting on Vector and Parallel Processing (VECPAR'96), Porto, Portugal, September 1996

[5] Texas Instruments *TMS320C40 User's Guide*, Texas Instruments, USA.

[6] Inmos Ltd., *The transputer Databook*, 2nd ed. Inmos Ltd., 1989

[7] Ruano, A., Fleming, P., Jones, D., *An Efficient Parallel Implementation of a Least Squares Problem*, Computing Systems in Engineering, Vol. 6, N° 4/5, 1995, pp. 313-318

THE WAY OF DISTRICT HEATING OUTPUT CONTROL
BY MEANS OF HOT WATER PIPING

Jaroslav Balátě, Tomáš Sysala

Department of Automatic Control, Faculty of Technology Zlin
Nam.TGM 275, 762 72 Zlin, Czech Republic
Tel./Fax: ++ 420 67 721 1521
E-mail: balate@zlin.vutbr.cz

Abstract: The presented algorithm design of the control of district heating output by means of the hot water piping eliminates the transport delay. That transport delay occurs in the case in which is used the qualitative way of control only and is depending upon the velocity of hot water and upon the length of hot water piping. The new qualitative-quantitative method of control of district heating output by means of hot-water piping which is simultaneous and continuous action of two manipulated variables i.e. qualitative part (control of temperature difference on the heating plant exchanger) and quantitative part (control of rate of flow through the circulating pumps) enable to eliminate the transport delay between the heating plant exchanger and a remote locality of relatively concentrated consumers. *Copyright © 1998 IFAC*

Key words: District heating, heat supply, hot-water system, heat feeder, control system, transport delay, heat supply prediction.

1. INTRODUCTION

The system of district heating has to ensure the heat energy supply to all consumers in sufficient quantity and according to their needs which varies in time. The energy supply must always be in accordance with the specified quality standards. In case of hot-water piping it concerns maintaining the specified temperature of water in the inlet piping.

Algorithm of so called qualitative-quantitative method of control using prediction of the course of daily diagram of heat supply in hot-water systems of updated heat supply enables to eliminate the influence of transport delay which arise in long feeders between the heating plant exchanger in the heat source and relatively concentrated consumer's heat consumption. The transport delay depends on the speed of heat-carrying medium flow (hot water) and on the length of feeder piping. New method of hot-water output control consist in simultaneous and

continuous action of two manipulated variables influencing the transmitted heat output and in utilizing required heat output prediction in the given locality. Newly designed control method was considered for a concrete case when the transport delay was supposed to be within the range of six up to twelve hours depending on the heat output withdrawn by consumers.

The designed method is the solution of the control method of heat output in the source of heat.

The present common heat output control method of heat supply by hot-water piping is using the dependence on the water temperature in return piping of the heat feeder or possibly in combination with the dependence on outside air temperature.

For the control of hot-water piping heat output from the source of heat there are available two manipulated variables:

- change of water temperature difference in the inlet and return piping of the hot water piping which is realized in practice by the change of heat input on the inlet to the heating plant exchanger-so called **qualitative method of heat output control;**

- change of rate of flow of hot water by means of changing speed of circulating pump - so called **quantitative method of heat output control.**

The above manipulated variables are usually used as acting separately and namely only of them. When both are used than it concerns a case when the qualitative method of control is the main control method and the qualitative method is used for starting and stopping operation of pumps with different transported rate of flow. Qualitative changes are carried out once in a season (summer, transition period, winter). Two or three sizes of circulating pumps are usually used for this purpose.

The disadvantage of the described control methods is the fact that they does not sufficiently include dynamic properties of the controlled plant (controlled subsystem). The transport delay in the inlet branch of the heat feeder and transport delay of the inertia members of heating plant exchanger remain forgotten. If there is a change of the output withdrawn in any point of hot-water network than the corresponding output of sources (production) controlled by classical qualitative method adjusts itself but with substantial delay, even when the change of rate of flow of hot water occurs due to self-regulation owing to self-regulative properties of static characteristic of circulating pump caused by the change of its working point position. The change of heat output withdrawal will be realized by the action of autonomous controllers of temperature in secondary networks of consumer's - transfer stations.

It happens that some of the requirements on the specified quality properties of heat-carrying medium will not be fulfilled.

2. ANALYSIS OF THE HOT-WATER PIPING DYNAMIC PROPERTIES

Technologic scheme of hot-water piping arrangement is illustrated in principle on the fig. 1. In the illustrated case the circulating (delivery) pump is arranged at the end of the return piping before the exchanger station.

As already mentioned in the chapter 1, there are variable two manipulated variables for the heat output control of hot - water piping for heat supply to the heat network:

- for *qualitative control method* it is the change of the difference between water temperature in inlet and return piping of the hot-water piping which is realized by the change of heat input of steam at the inlet to the heating plant exchanger,

- for *quantitative control method* it is the change of hot water rate of flow realized by means of the change of circulating pump speed.

The following relation is valid

$$P_T = M_v \cdot c \cdot \Delta\vartheta \qquad (1)$$

$\underset{\text{method}}{\text{quantitative control}} \uparrow \qquad \uparrow \underset{\text{method}}{\text{qualitative control}}$

where P_T (W) is the heat output of the hot-water piping, M_v (kg.s^{-1}) - rate of flow (mass flow) of the heat-carrying medium, $\Delta\vartheta$ (K) - temperature difference, c (J.kg^{-1}.K^{-1}) - specific heat capacity.

2. 1. Behaviour of Controlled System at the Qualitative Control Method

At the qualitative control method the hot-water piping behaves like a proportional system with inertia of higher order with transport delay. Expressed by transfer function $G_S{}^{qual}$(p) as follows:

$$G_S{}^{qual}(p) = \frac{\Delta\Theta_{SP}(p)}{\Delta M_p(p)} = \frac{k^{qual}}{1 + T_1 p + T_2{}^2 p^2 + T_3{}^3 p^3} e^{-pT_d}$$

$$(2)$$

where k^{qual} (-) is the amplification of the controlled system, T_1, T_2, T_3 (s) - time constants of the heat inertia members of the controlled plant, T_d (s) - transport delay, M_p (kg.s^{-1}) - steam rate of flow at the inlet of the heating plant exchanger.

Laplace transformations of the original function of the respective variables are marked by capital letters in the text.

Dynamic properties of the hot-water piping are defined by means of behaviour of heating plant exchanger and for this case they are approximated by the properties of proportional system having the inertia of the third order (n = 3). However the time contants of the transition process (theoretical transient time, time of rising, transit time) are gaining the values of tens minutes and depend on the rate of flow of the heated water through the exchanger. The step response on the fig. 2 expresses by parameters the time course of temperature in the inlet piping ϑ_P (rate of flow of circulating water M_v is the parameter).

The transport delay in the hot-water network is a function of M_v and is given by the following relation:

$$T_d = f(M_v) = \frac{l \, S \, \rho_v}{M_v} \qquad (3)$$

where l (m) is the length of the controlled part of the piping, S (m^2) - the cross section of the inlet branch piping of the heat feeder, ρ_v (kg.m^{-3}) - specific weight of the circulating water.

The described contemplation is justified in cases similar to that for which it was cerried out i. e. in localities where relatively concentrated withdrawal of heat is considerably remote from the source of heat (e.g. heat feeder from the Nuclear Power Plant Dukovany to the system of centralized heat supply in the city of Brno).

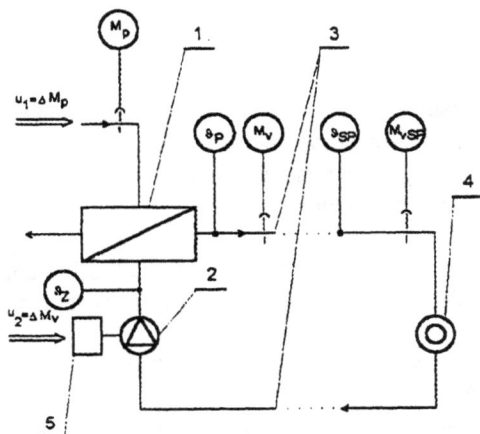

Fig. 1. Principal scheme of hot-water piping.

1 - Heating plant heat exchanger in the source of heat, 2 - Circulating pump, 3 - Hot-water piping (pipeline of the heat inlet), 4 - Consumers (exchangers in the consumers'- transfer stations), 5 - Speed-changing device of the pumps.

Used designation:
ϑ_Z (^{0}C) is the temperature in the return branch of hot-water piping,
ϑ_P (^{0}C) - temperature in the inlet branch of hot-water piping,
ϑ_{SP} (^{0}C) - temperature of hot water in the inlet branch at the place of consumers,
M_v (kg.s^{-1}) - rate of flow of heat-carrying medium at the outlet of exchanger,
M_p (kg.s^{-1}) - rate of flow of steam at the inlet to the heating plant exchanger,

M_{vSP} (kg.s^{-1}) - rate of flow of heat-carrying medium in the inlet branch at the place of consumers,
u_1 (-) - manipulated variable 1 (change of rate of flow of steam),
u_2 (-) - manipulated variable 2 (change of speed of circulating pumps).

It is obvious from the above that the qualitative method of control becomes evident at the consumer of heat with a time delay corresponding to the transport delay. Therefore it is valid:

$$\vartheta_{SP}(t) = \vartheta_P\left(t + T_d\right) \qquad (4)$$

Acting of the manipulated variable u_1 adjusts the corresponding difference of water temperature on the inlet and outlet of the heating plant exchanger, which is determined by the following relation:

$$\Delta\vartheta = \vartheta_P - \vartheta_Z = \frac{P_T}{M_v \, c} \qquad (5)$$

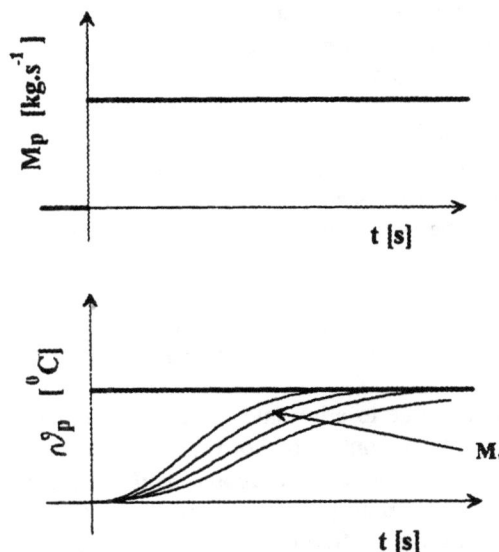

Fig. 2. Step response of the time course of temperature in the inlet piping ϑ_P

2.2 Behaviour of controlled system at the quantitative control method

The quantitative control method realizes by means of speed-changing device of the circulating pump the change of rate of flow of the circulating water and by means of that also the change of delivered heat output (1). It encloses the inertia delay of speed-changing device and includes also the time constant of the piping which covers the time necessary for acceleration or possibly deceleration of circulating mass of incompressible heat-carrying medium. The hot-water piping itself (pipeline) behaves as a proportional system without inertia delay. The mentioned properties can be transfer $G_S^{quant}(p)$:

$$G_S^{quant}(p) = \frac{\Delta M_{vSP}(p)}{\Delta M_v(p)} = \frac{k^{quant}}{1 + T_1' p + T_2'^2 p^2 + T_3'^3 p^3} = 1$$

(6)

where time constants of inertia delay of speed-changing devices T_1', T_2' are defined by the kind of speed of speed-changing devices (hydraulic clutch, electric speed-changing device); time constant of piping T_3' is defined by the length of piping, speed of the heat-carrying medium and transport height of the circulating pump. They are generally much smaller (the order of seconds, tens seconds) than time constants in the relation (2) i.e. than time constants of heating plant exchanger (the order of tens minutes).

However it is very important that the transfer function (6) does not include the transport delay which in the transfer function (2) represents the order of hours.

It is obvious from the above that from the point of view of dynamic properties the quantitative control method of hot-water piping output has its advantage. Nevertheless it is used very seldom till this time.

2. 3. Elimination of transport delay at the control of hot-water piping heat output.

It is possible to eliminate the influence of transport delay at the control of hot-water feeder heat output by means of **simultaneous and uninterrupted control by manipulated variables** i.e. control of temperature difference at the heating plant exchanger and control of rate of flow of circulating water - heat - carrying medium.

The basis of the method is in correction of the deviation arising at the use of qualitative control method influenced by transport delay (2) by means of using quantitative control method which is not influenced by transport delay and can act almost immediately.

The algorithm for such control method was designed and verified by simulation. It has been named **qualitative-quantitative control method of hot-water piping heat output using the prediction of the course of heat supply daily diagram.**

The algorithm of the above described and named control method i.e. using two manipulated variables namely separately for qualitative and also quantitative control method is on fig. 3.

The course of qualitative control method is as follows:

- measuring the rate of heat -carrying medium (hot water) (step 1),
- determination of transport delay (step 2),
- determination of the time in which the action (intervention) of the qualitative control method displays at consumers (step 3),
- calculation of heat output which is adjusted by qualitative control method including also correction of heat content in the inlet branch of the feeder (step 4), (step 5),
- change of control signal into the manipulated variable i.e. position of control valve of inlet steam at the inlet into the heating plant exchanger (step 6).

The course of quantitative control method is as follows:

- measuring real (actual) parameters necessary for further calculation (step 1),
- calculation of real heat output withdrawn at the place of consumers (step 2),
- calculation of deviation between heat output adjusted in time of action of qualitative control method and real (actual) output withdrawn at consumers (step 3),
- calculation of quantitative correction of heat output (step 4),
- change of control signal into the manipulated variable i.e. to the value of speed of circulating pump (step 5).

Designation to the fig. 3:
c - specific heat capacity,
l - length of the inlet branch of the heat feeder,
RT - real time (time in which the manipulated variable of the qualitative control method acts on the heating plant exchanger),
S - cross section of the inlet branch of the feeder,
T - time in which the action of the manipulated variable of qualitative control method becomes evident at the locally concentrated consumers,
T_d - transport delay,
T_d^p - anticipated transport delay,
$T_{př}$ - time advance,
$T_{přech}$ - time for transition of heating plant exchanger at the action of manipulated variable,
T_{vz} - period of sampling (approx. 15 minutes),
M_v - rate of flow of the circulating water,

$M_{v,RT}^s$ - real rate of flow of circulating water during the time RT,

$M_{v,T}^s$ - real rate of flow of circulating water during the time T,

P_T - heat output of the hot-water piping,

P_T^p - anticipated heat output taken from the prediction of heat supply daily diagram (DDDT),

$P_{T.T}^p$ - anticipated heat output in the time T,

$P_{T.T}^s$ - real measured output in the time T,

$\vartheta_{P.T}^s$ – real temperature in the inlet branch of the feeder at consumer's in the time T ,

$\vartheta_{Z.T}^s$ - real temperature in the return branch of the feeder at consumers in the time T,

ΔP_T - deviation between the anticipated and real withdrawn output in the time T,

$\Delta M_{v.T}$ - quantitative correction, i.e. the change of circulating water rate of flow,

ΔQ - change of heat content in the inlet branch of the feeder caused by quantitative correction,

$\Delta\vartheta_T^s$ - real temperature difference at consumers in the time T,

$\Delta\vartheta_T^p$ - anticipated temperature difference on the heating plant exchanger in the time T, which is calculated from $P_{T.T}^s$ and is an manipulated variable of the qualitative control method,

$\Delta\vartheta_T^{p.Q}$ - anticipated temperature difference on the heating plant exchanger in the time T including correction of the heat content in the inlet branch of feeder ΔQ.
This heat has to be supplied or possibly to decrease by it the supply of heat depending on the sense (sign) of quantitative correction $\Delta M_{v,T}$,

ρ_1 - specific weight of circulating water in inlet branch of the feeder.

3. CONCLUSION

The described way of control enables following advantage:

- the elimination of transportation delay in the case of the qualitative output control in hot water piping,
- the minimum of the heat losses along the hot water piping,
- the minimum of the power consumption of network-water pumps,
- modification of the algorithm of control (fig. 3) when using part of the heat feeder for heat accumulation.

The described method is meaningful in situations when the heat consumers are concentrated in location which is situated in long distance from the heating plant with heat exchanger. In the concrete respective case this transportation delay is supposed to be in the range from six to twelve hours. In these cases the elimination of transportation delay of heat output control by means of water piping is necessary.

REFERENCES

Baláté, J.: *Patent No. 2792532*, Bulletin of Central Patent Office (in Czech), Praha, 15. 2. 1995.

Baláté, J.: *Controle de la production thermique dans des circuits de chaleur et méthode utilisée*, BRUSSELS EUREKA '97 - Médaille d'or, Bruxelles, 11.11.1997

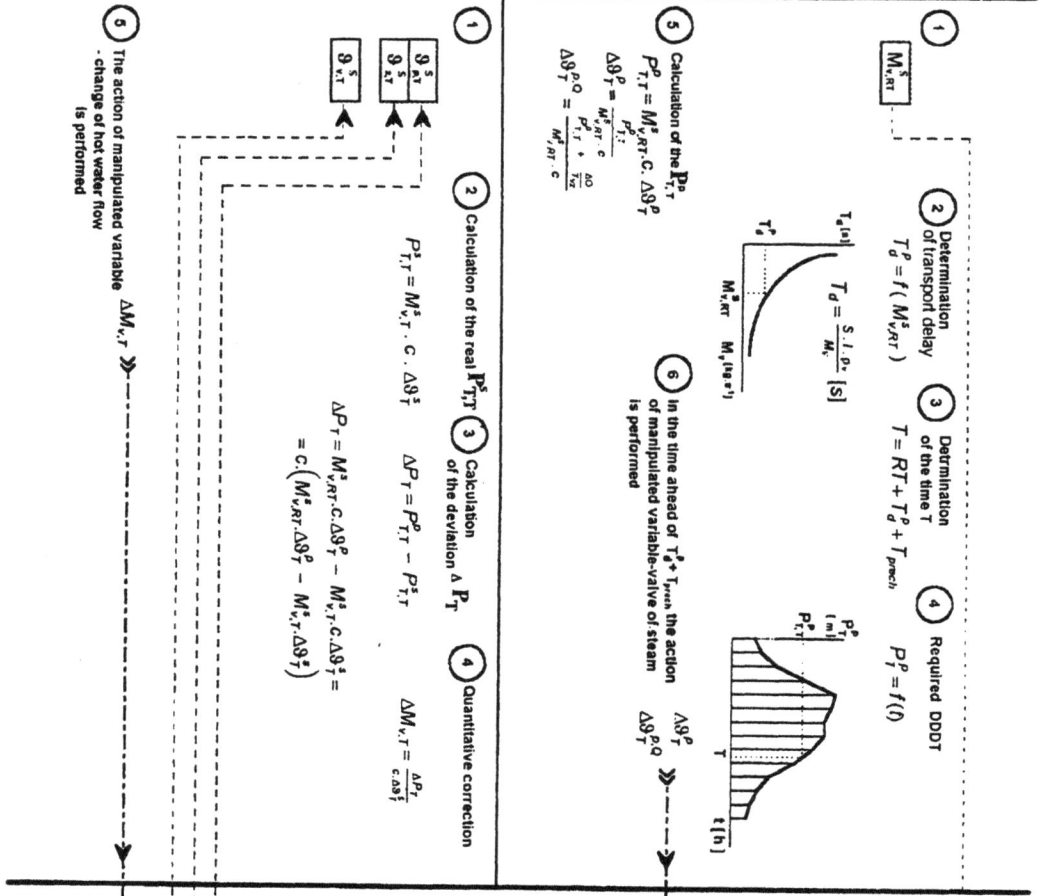

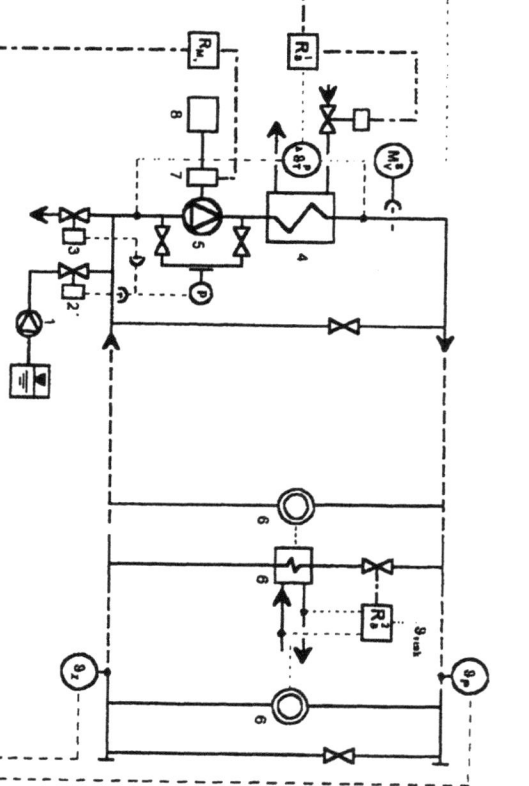

Fig. 3 Algorithm of the qualitative-quantitative control method of heat suply by hot-water piping

PROCESS DYNAMIC MODELLING USING CONTINUOUS TIME LOCAL MODEL NETWORKS

Séamus Mc Loone and George Irwin

Advanced Control Engineering Research Group,
Dept. of Electrical and Electronic Engineering,
The Queen's University of Belfast,
Belfast, UK, BT9 5AH

seamus.mcloone@qub.ac.uk, g.irwin@qub.ac.uk

Abstract: The capabilities of the continuous-time Local Model (LM) network for representing a non-linear dynamic process are studied by simulation of a coupled tank system. The paper shows how normalisation of the network weighting functions limits the modelling accuracy and proposes a solution via the addition of constant bias terms to each of the individual local models. *Copyright © 1998 IFAC*

Keywords: Multiple model, continuous-time modelling, non-linear modelling, Local Model Networks

1. INTRODUCTION

There has been considerable research interest in neural networks for the identification and control of non-linear dynamic systems in recent years, (Hunt, Irwin and Warwick, 1995) and (Hunt et al., 1992). Although there has been significant progress, including industrial applications in the fields of aerospace (Morita, 1993), robotics (Pham et al., 1994), power generation (Brown et al., 1995) and chemical (Lightbody et al., 1994), there remain a number of significant disadvantages to be addressed. These include a lack of transparency in the identified models which are inherently 'black-box' in nature, the difficulty in incorporating 'a-priori' plant knowledge which is usually available in practice and the weakness of theoretical support which is particularly evident when neural models are employed for control purposes.

An alternative modelling strategy, which addresses many of these limitations, is the Local Model (LM) network (Murray-Smith and Johansen, 1997), first studied by (Johansen and Foss, 1992, 1993). A LM network comprises a set of local models and

associated validity functions. The non-linear representation can be either discrete- or continuous-time, depending on the nature of the local model employed. Although the majority of the work to date has concentrated on the former, (Hunt, Johansen et al., 1996) and (Johansen and Foss, 1995), there has also been some progress on continuous-time LM networks, (Gawthrop, 1995).

In this paper the capabilities of the continuous-time Local Model (LM) network for representing a non-linear dynamic process is studied by simulation of a coupled tank system. The paper shows how normalisation of the network weighting functions limits the modelling accuracy and proposes a solution via the addition of constant bias terms to each of the individual local models.

2. LOCAL MODEL NETWORKS

Local Model (LM) networks were first described by (Johansen and Foss, 1992, 1993). A local model network (see Fig. 1) is a set of models, each valid for a specific regime in the operating space, weighted by

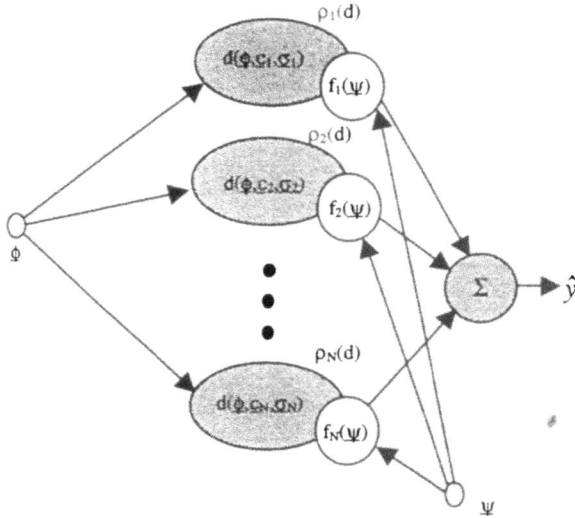

Fig. 1: General Architecture of a LM Network

some activation function. The same inputs, $\underline{\psi}$, are fed to all the models and the outputs are weighted according to some scheduling variable or variables, $\underline{\phi}$. The underlying local models can be either linear or nonlinear. The LM network output is given by:

$$\hat{y} = \sum_{i=1}^{N} \rho_i (d(\underline{\phi}, \underline{c_i}, \underline{\sigma_i})) f_i (\underline{\psi}) \qquad (1)$$

where $\rho_i(d(\underline{\phi}, c_i, \underline{\sigma_i}))$ is the basis function (in this case, a Gaussian function) of the ith model, $\underline{\phi}$ is a vector of scheduling variables, N is the number of models in the network and $f_i(\underline{\psi})$ is the ith local model output.

The LM network can be viewed as a generalisation of the RBF neural network. In the latter case, the weights associated with each basis function are constant parameters. In the LM network, these coefficents have been generalised to include more powerful functions of the inputs, $\underline{\psi}$. This means that a smaller number of local models can cover larger operating regimes of the input domain as illustrated in Fig. 2.

Obviously, there is a trade-off between the number and size of the operating regimes on the one hand, and the complexities of the local models on the other (Murray-Smith and Johansen, 1997). For instance, at one extreme one can have only one large operating regime that covers the full range of operation and, therefore, the local model must typically be complex since it is actually the global model. In general, a decomposition into a few *large* operating regimes will require more complex local models than a decomposition into numerous *small* operating regimes, see Fig. 2 (b) and (c). On the other extreme, one can partition the input domain into a large number of operating regimes, so that the function to be approximated can be represented by constant values locally, see Fig. 2 (a). The latter function

approximation is the principle underlying RBF networks.

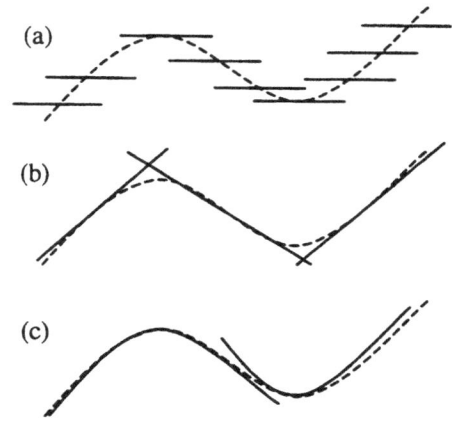

Fig. 2: Function Approximation by Local Models: (a) local constant functions, (b) local linear functions, (c) local quadratic functions.

2.1 Weighting Functions

In the local model network, the weighting of the local models is calculated using weighting or *activation* functions. The most commonly used weighting functions are the Gaussian basis functions, given by:

$$\rho_i = d(\underline{\phi}, \underline{c_i}, \underline{\sigma_i}) = \exp\left[-\left\| \frac{\underline{\phi} - \underline{c_i}}{\underline{\sigma_i}} \right\|^2 \right] \quad for \ i = 1,2,...,N \qquad (2)$$

where $\underline{c_i}$ defines the Gaussian centre, $\underline{\sigma_i}$ defines the Gaussian width and ρ_i is the ith activation function output. The scheduling variables, $\underline{\phi}$, can be a function of a system state, an input, and/or some other system parameter. The number of centres and widths defining the Gaussian function depends on the number of scheduling variables used in the local model network. For one scheduling variable the Gaussian activation function is simply a 2-D bell-shaped curve while for two scheduling variables it is a 3-D dome-shaped surface.

It is common practice to use normalised Gaussian basis functions, such that every point in the input space is covered by the basis functions to the same degree, i.e. the basis functions form a *partition of unity* across the input space. The normalised Gaussian function output is given by:

$$\bar{\rho}_i = \frac{\rho_i}{\sum_{j=1}^{N} \rho_i} \qquad (3)$$

150

3. THE COUPLED TANK SYSTEM

The simulated application is a standard one for control studies and is a highly complex, nonlinear plant. Fig. 3 shows a simplified diagram of the coupled tank system.

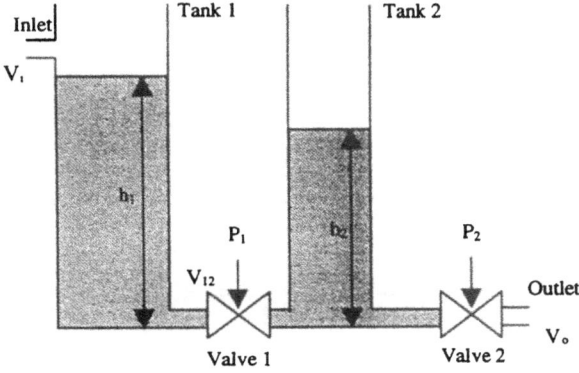

Fig. 3 Coupled Tank System

V_i, V_o and V_{12} represent the input flow rate, the output flow rate and the flow rate between the tanks respectively. The height of water in tank 1 is given by h_1, while the height of the water in tank 2 is h_2. If A_1 is the cross-sectional area of tank 1 and A_2 the cross-sectional area of tank 2 then the set of equations representing the coupled tank system is as follows:

$$\dot{h_1} = (V_i - V_{12})/A_1$$

$$\dot{h_2} = (V_{12} - V_o)/A_2 \qquad (4)$$

$$V_{12} = K_1(P_1)\sqrt{h_1 - h_2}$$

$$V_o = K_2(P_2)\sqrt{h_2}$$

where K_1 and K_2 are functions of the valve positions, P_1 and P_2 respectively.

Fig. 4 shows a graph of the flow characteristics for the valves in the coupled tank system, (Heckenthaler and Engell, 1994). This graph, along with the square roots in equations (4), illustrates the nonlinearity of this system.

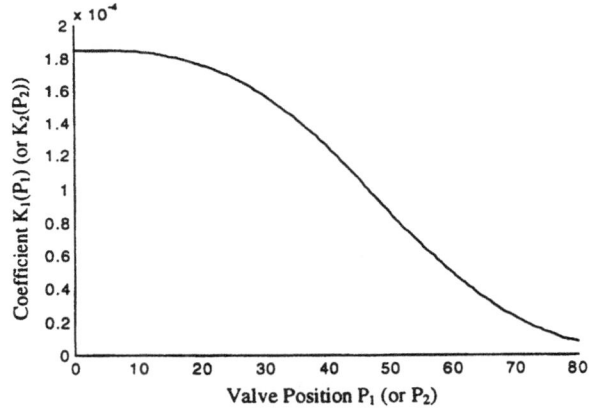

Fig. 4 Valve Flow Characteristics

4. LM NETWORK MODELLING OF THE TANK SYSTEM

It was decided to model the response of h_1 to changes in P_1 and P_2, as h_2 was not affected by P_1 in the steady-state. The steady-state output of h_1 is shown in Fig. 5. Nine local second order linear state space models were developed at various operating points throughout the operating range. Normalised Gaussian functions were used to interpolate between the models with the centres and widths optimised using the BFGS (Broyden, Fletcher, Goldfarb and Shanno) Quasi-Newton algorithm, (Gill, Murray and Wright, 1981).

4.1 Initial Modelling Results

Fig. 6 shows the steady-state output of the LM network, while the error between the actual steady-state output of the plant and that of the LM network is shown in Fig. 7. It can be seen that the LM network output is reasonably inaccurate. This is due to the fact that normalisation, defined in (3), limits the modelling accuracy of the LM network.

Normalisation causes the Gaussian functions to sum to unity throughout the operating region. In doing so, the steady-state output of the LM network is forced to lie within the space bounded by the local models chosen. Fig.8 illustrates this using a simple example, whereby two linear models are used to represent the steady-state curve shown. The steady-state output of the LM network in this particular case is forced to lie within the shaded area. It can be clearly seen that, despite any improvement in the weighting functions or change in the positioning of the linear models, the LM network will not accurately represent the nonlinear plant in the region between the operating points and, indeed, in the regions outside the operating points also.

151

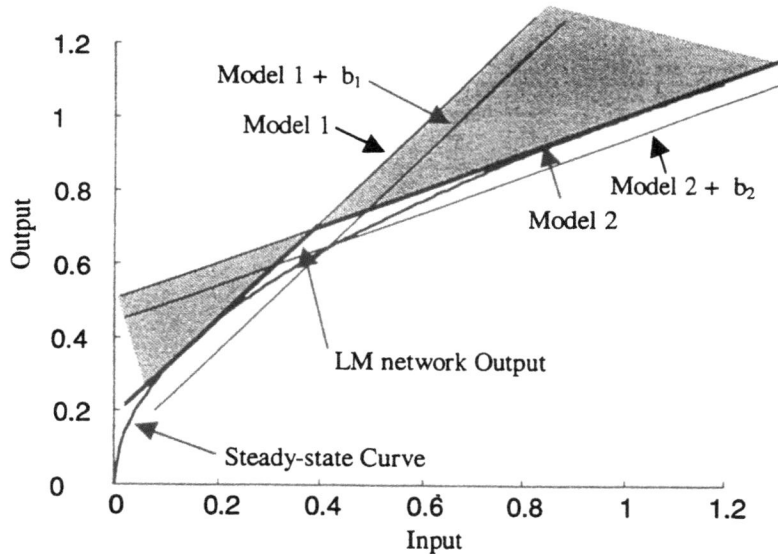

Fig. 8 Effect of Adding Constant Bias to each Model

Introducing more models will reduce the actual steady-state error but there will always exist some residual error that cannot be eliminated using this approach.

4.2 Bias Term Modelling

By adding constant bias terms, b_1 and b_2, to the two models in Fig. 8, they can be shifted upwards or downwards accordingly. An accurate representation may be achieved by optimising these bias terms. It is worth noting that, in order to achieve a completely accurate representation, the point of intersection between the two locally valid models needs to lie on the actual steady-state curve of the nonlinear plant.

This concept was applied to the modelling of the coupled tank system, i.e. a constant bias was added to each of the nine models. These bias terms, along with the centres and widths of the weighting functions, were optimised to give a minimal modelling error. The steady-state output is shown in Fig. 9 with the error displayed in Fig. 10. These results clearly show a remarkable improvement in the modelling accuracy of the LM network.

5 RESULTS

A series of step changes in P_1, with P_2 held constant, was used to test the dynamic response of the LM network, see Fig. 11. The LM network with biased local models produces a better response than the original LM network.

In the latter case, 'jumps' in the response are apparent. These 'jumps' occur when the network is traversing between local models and the reason for this again lies with normalisation. The LM network is optimised so as to give a minimum error between the LM network output and the plant output. From Fig. 8, it is plain to see that the minimum error is achieved when the output of the LM network is as indicated. The transition between the local models is, in this case, not smooth and hence the 'jump' in the step response.

The addition of the bias terms results in a smooth and continuous steady-state LM network output and, hence, no 'jumps' in the relevant output in Fig. 11.

6 CONCLUDING DISCUSSION

The LM network is simply a set of local models, based at various operating points throughout the operating range, and a set of basis functions that interpolates between these models. In this paper local linear models were used along with normalised Gaussian activation functions to model a Coupled Tank System.

Normalisation of the Gaussian functions has the advantage of ensuring that every point in the operating region is covered by the basis function to the same degree. However, it can limit the modelling accuracy of the LM network and can sometimes result in undesirable dynamical responses.

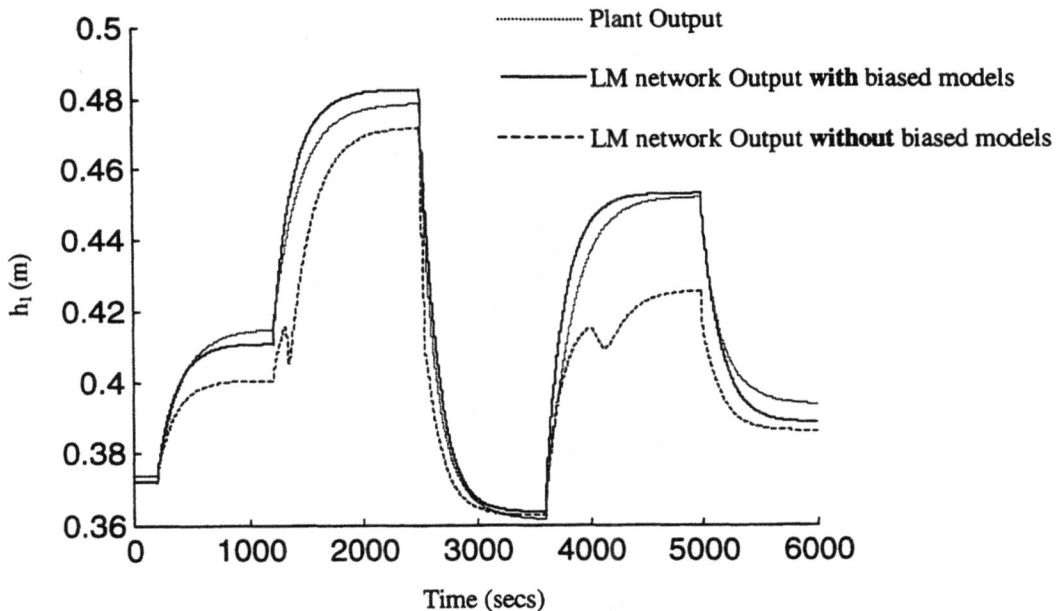

Fig. 11 Dynamical Responses of Various Systems

The steady-state output of the local models needs to be changed in order to obtain a LM network with a smooth continuous steady-state response. This was achieved by adding a constant bias to each of the local linear models. With the incorporation of these bias terms, it was possible to develop a local model network that accurately modelled the example system. In doing so, the disadvantages due to normalisation were overcome, while the associated advantages were maintained.

In this paper the input operating range was decomposed manually and suitable operating points chosen accordingly. Future work includes investigating operating regime decomposition and locating the optimal operating points for a nonlinear plant.

7 REFERENCES

Brown, M. D., Irwin, G. W., Hogg, B. W. and Swidenbank, K. E. (1995). Neural network modelling of a 200 MW boiler system, *IEE Proc. Control Theory & Applications*, **142**, No. 6, 529-536.

Gawthrop, P. J. (1995). Continuous-time local state local model networks. *Proc. IEEE Conf. Systems, Man and Cybernetics, Vancouver, Canada*, 852-857.

Gill, P. E., Murray, W. and Wright, M. H. (1981). *Practical Optimisation*, Academic Press, London.

Heckenthaler, T. and Engell, S., (1994). Approximately time-optimal fuzzy control of a two-tank system. *IEEE Control Systems*, 24-30.

Hunt, K. J., Sbarbaro, D., Zbikowski, R. and Gawthrop, P. J. (1992). Neural networks for control systems – A survey, *Automatica*, **28**, No.6, 1083-1112.

Hunt, K.J., Irwin, G. W. and Warwick, K. (1995). *Neural Network Engineering in Dynamic Control Systems*, Springer.

Hunt, K. J., Kalkkuhl, J. C., Fritz, H. and Johansen, T. A. (1996). Constructive empirical modelling of longitudinal vehicle dynamics using local model networks. *Control Engineering Practice*, **4**, 167-178.

Johansen, T. A. and Foss, B. A. (1992) A NARMAX model representation for adaptive control based on local models. *Modelling, Identification and Control 13*, 25-39.

Johansen, T. A. and Foss, B. A. (1993). Constructing NARMAX models using ARMAX models. *Int. J. Control 58*, 1125-1153.

Johansen, T. A. and Foss, B. A. (1995). Empirical modelling of a heat transfer process using local models and interpolation. *Proc. American Conference, Seattle, WA*, 3654-3658.

Lightbody, G., Irwin, G. W., Taylor, A., Kelly, K. and McCormick, J. (1994). Neural network modelling of a polymerisation reactor, *Proc. IEE Int. Conf. Control '94*, **1**, 237-242.

Morita, S. (1993). Optimisation control for combustion parameters of petrol engines using neural networks – in the case of on-line control, *Int. J. of Vehicle Design*, **14**, No. 5/5, 552-563.

Murray-Smith, R. and Johansen, T. A. (1997). *Multiple Model Approaches to Modelling and Control*, Taylor & Francis, Chapter 1, 3-72.

Pham, D. T. and Oh, S. J. (1994). Adaptive control of a robot using neural network controllers and filters, *IEEE Trans. Neural Networks*, **5**, No.2, 198-212.

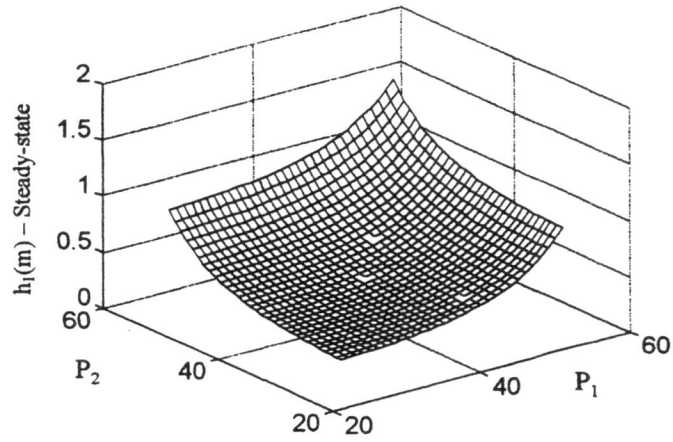

Fig. 5 Variation in steady-state height h_1 with
valve positions P_1 and P_2

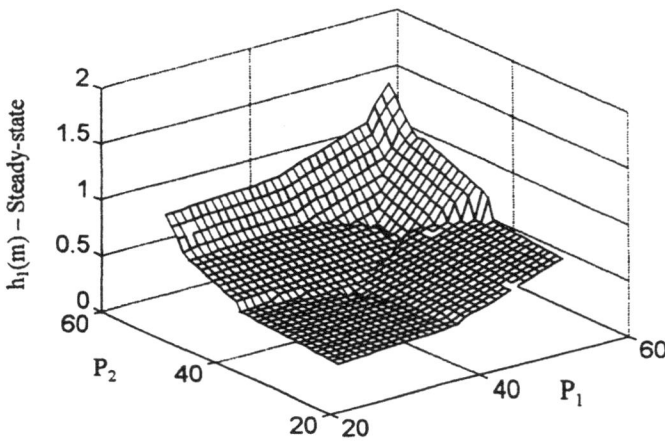

Fig. 6 LM network Output

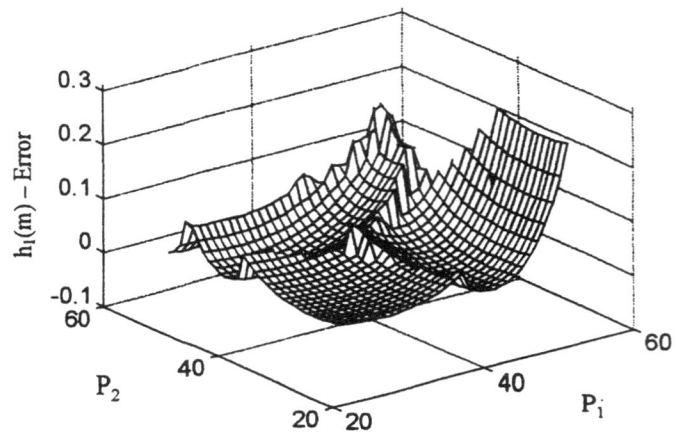

Fig. 7 Modelling Error in LM network Output

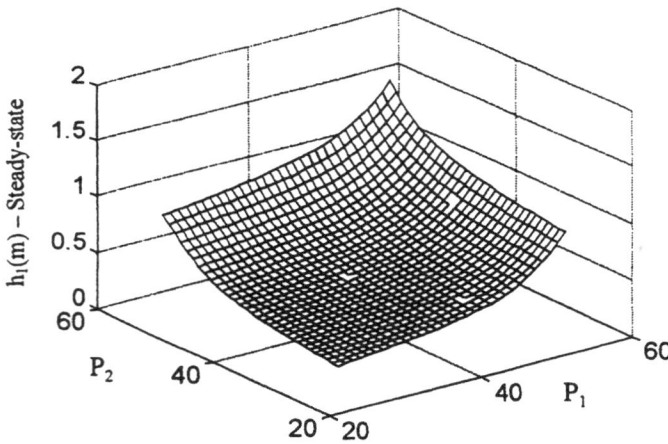

Fig. 9 Output of LM network with biased
local models

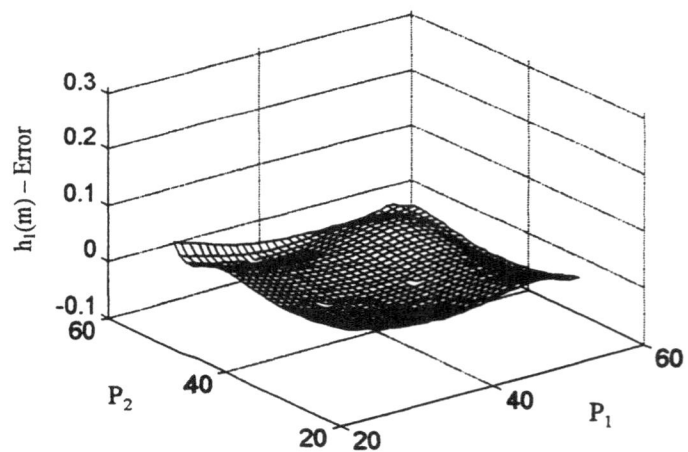

Fig. 10 Modelling Error in output of LM
network with biased local models

154

FUZZY LINEAR QUADRATIC REGULATOR APPLIED
TO THE REAL TIME CONTROL OF AN INVERTED PENDULUM

Doris Sáez and Aldo Cipriano

Faculty of Engineering, Catholic University of Chile
P.O.Box 306, Santiago 22, Chile
Phone: 56-2-6864286; fax: 56-2-5522563; e-mail: dsaez@ing.puc.cl

Abstract: The paper presents the design and real time implementation of fuzzy model based
control algorithms, specially fuzzy LQR ("Linear Quadratic Regulator"). Fuzzy LQR controller is
favorably compared with a conventional LQR to the angular stabilization of an inverted
pendulum. The Fuzzy LQR is also applied to the real time control of an inverted pendulum.
Copyright © 1998 IFAC

Keywords: Models based control, fuzzy models, fuzzy LQR, inverted pendulum.

1. INTRODUCTION

During the last years, various works have been
published describing successful application of fuzzy
logic to the control of non-linear dynamic processes
(Lee, 1990). Nonetheless, in most of the cases, the
design of fuzzy controllers doesn't follow a formal
procedure, as is the case in more conventional
controllers. As a contribution to the fuzzy controller
design, in this work we propose a model based fuzzy
control algorithm for non-linear dynamic systems.
The design method is illustrated with the control of
an inverted pendulum.

The inverted pendulum is a classic problem of
nonlinear unstable behavior (Mori, *et al.*, 1976). In
specialized literature, various techniques have been
introduced to solve the control problem, using
different manipulated and controlled variables,
together with different control objectives. This study
considers the problems of angle stabilization and the
angle stabilization with distance control, for which
different solutions have been developed.

First, for the angle stabilization Dorf (1986) proposed
to use state feedback controllers, such as linear
quadratic regulators and optimal controllers based on
pole assignment. On the other hand, Yamakawa
(1989) developed an expert fuzzy controller with
only seven control rules. To solve this problem, also
more complex controllers have been developed. For
example, Jang (1992) implemented a self-learning

fuzzy controller based on the temporal
backpropagation algorithm, Burkhardt and Bonissone
(1992) designed a modified self-organizing fuzzy
control architecture to generate the rules base, Wang
(1994) developed a supervisory controller for a fuzzy
expert control system that assures global stability and
Lin and Chen (1994) used an adaptive fuzzy sliding
mode controller.

For the angle stabilization with distance control,
Mori, *et al.* (1976) implemented a linear feedback
controller combined with a state observer. Another
control proposal was presented by Kyung and Lee
(1993) that designed a self-learning fuzzy controller
based on neural networks. Finally, Kun, *et al.* (1994)
developed an adaptive fuzzy controller with a
cascade architecture.

The work starts with the description of fuzzy models
proposed by Takagi and Sugeno. Then, the derivation
of control algorithm based on fuzzy models, specially
fuzzy LQR, is presented. The following section
presents the tests made with fuzzy LQR to stabilize
the angle, together with the results from the
comparison with the conventional LQR. Finally, the
real time control of an inverted pendulum is
presented.

2. FUZZY MODELING

In the fuzzy model proposed by Takagi and Sugeno (1985), the input variables of the premises of each rule are combined by AND operators and the output variables represent linear models in the process' state variables. Thus, the fuzzy model rules are the following:

$$R_i: \text{If } Z1 \text{ is } P1_i \text{ and} \ldots \text{ and } Zk \text{ is } Pk_i$$
$$\text{then } Y_i = p_o^{\,i} + p_1^{\,i} X1 + \ldots + p_k^{\,i} Xk \qquad (1)$$

where $Z1, \ldots, Zk$ are the model's input variables, $P1_i, \ldots, Pk_i$ are the fuzzy sets associated to the input variables, $X1, \ldots, Xk$ are the state variables, $p_o^{\,i}, \ldots, p_k^{\,i}$ are the parameters of the i rule and Y_i is the output of linear model of rule i.

Thus, the output of the model, Y, is obtained weighting the output of each rule by their respective degree of activation, W_i, that is:

$$Y = \frac{\sum\limits_{i=1}^{M} W_i Y_i}{\sum\limits_{i=1}^{M} W_i} \qquad (2)$$

where M is the number of rules of the model. W_i is calculated as the product of the membership grades involved in the same rule.

3. FUZZY CONTROL ALGORITHM

The model based fuzzy control here proposed consists of rules with the following structure:

$$R_i: \text{If } Z1 \text{ is } P1_i \text{ and} \ldots \text{ and } Zk \text{ is } Pk_i$$
$$\text{then } u_i = f_i (X1, \ldots, Xk) \qquad (3)$$

where u_i is the manipulated variable of rule i and the control law of each rule, f_i, corresponds, for example, to a LQR obtained from a linear model of rule i (see equation (1)). Note the premises of the fuzzy controller are the same of those of the fuzzy model rules.

The final control action is obtained as:

$$u = \frac{\sum\limits_{i=1}^{M} (W_i \, u_i)}{\sum\limits_{i=1}^{M} W_i} \qquad (4)$$

where M is the number of rules of the controller and W_i is the degree of activation of rule i calculated as the product of the membership grades involved in the same rule.

4. PROCESS DESCRIPTION

The experimental assembly consists on a Quanser Consulting system that includes an inverted pendulum with a linear sensor, an angle sensor and a DC motor that acts as an actuator, connected to a PC-486 DX personal computer through a 12-bit data acquisition Data Translation DT-2811 board.

The inverted pendulum is composed of a beam mounted on a cart that slides on a rail as shown in Fig. 1. The cart is equipped with a gear mechanism that allows to exert force on the system and allows to measure the cart's position. Also, a potentiometer mounted on the rotating axis allows to measure the angle of the pendulum with respect to the vertical axis.

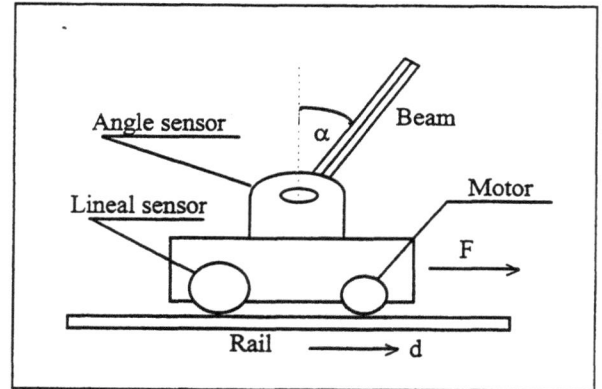

Fig. 1. Experimental platform.

The state of the process is determined by four variables: the angle between the pendulum and the vertical axis (α), the angular velocity (α'), the horizontal position (d) and the cart's horizontal velocity (d').

The movement of the inverted pendulum can be modeled through the following non-linear equations (Anderson, 1989):

$$\alpha'' = \frac{(M + m)g \sin\alpha - F\cos\alpha - \dfrac{ml}{2}\alpha'^2 \sin\alpha \cos\alpha}{\dfrac{2(M+m)l}{3} - \dfrac{ml \cos^2\alpha}{2}} \qquad (5)$$

$$d'' = \frac{\dfrac{2}{3}F + \dfrac{ml}{3}\alpha'^2 \sin\alpha - \dfrac{mg}{2}\cos\alpha \sin\alpha}{\dfrac{2(M+m)}{3} - \dfrac{m\cos^2\alpha}{2}} \qquad (6)$$

where α and d are the output variables and F is the manipulated variable.

5. SIMULATION TEST

5.1 Basis of evaluation

For testing the controllers, an inverted pendulum simulator programmed in Simulink-Matlab is used, with the parameters shown in Table 1.

Table 1 Process parameters

Parameter	Value	Unit
Cart mass (M)	1.0	Kg.
Pendulum mass (m)	0.1	Kg.
Pendulum length (l)	1.0	m.
Gravity acceleration (g)	9.8	m./sec^2

These test considers the control objective of stabilizing the angle; so the controlled variable is angle α and the manipulated variable is the force F. The desired final state is given by $\alpha_f = 0$, $\alpha'_f = 0$. The horizontal position d is not controlled.

To design and evaluate fuzzy LQR and conventional LQR, the following cost function is defined (Jang, 1992):

$$J = \sum_{j=1}^{100} \alpha^2(j) + 10 \sum_{j=0}^{99} F^2(j) \qquad (7)$$

5.2 LQR controller

The LQR controller is derived from the Ricatti equation (Astrom and Wittenmark, 1984) using a linearized model of equation (5), at the upright position point ($\alpha = \alpha_f$, $\alpha' = \alpha'_f$).

Then, the control law is:

$$F = 21.5646\,\alpha + 5.4288\,\alpha' \qquad (8)$$

5.3 Fuzzy LQR controller

Due to two fuzzy sets are associated to inputs variables, the designed Fuzzy LQR controller has four rules which take the form:

R_i: If α is A_i and α' is B_i then $F_i = -k_1^i\,\alpha - k_2^i\,\alpha' + k_o^i$ (9)

where k_j^i are the coefficient of rule i for the state variable j, ($j = 1, 2$) and k_o^i is a bias given by the difference between the operation points (see Table 2) and the equilibrium point ($\alpha = \alpha_f$, $\alpha' = \alpha'_f$).

The membership functions are shown in Fig. 2. The parameters of membership functions has been determined by a trial and error procedure and its

values are presented in Table 3. The consequences coefficients, corresponding to a LQR controller for each rule, are shown in Table 4.

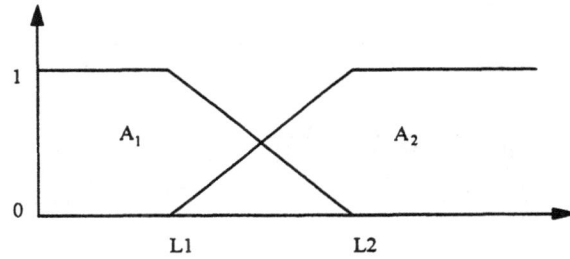

Fig. 2. Membership functions.

Table 2 Operation points

Rule i	α (rad)		α' (rad/sec)	
1	-0.698	(-40°)	-1.745	(-100°/sec)
2	-0.698	(-40°)	1.745	(100°/sec)
3	0.698	(40°)	-1.745	(-100°/sec)
4	0.698	(40°)	1.745	(100°/sec)

Table 3 Parameters of membership functions

	L1	L2
α (rad)	-0.349 (-20°)	0.349 (20°)
α' (rad/sec)	-0.873 (-50°/sec)	0.873 (50°/sec)

Table 4 Coefficients of consequences

Rule i	k_1	k_2	k_o
1	-20.24	-5.989	-2.079
2	-20.24	-6.213	-2.079
3	-20.24	-6.213	2.079
4	-20.24	-5.989	2.079

5.4 Comparative analysis

Fig. 4 shows the responses of the inverted pendulum, for initial conditions $\alpha_o = 0.349$ rad, $\alpha'_o = 0.698$ rad / seg, using LQR and Fuzzy LQR controllers. Fig. 4 (a) (d) shows that the responses of the Fuzzy LQR is better than the conventional LQR.

Fig. 5 shows the degree of activation of each rule for the same initial condition. It must be noted that four rules are activated during the complete simulation, because of the shape of membership functions (see Fig. 2).

Table 5 presents the values of the cost function, J (see equation (7)), for all the test performed with different initial conditions.

Taking account the graphics in Fig. 4 and the cost function values in Table 5 it can say that the Fuzzy LQR has the best performance in the simulation tests.

Table 5 Cost function values

α_0 (rad)	0.175 (10°)	0.175 (10°)	0.262 (15°)	0.349 (20°)
α'_0 (rad/sec)	0	0.349 (20°/sec)	0.522 (30°/sec)	0.698 (40°/sec)
LQR	4956	8437	19574	37240
Fuzzy LQR	4578	7747	17805	34635

6. EXPERIMENTAL TEST

6.1 Basis of evaluation

For testing the fuzzy LQR, the experimental platform of an inverted pendulum is used, with the parameters shown in Table 6.

Table 6 Experimental platform parameters

Parameter	Value	Unit
Cart mass (M)	0.455	Kg.
Pendulum mass (m)	0.210	Kg.
Pendulum length (l)	0.61	m.
Rail length (d_f)	1.0	m.
Gravity acceleration (g)	9.8	m./sec^2

To perform the real time tests of the fuzzy LQR controller, the blocks configuration programmed in Simulink shown in Figure 6 is used. The ADC and DAC data acquisition blocks sense the angle α voltage signals and distance d, and generate the voltage V to the motor at a frequency of 200 Hz. Calib 1 and Calib 2 blocks calibrate the measurement of the angle and the distance respectively, during the first two seconds of running. Factor 1 and Factor 2 blocks make the conversion of voltage to degrees and centimeters for the angle and distance respectively. The Filter 1 block is a set of lowpass filters at 3 Hz, with the function of eliminating the noise components of the signals. The Filter 2 block is simultaneously a set of state observers and lowpass filters for the α' angle velocity and the horizontal velocity of the cart d'. The LQR block is the fuzzy LQR controller. The Saturation for V block limits the voltage around ±5 volts. Lastly, the Zero-Order Hold blocks sample the signal at 20 Hz to subsequently analyzing them.

6.2 Implementation and results

Due to the rail of our experimental pendulum has a limited length of 1 m., the horizontal position d should be considered as a controlled variable too. The rules of the Fuzzy LQR controller takes then the following structure:

$$R_i: \text{If } \alpha \text{ is } P_i \text{ then}$$

$$V_i = -k_1{}^i \alpha - k_2{}^i \alpha' - k_3{}^i d - k_4{}^i d' + k_o{}^i \qquad (10)$$

where $k_j{}^i$ are the coefficients of rule i for the state variable x_j, (j = 1, 2, 3, 4) and $k_o{}^i$ is a bias given by the difference between the operational point used for the linearization corresponding to rule i and the equilibrium state defined by $\alpha = \alpha_f$, $d = d_f$, $\alpha' = \alpha'_f$, $d = d'_f$. The manipulated variable is now a voltage V which is applied to the motor and generates the appropriate force F. For experimental conditions this voltage is restricted to be inside the limits ± 5 volts.

The values of the weighting matrices Q and R are the ones proposed in the User's Manual of the experimental inverted pendulum. They are:

$$Q = \begin{bmatrix} 4 & 0 & 0 & 0 \\ 0 & 0.25 & 0 & 0 \\ 0 & 0 & 0 & 0 \\ 0 & 0 & 0 & 0 \end{bmatrix} \qquad R = 0.0003 \qquad (11)$$

The operational points for the linearization are $\alpha = \alpha_f$, $\alpha' = \alpha'_f$, $\alpha = 10°$, $\alpha' = 0$ and $\alpha = -10°$, $\alpha' = 0$. The parameters of the membership functions (see Fig. 3) of the P_i fuzzy sets are L1 = 2.5° and L2 = 7.5°. The coefficients of the consequences, corresponding to the ones of a LQR controller for each rule, are shown in Table 7.

Fig. 7 shows the response of the inverted pendulum for the tests made with the experimental platform, using the fuzzy LQR controller.

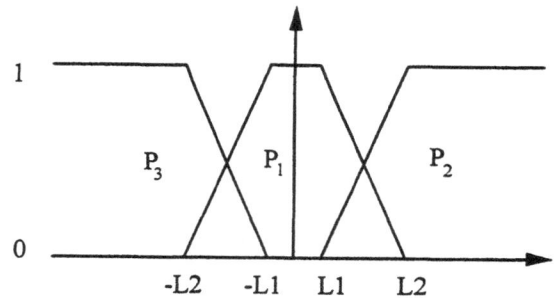

Fig. 3. Membership functions.

Table 7 Coefficients of consequences

Rule i	$k_1{}^i$	$k_2{}^i$	$k_3{}^i$	$k_4{}^i$	$k_o{}^i$
1	2.441	0.289	0.236	0.331	0
2	2.443	0.289	0.241	0.333	0.025
3	2.443	0.289	0.241	0.333	-0.025

7. CONCLUSIONS

The fuzzy model based controller has been designed and implemented in this paper. The simulation tests show the superiority of the fuzzy LQR over conventional LQR.

The model based fuzzy controllers, fuzzy LQR, have also the advantage that the design procedure and the tuning of the controller parameters is simple to understand and to implement.

Finally, the real time tests show that the fuzzy LQR responds appropriately in the presence of disturbances.

ACKNOWLEDGMENTS

The authors wish to thank FONDECYT for the financial support given to the projects 1960394, Predictive Optimal Control Based on Fuzzy Models and 2980029 Design of Predictive Control Strategies Based on Non Linear Models and its Application to the Control of Power Plants.

REFERENCES

Anderson C. (1989). Learning to control an inverted pendulum using neural networks. *IEEE Control Systems Magazine*, April, 31-36.

Astrom K. and B. Wittenmark (1984). *Computer Controller Systems, Theory and Design.* Prentice-Hall, Inc.

Burkhart D. and P. Bonissone (1992). Automated fuzzy knowledge base generation and tunning. *IEEE Int. Conf. Fuzzy Syst.*, San Diego, California, March 8-12, 179-188.

Dorf R. (1986). *Modern control systems.* Addison-Wesley.

Jang J. (1992). Self-learning fuzzy controllers based on temporal back propagation. *IEEE Trans. Neural Networks*, **Volume No 3**, 714-723.

Kun M., H. Chung, T. Hwee and P. Zhuang. (1994). A cascade architecture of adaptive fuzzy controllers inverted pendulums. *IEEE Int. Conf. Fuzzy Syst.*, Orlando, June 26-29, 1514-1519.

Kyung K. and B. Lee. (1993). Fuzzy rule base derivation using neural network-based fuzzy logic controller by self-learning. *IEEE Conf. Ind. Elect., Contr. and Instr.*, Hawai, Nov. 15-19, 435-440.

Lee, Ch. (1990). Fuzzy logic in control systems: Fuzzy logic controller. *IEEE Trans. Syst., Man and Cybern.*, **Volume No SMC-20**, 404-435.

Lin S. and Y. Chen (1994). Design of adaptive fuzzy sliding mode for nonlinear system control. *IEEE .Int. Conf. Fuzzy Syst.*, Orlando, Florida, June 26-29, 35-39.

Mori S., H. Nishihara and K. Furuta (1976). Control of unstable mechanical system control of pendulum. *Int. J. Contr.* **Volume No 23**, 673-692.

Takagi T. and M. Sugeno (1985). Fuzzy identification of systems and its applications to modeling and control. *IEEE Trans. Syst., Man and Cybern.*, **Volume No SMC-15**, 116-132.

Wang L. (1994). A supervisory controller for fuzzy control systems that guarantees stability. *IEEE Trans. Automat. Contr.*, **Volume No 39**, 1845-1847.

Yamakawa T. (1989). Stabilization of inverted pendulum by a high-speed fuzzy logic controller hardware system. *Fuzzy Sets and Systems*, **Volume No 32**, 161-180.

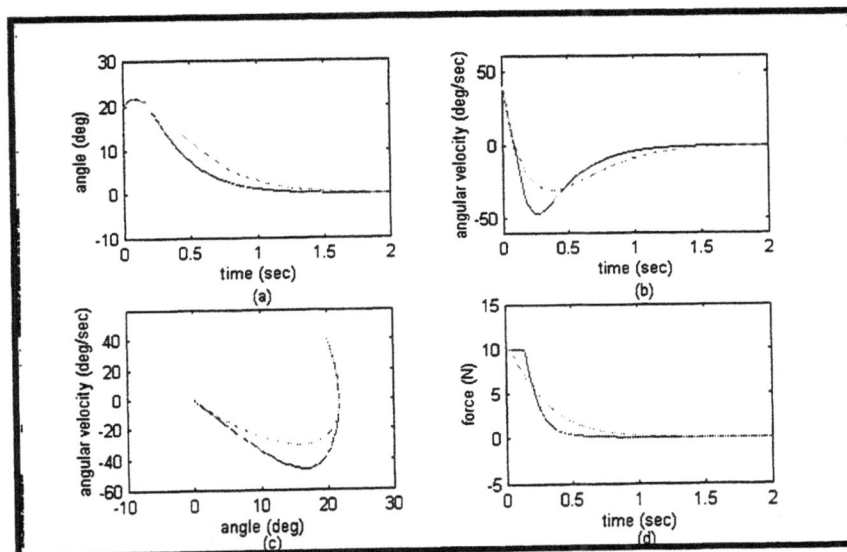

Fig. 4. Simulation results.
Initial conditions: $\alpha_0 = 0.349$ rad (20°), $\alpha'_0 = 0.698$ rad/sec (40°/sec). ——— : Fuzzy LQR, ----: LQR.

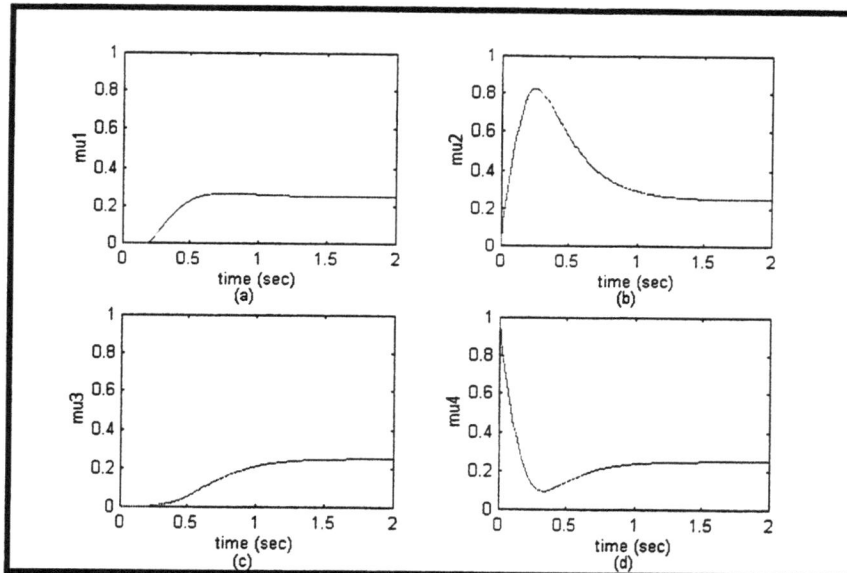

Fig. 5. Degrees of activation of the rules.
Initial conditions: $\alpha_0 = 0.349$ rad (20°), $\alpha'_0 = 0.698$ rad/sec (40°/sec).

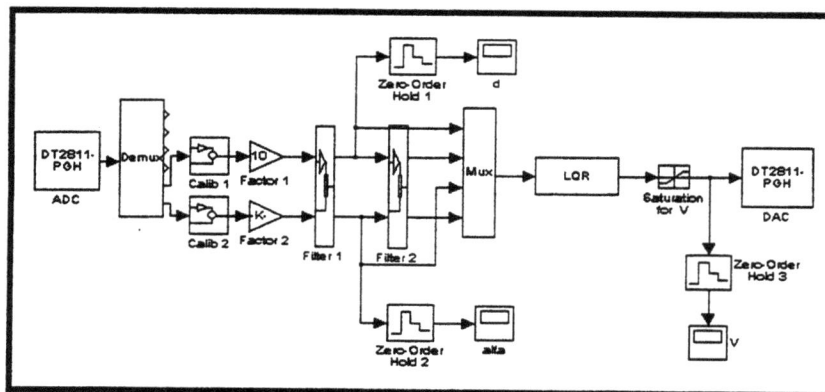

Fig. 6. Fuzzy LQR implementation.

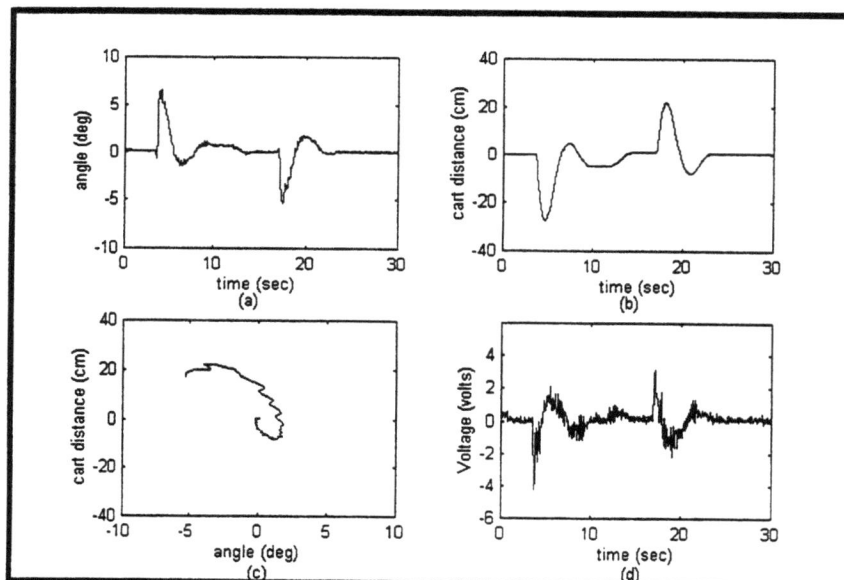

Fig. 7. Real time test with fuzzy LQR controller.

ON REAL-TIME MARKOV SIMULATION FOR GAS TURBINE ENGINE CONDITION MONITORING

G.G.Kulikov*, T.V.Breikin*, V.Y.Arkov** and P.J.Fleming***

* Department of Automated Control Systems
Ufa State Aviation Technical University
K.Marx Street 12, Ufa, 450000, Russia.

** Institute of Mechanics,
Russian Academy of Science
K.Marx Street 12, Ufa, 450000, Russia.

*** Department of Automatic Control and Systems Engineering
University of Sheffield, Mappin Street
Sheffield S1 3JD, UK

Abstract: This paper focuses on real-time simulation for gas turbine engine condition monitoring. The problem under investigation is to design a method for non-parametric modelling of a gas turbine engine as a non-linear stochastic system using input-output realisations via Markov models. The analysis of real engine data obtained from an engine test-bed and identification of the engine Markov model have been done. A technique of Markov model based engine condition monitoring is suggested. The results of implementation of obtained models for engine real-time simulation are considered *Copyright © 1998 IFAC*

Keywords: signal processing, aerospace, intelligent control.

1. INTRODUCTION

Gas turbine engines are now widely used in different fields of human activity and creating of engine condition monitoring system can results in many benefits. A condition monitoring system is especially important for an aircraft gas turbine engine because a fault of the engine can lead to disastrous results. Real-time condition monitoring of such complex non-linear dynamical systems as aircraft gas turbine engines is a very complicated problem. This problem getting even more complex when taking into accounts a non-linear and stochastic nature of a gas turbine engine.

The presence of a fault in a system changes the general characteristics of the system and detection of these changes can help in detection of the fault. This assumption leads to development of different kinds of model-based fault detection and diagnosis methods (Isermann, 1984; Basseville, 1988; Isermann, 1993). This approach has been investigated for gas turbine engine (Breikin, 1997b; Green, 1997).

Utilisation of model-based approaches for condition monitoring purposes faces a problem of finding a compromise between model complexity and accuracy of engine parameters prediction. A very complex model is difficult for identification and simulation in

real time. A low accurate model is useless for a condition monitoring system. Also system identification for stochastic systems is a very complex problem because types of noises and their point of occurrence in the system under investigation are usually unknown. The problem of parameter estimation for linear stochastic systems with different kinds of assumptions has been considered (Soderstrom and Stoica, 1989; Ljung, 1987; Goodwin and Sin, 1984; Tugnait, 1995). The creation of a parametric analytical model for non-linear stochastic systems is a far more complicated problem. A non-parametric state-space Markov models approach has been suggested by Breikin *et al.* (1997b) for stochastic system identification.

In this paper, an application of Markov model based fault detection method for condition monitoring of a gas turbine engine is considered. The engine Markov model is built using real data obtained from an engine test-bed. The obtained model is applied for real-time gas turbine engine simulation. The comparison of Markov model performance with other non-linear engine models is done. The paper is organised as follows: In section 2, a condition monitoring problem for a gas turbine engine is formulated. Section 3 describes identification of non-linear systems via Markov modelling approach. In Section 4 results of simulation are presented and Section 5 gives the summary of results and conclusions.

2. PROBLEM FORMULATION

The gas turbine engine data are obtained from two information channels, which measure the same input and output. These measurements were made possible by the use of a dual control system. The measurement data are the product of engine testing on an engine test-bed. Amongst the variables measured, the fuel flow (denoted by $u(t)$) and the engine high and low pressure shaft speeds (denoted by $x_h(t)$, $x_l(t)$ respectively) are of interest here. The measurement scheme is given in Figure 1, where the feedback signals from the engine to the controller are not shown.

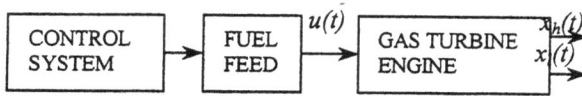

Fig.1. The engine signal measurement scheme, without the feedback shown.

Since two independent measurement sensors measure the same variable, redundancy in the available information exists. Figure 2 shows the high pressure shaft speed measurements from the two channels and the difference between them.

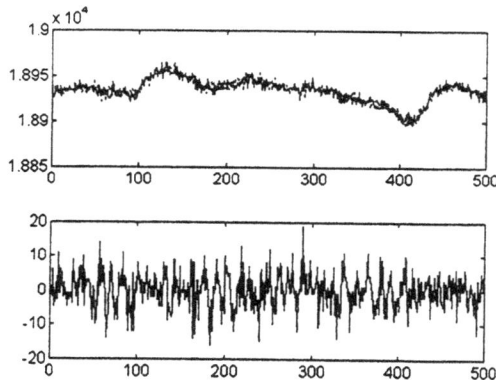

Fig. 2. The data from different measurement channels (top) and difference between them (bottom).

Clearly, sensor faults such as drift and malfunction in one can be detected by considering the maximum allowed measurement difference in the two channels, and this has been demonstrated by Kulikov *et al.* (1995). However, this method does not provide information as to which channel is faulty and whether there are any faults in the engine itself. This problem can be overcome with the help of a high accuracy engine model.

The presence of a fault in a gas turbine engine leads to behaviour that deviates from normal operation.

This deviation is best extracted if an accurate model of the engine is available. Explicit modelling of a gas turbine engine is rather complex and analysing its behaviour in real-time is extremely computationally demanding. An alternative approach is to use system identification methods to model the normal engine behaviour, as in model-based approaches (Breikin, 1997b).

3. MARKOV MODELS APPROACH

There are two types of stochastic systems:

- **Case I:** The system does not generate internal noises but operates under environmental noise conditions. In this case, a model of the system and a model of additional noises can be obtained separately.

- **Case II:** The system generates internal noises. In this case, the problem of parametric model identification becomes very complicated. A gas turbine engine is such a system.

Let us consider for simplicity a SISO stochastic system with an input, denoted $u(t)$ and an output, denoted $x(t)$. It is clear that $x(t)$ is a stochastic process. The problem to be solved is to create such a stochastic system model, whose output, denoted $\hat{x}(t)$, is the same stochastic process as $x(t)$ under condition that inputs are the same

Markov models are an old and well studied field by modern standards. They find implementation in different kinds of applications. The subject of this paper is controlled Markov chain in a finite state space.

In the stochastic dynamical system,

$$x(t)=F(x(t-1),u(t-1),x(t-2),u(t-2),..., e(t),e(t-1),...),$$

where $e(t)$, $e(t-1)$, ... are noise terms, the states $x(t)$ are assumed to follow internal dynamics which can generally be described by an r-order Markov process. In the simple case, assuming a first order controlled Markov process,

$$P\{x(t)|(x(t-1),u(t-1),x(t-2),u(t-2),...)\} =$$

$$= P\{x(t)|(x(t-1),u(t-1))\}.$$

For a discrete time - discrete space dynamic system it will be a first order controlled Markov chain,

$$P\{x(t_n)=q|(x(t_{n-1})=q_1,u(t_{n-1})=p_1,x(t_{n-2})=q_2,...)\} =$$

$$= P\{x(t_n)=q|(x(t_{n-1})=q_1,u(t_{n-1})=p_1)\}.$$

If stochastic processes within a dynamic system are assumed to be stationary and ergodic then the Markov chain is homogeneous,

$$P\{x(t_n)=i|(x(t_{n-1})=j, u(t_{n-1})=k)\} = P_{ijk}.$$

Thus the problem of stochastic system identification is to estimate the Markov chain transition probabilities P_{ijk}.

Two approaches for the estimation of $x(t)$ probability density can be considered:

Parametric approach: If there is any prior information about the type of $x(t)$ probability distribution, it is possible to estimate its parameters (such as mean value $\bar{u}_{jk}$ and standard deviation σ_{jk} for Gaussian distribution) for every (x_j, u_k) domain using experimental data.

Non-parametric (histogram) approach: In this case, every transition probability P_{ijk} is to be estimated separately via the formula

$$P_{ijk} = \frac{N^i_{jk}}{N_{jk}},$$

where N_{jk} is the number of system output values in the jth interval of discretisation on x and N^i_{jk} is the number of output transitions from the jth interval to the ith interval, with the condition that the input value is in the kth interval of discretisation on u. A more complex formula for estimating transition probabilities P_{ijk} is following

$$P_{ijk} = \frac{\sum_n x(t_n)\varphi_j u(t_n)\psi_k x(t_{n+1})\phi_i}{\sum_n x(t_n)\varphi_j u(t_n)\psi_k},$$

where $x(t_n)$, $u(t_n)$ are current values on x and on u correspondingly, φ_j is the probability density function for j^{th} state on x, ϕ_i is the probability density function for i^{th} state on x, ψ_k is the probability density function for k^{th} state on u.

The investigation of the noise signals generated during the gas turbine engine operation has shown that these noise terms can be considered as normal processes. In this case, the stochastic difference equation describing the engine behaviour can be considered as a controlled Markov process. On the other hand, at the steady state conditions of the engine the size of the input and output variation is rather small. This allows producing input and outputting discretisation and thus considering the

Markov process as a Markov chain. In assumption that the stochastic process $e(t)$ is stationar the Markov chain can be considered as homogeneous. In this case it is possible to obtain a Markov chain, which has a reasonable number of elements in the transition probabilities matrix. It is clear that the matrix dimension depends on the order of the stochastic difference equation describing the process.

The homogeneous Markov chain can be described with the help of the stochastic matrix of transitions probabilities π with dimension $m \times m$, where m is the number of states in the chain. Each element of this matrix is the probability of system transition from the state X_j at $(n-1)^{th}$ time moment into the state X_i at n^{th} time moment.

$$P_{ji} = P\{X(t_{n-1}) = X_j \rightarrow X(t_n) = X_i\},$$
$$\sum_j P_{ji} = 1.$$

The probability for the system co-ordinate $x(t)$ to be in the state X_i can be calculated by the following formula

$$P_i = \int_{x_i - \Delta x/2}^{x_i + \Delta x/2} P(x)dx = P\{x \in [x_i - \frac{\Delta x}{2}, x_i + \frac{\Delta x}{2}]\},$$

where x_i is the mean value of ith interval of discretisation on x, and Δx is a step of discretisation on x.

In the case of a controlled Markov chain it can be represented by an array of stochastic matrixes of transition probabilities $\pi(u_k)$, where u_k is the mean value of kth interval of discretisation on u.

Such representation of stochastic dynamic systems in the form of homogeneous Markov chains made it possible to obtain via Markov formula the vector of system state probabilities at n^{th} time moment if initial system states probabilities vector P_0 is given. For controlled dynamic systems this formula is the following

$$P_i = P_0 \prod_{n=1}^{i} \pi(u(t_n) = u_k) = P_{i-1}\pi(u(t_n) = u_k).$$

The problem of simulation can be easily solved with the help of Monte Carlo methods.

4. SIMULATION RESULTS

Markov models of different channels of a modern gas turbine engine and hydro-mechanical part of its control system have been identified for condition monitoring purposes. The results of simulation of fuel pomp work are shown in the Figure 3 compared with real data. A dotted line defines a 99% confidence intervals for the simulated parameter to be in the co-ordinate space divided by states of the Markov model.

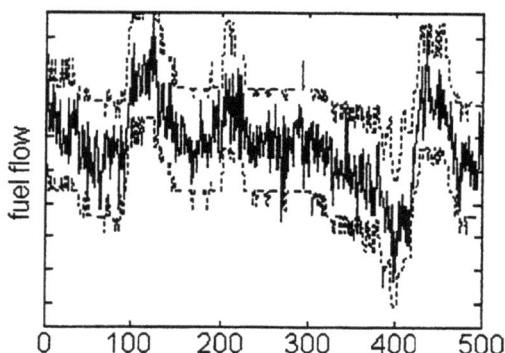

Fig.3. Markov simulation of fuel pomp compared with real data (99% confidence intervals are shown by dotted line).

This Markov modelling technique has also been applied for such engine outputs as shafts speed, pressure, and temperature. The results of Markov simulation of high pressure shaft speed with maximal probability value compared with real data and the error of simulation are shown in Figure 4.

For estimating of Markov model performance the possibility of RBF neural network utilising for this problem has also been investigated. The results of RBF network simulation of high pressure shaft speed compared with real data and the error of simulation are shown in Figure 5. The mean squared error of prediction is 3.98 rpm for the Markov model approach and 4.55 rpm for RBF neural network simulation.

An important advantage of the Markov simulation technique is the possibility to implement it in real-time applications. In this practical example the simulation of engine work using Markov model was approximately 20 times faster then the same simulation using a RBF neural network.

5. CONCLUSIONS

In this paper a Markov model-based aircraft engine condition monitoring technique has been considered. A stochastic system identification approach based on Markov models has been described. Gas turbine engine identification was carried out on input-output realisations. It was shown that this approach gives good results for identification of nonlinear systems.

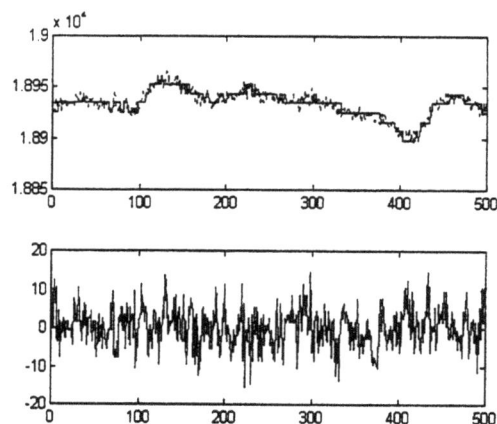

Fig.4. Markov simulation of high pressure shaft speed with maximal probability value (top) compared with real data (dotted line) and the error of simulation (bottom).

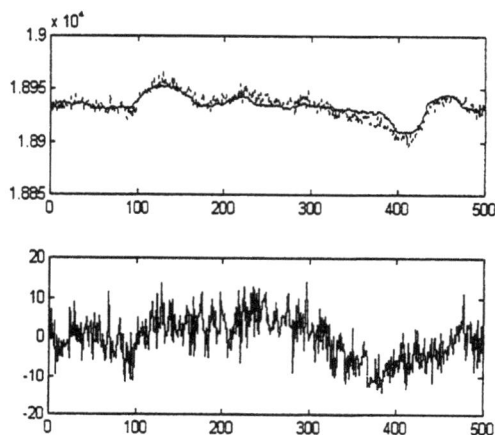

Fig.5. RBF-network simulation of high pressure shaft speed (top) compared with real data (dotted line) and the error of simulation (bottom).

The possibility of the engine real-time simulation on its Markov model was investigated. It was shown that the engine Markov model can be used as a third informational channel for detecting of which of the engine control system sensor is faulty or about a fault within the engine itself. An advantage of the proposed method is its simplicity. This technique can be applied within present condition monitoring systems without any modification to the equipment.

Acknowledgement. This work was partially supported by NATO Linkage Grant HTECH.LG 970611.

6. REFERENCES

Basseville, M. (1988). Detecting changes in signals and systems – a survey. *Automatica*, 24 (3), 309-326.

Breikin, T.V., V.Y.Arkov and G.G.Kulikov (1997a). On stochastic system identification: Markov models approach. *Proc. 2nd Asian Control Conf. ASCC'97*, Vol.2, pp. 775-778.

Breikin, T.V., V.Y.Arkov, G.G.Kulikov, V.Kadirkamanathan and V.C.Patel (1997b). On gas turbine engine and control system condition monitoring, *Prepr. IFAC Symposium on Fault Detection, Supervision and Safety for Technical Processes SAFEPROCESS'97*, Vol. 1, pp. 66-70.

Goodwin, G.C. and K.S.Sin (1984). Adaptive Filtering, Prediction and Control. *Prentice-Hall*: Englewood Cliffs, N.J.

Green, M.D., A.Duyar and J.S.Litt (1997). Model-based fault diagnosis for turboshaft engine. *Prepr. IFAC Symposium on Fault Detection, Supervision and Safety for Technical Processes SAFEPROCESS'97*, Vol. 1, pp. 71-76.

Isermann, R. (1984). Process fault detection based on modelling and estimation methods – a survey. *Automatica*, 20 (4), 387-404.

Isermann, R. (1993). Fault diagnosis of machines via parameter estimation and knowledge processing – tutorial paper. *Automatica*, 29 (4), 815-835.

Kulikov, G.G., V.Y.Arkov and T.V.Breikin (1997a). On condition monitoring of FADEC information channels (in Russian). *Izvestiya vuzov. Aviatsionnaya technika*, 4, 75-79.

Ljung, L. (1987). System Identification: Theory for User. *Prentice-Hall:* Englewood Cliffs, N.J.

Soderstrom, T. and P.Stoica (1989). System Identification. *Prentice Hall Intern.*: London.

Tugnait, J.K. (1995). Techniques for stochastic system identification with noisy input and output system measurements. *Control and dynamic systems*, 73, 41-88.

PARALLEL IMPLEMENTATION OF CONTROL ALGORITHMS

D. N. Ramos-Hernandez ‡ M. O. Tokhi ‡ J. M. Bass 1 and A. R. Browne ‡

‡ *The University of Sheffield, UK. Tel: +44 (0)114 222 5236.*
email: ramos@acse.shef.ac.uk
1 *The University of Hertfordshire, UK. Tel: +44(01707) 284172.*

Abstract: This paper presents an investigation of two approaches to parallel
implementation of an LMS adaptive algorithm using several development platforms.
The first approach is a modification of previously proposed parallel implementation of
the algorithm and the second is the parallel implementation using the Development
Framework. The target architecture consists of a two processor message-passing
system. A comparative evaluation of the two approaches is presented on the basis of
the execution time of the algorithm. *Copyright © 1998 IFAC*

Keywords: Parallel processing, adaptive digital filters, multiprocessor, control
systems.

1. INTRODUCTION

In control systems domain, algorithms are often
constructed from a large number of computationally
undemanding tasks with, in comparison, significant
communication demand. The combination of these
characteristics and the fast cycle times required often
lead to parallel processing being intractable for
algorithms in this form. In practice single processor
solutions are often favourable for purposes of
simplicity and for the elimination of inter-processor
communication. Addressing the imbalance of
computation and communication demand has been
found to be challenging as a number of highly
interacting issues are responsible for this
performance degradation. However, tractable parallel
solutions for control algorithms can be afforded with
careful consideration during the development of a
parallel implementation. For example, the use of
well-understood modern control theory permits the
restructuring and tuning of an algorithm to the
hardware (Baxter et. al, 1994; Baxter et. al, 1995a).

To effectively exploit parallel architectures,
hardware demands of the algorithms and the
available hardware resources must be closely
matched. An approach to this involves the profiling
of the application algorithm to identify the demands,
and the characterisation of each of the processors
available to build the architecture (Ghafoor, 1993;
Baxter et. al, 1995b). The above approach, however,
can be simplified providing the application
algorithms are primarily constructed from a small
number of computational block types. For example,
transfer functions, state-space equations,
multiplexes, unit-delays, integrators, gains and
adders are all commonly found in control
algorithms. Digital signal processing algorithms are
also, in general, built from a small number of blocks
and this characteristic can be found in a broad range
of application domains.

Adaptive digital filters have achieved a widespread
acceptance in the past decade and are now included

167

in many application areas. One of the most popular and widely used of these algorithms is the least mean squares (LMS) algorithm. Due to its simplicity, the LMS algorithm has been applied to such diverse areas as channel equalisation, noise cancellation, narrow-band information signal enhancement, and echo cancellation, as well as many others (Miller et al., 1986). Whereas the concept of applying LMS to a particular application is often straightforward, in many cases the hardware or software implementation is somewhat more difficult. In the interest of distributing the computational load and increasing the computational speed, one is often interested in applying multiprocessor configurations and arithmetical speedup techniques. However, when multiprocessors are considered, one of the more difficult tasks is solving the timing, or latency problem, in which the filter prediction error must be fed back to all previous filter stages so that the LMS adaptive filter coefficients may be updated (Lawrence and Tewksbury, 1983; Miller et al., 1986). This requirement complicates the multiprocessor implementation of the LMS adaptive filter, since the output must be fed back to the processors responsible for updating the filter coefficients. Therefore, LMS architectures which achieve this output feedback in an efficient manner are very desirable. An SIMD architecture which solves the output feedback problem inherent in the LMS adaptive algorithm has previously been reported (Miller et al., 1986). Recently, it has been shown that it is possible to introduce a fixed delay in the coefficient update equation of the LMS algorithm. In 1989, Long et al. presented the delayed least mean squares (DLMS) algorithm, which uses a delayed prediction error signal to update the filter weights. Later, in 1993 Meyer and Agrawal, proposed a systolic implementation of the DLMS algorithm, which provides a significant computational speedup over single-processor LMS filters. This paper will continue with the description of the LMS algorithm. Subsequently, the parallel approaches will be described. Finally, the results of implementation of the algorithm using these approaches will be given, and the concluding remarks will be presented.

2. APPLICATION ALGORITHM

The LMS adaptive filter algorithm was developed by Widrow and his co-workers (Widrow et al., 1975). It is based on the steepest descent method where the weight vector is updated according to

$$W_{j+1} = W_j + 2e_j \mu X_j \qquad (1)$$

where:

W_j = weight vector.

X_j = input signal vector.

μ = constant that controls the stability and rate of convergence.

and the error is given by

$$e_j = y_j - W_j^T X_j$$

y_j = current contaminated signal sample.

A block diagram representation of the algorithm is shown in Fig. 1.

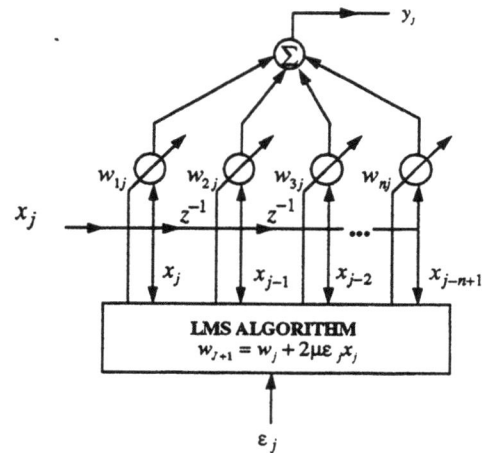

Fig. 1 The LMS adaptive filter (Block diagram)

3. PARALLEL IMPLEMENTATIONS

This research has investigated two parallel approaches of the LMS algorithm. The first approach consists of the implementation of the algorithm on two INMOS T805 (T8) processors. This approach is based on the work proposed by Miller et. al, (1986) with a modification that decreases the execution time of the parallel implementation of the algorithm. In the second approach, the LMS algorithm was implemented in Simulink (The MathWorks, 1992) and later in the Development Framework, which translated the LMS algorithm into several tasks.

3.1 Two processor implementation.

For implementing the LMS algorithm on two processors ($M = 2$), the N-vectors $x(j)$ and $w(j)$ are partitioned among the processors (N / M). This partitioning requires interprocessor communication

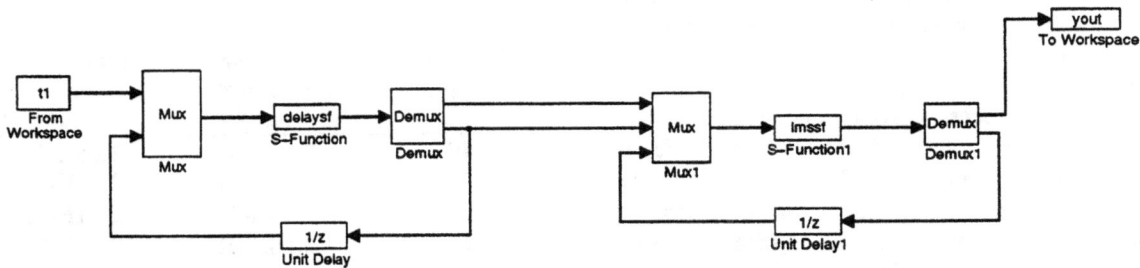

Fig. 3 Simulink diagram of the LMS algorithm

at three levels: (1) computation of the inner product $w(j)*x(j)$; (2) feedback of the output error signal $e(j)$; (3) update of the input element $x\left[\dfrac{N}{2}-1\right]$. This approach was previously implemented by Miller et. al, (1986) for an SIMD implementation, where the input vector $x(j)$ is updated. In this paper only one element of this vector is updated which makes the implementation of the algorithm faster. This is shown in Fig. 2.

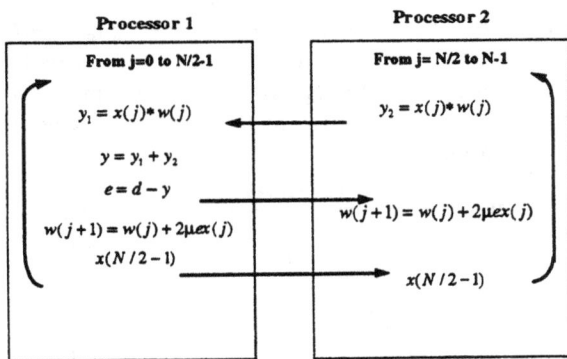

Fig. 2. Implementation with two processors (Block diagram).

3.2 Framework implementation.

Control system algorithms are typically constructed out of a number of computational blocks with data flowing between them. This can be viewed as a data-flow diagram. The usual approach to implementing such an algorithm on a parallel architecture is through algorithmic parallelism. Each computational block is assigned to a processor of the architecture and the communications between blocks are routed via communications channels, either on chip- or through hardware links to other processors.

Firstly, the application algorithm is implemented in Simulink (The Mathworks, 1992). This is shown in Fig. 3. Then, the algorithm is imported into the Development Framework (Bass et. al, 1994; Hajji et. al, 1997) where it is represented as a data-flow diagram (DFD), see in Fig. 4. The algorithm in this form is architecture independent and each node in the DFD represents a block of the algorithm. The DFD is generated using the Framework Information Interchange (FII) which provides data storage and could be seen as a graphical structure, where it is possible to identify three types of elements: nodes that represent processes and states; arcs that represent data flows and transitions; and diagrams that are combination of the above. The mapper subsequently appends an annotation to the DFD, of the task to processor mapping, and determines the task execution order. From this information the code generator creates the source code for the implementation which is then compiled. There are several code generators which exist for the different forms of hardware architectures. These code generators have been designed to produce code for different processes. This code is generated in C language. In addition, code generators exist for both single processor UNIX workstations and heterogeneous architectures consisting of mixed networks of transputers, Texas Instruments C40 DSPs, and Intel i860 purposes. The code generator for transputer-based architectures produces code which works in conjunction with a micro-kernel specially developed to allow the implementation of systems using the semantics of the Development Framework (Browne, 1996). This micro-kernel provides interprocess communication service and a priority based scheduler, which allow processes to be scheduled. The actual code produced for each system is split into two categories, application code and harness code.

169

Application code is used to implement the specified functionality of each process in the system and it is architecturally independent. Harness code is used to specify the scheduling, timing, control, and synchronisation properties of the processes and to handle inter-process communication. Unlike application code, much of the harness code is architecturally dependent. The code generator utilises a library of code fragments, described as templates, which represent the typical building blocks of the application domain for the generation of the source code (Browne et. al, 1997). Due to the template selection library it is still in an early stage of development. It was extended to support modules such as multiplex, demultiplex and unit-delays with several inputs and outputs. Also, this library was extended with S-Functions which has the main code for the LMS algorithm. The code generated was compiled and linked with ANSI C and the execution time was taken for each process.

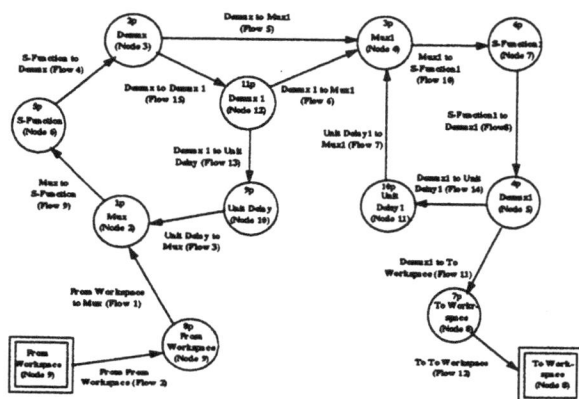

Fig. 4. Data flow diagram of the converted LMS algorithm

4. RESULTS

In implementing the LMS algorithm in this investigation the number of weights were varied from 10 to 200, in increments of 10. The operation at each step involved 1000 data points. The (convergence rate parameter) μ = 0.05, was utilised throughout.

Table 1 shows the execution times for the LMS algorithm with the Development Framework. The second column shows the execution time without considering communication and scheduling overhead, and in the third column these are considered. The time was obtained with increasing the number of weights from 10 to 200, in increments of 10.

Table 1 The execution times for the LMS algorithm (Framework implementation).

Number of weights	Processes execution time (sec)	Total execution time (sec)
10	5.58E-02	1.63E-01
20	1.11E-01	2.18E-01
30	1.64E-01	2.71E-01
40	2.18E-01	3.26E-01
50	2.70E-01	3.78E-01
60	3.25E-01	4.32E-01
70	3.82E-01	4.89E-01
80	4.36E-01	5.43E-01
90	4.89E-01	5.97E-01
100	5.44E-01	6.51E-01
110	5.97E-01	7.05E-01
120	6.51E-01	7.59E-01
130	7.12E-01	8.19E-01
140	7.67E-01	8.74E-01
150	8.21E-01	9.28E-01
160	8.76E-01	9.83E-01
170	9.29E-01	1.04E+00
180	9.75E-01	1.08E+00
190	1.04E+00	1.14E+00
200	1.09E+00	1.20E+00

Fig. 5 shows the execution times of the parallel implementation of the LMS algorithm on two T8 processors. The implementation on one transputer is also shown. It is noted that the parallel implementation of the algorithm is faster than single processor implementation, as the number of weights are increased.

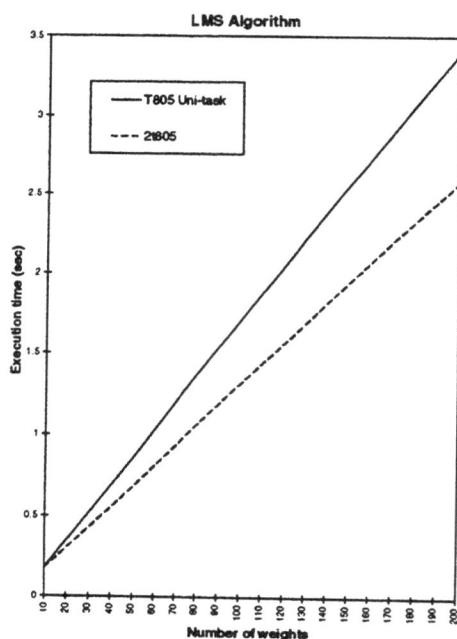

Fig. 5. Execution times of the LMS algorithm with two T8 processors.

Fig. 6 shows the implementation of the LMS algorithm with one and two transputers using the Framework approach. The execution times in these implementations are compared with the parallel representation of the algorithm obtained with the Development Framework. It is noted that the framework implementation is considerably faster than the two transputers implementation and than the single transputer implementation, as the number of weights are increased. This indicates that better performance can be achieved if an algorithm is partitioned in this form.

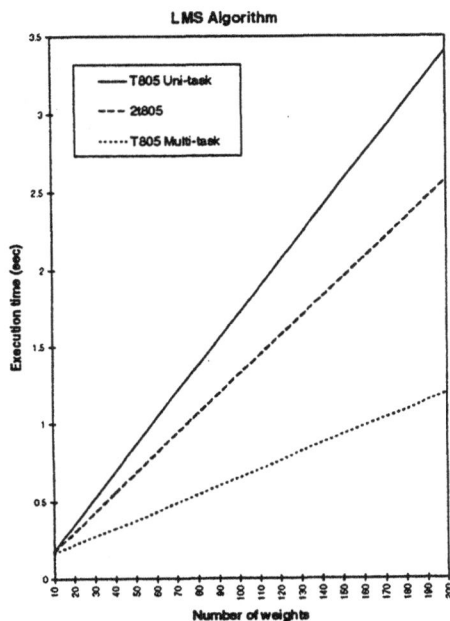

Fig. 6. Execution times of the LMS algorithm with one and two T8s, and the Framework implementation.

5. CONCLUDING REMARKS

Parallel implementations of an LMS adaptive algorithm have been presented. Two approaches, namely two processor and the Development Framework implementation have been used. The first approach can be seen as a direct method, where the task size of the algorithm is distributed onto all the processors. The second approach is not a straightforward method. Firstly it is necessary to have the representation of the algorithm in Simulink and then follow several steps to produce the code which represent the algorithm. However, the results obtained using the Development Framework are faster. Therefore, it should be worthwhile to experiment with other algorithms and with different types of processors including heterogeneous architectures.

ACKNOWLEDGEMENTS

The authors acknowledge the financial support of CONACYT-MEXICO and the UK EPSRC (Grant No. GR/K 64310).

REFERENCES

Bass, J.M., Browne, A.R., Hajji, M.S., Marriott, D.G., Croll, P.R., Fleming, P.J. (1994). Automating the development of distributed control software, *IEEE Parallel & Distribuited Technology*, **2** (4), Winter 1994, pp. 9-19.

Baxter, M.J., Tokhi, M.O., Fleming, P.J. (1994). Parallelising algorithms to exploit heterogeneous architectures for real-time control systems, *Proceedings of the IEE Control'94 Conference*, **2**, pp. 1266-1271.

Baxter, M.J., Tokhi, M.O., Fleming, P.J. (1995a). Parallelising and developing control algorithms for heterogeneous architectures. *Research report N. 597. Department of Automatic Control and Systems Engineering, The University of Sheffield*, August 1995.

Baxter, M.J., Tokhi, M.O., Fleming, P.J. (1995b). Mapping real-time control algorithms onto heterogeneous architectures, *Preprints of the IFAC Workshop on Algorithms and Architectures for Real-Time Control*, pp. 563-568.

Browne, A.R. (1996). PhD Thesis: *Automating the Development of Real-Time Control System Sotfware. Departement of Automatic Control and Systems Engineering, University of Sheffield.*

Browne, A. R., Bass, J. M., and Fleming, P. J., (1997). A building block approach to the temporal modeling of control software, *IFAC 4th Workshop on Algorithms and Architectures for Real-Time Control*, Vilamoura, Portugal, April 1997, pp. 433-438.

Ghafoor, A. (1993). A distributed heterogeneous supercomputing management system, *IEEE Computer*, June, pp. 78-86.

Hajji, M. S., Bass, J. M., Browne, A. R. and Fleming, P. J., (1997). Design tools for hybrid control systems, *Int. Workshop on Hybrid & Real-Time Systems, HART'97*, Oded Maler (ed.), Grenoble, France, LNCS 1201, Springer-Verlag, March 1997, pp. 87-92.

Lawrence, V. B. and Tewksbury, S. K. (1983). Multiprocessor implementation of adaptive digital filters, *IEEE Trans. Commun.*, **31** (1), pp. 826-835, June 1983.

Long, G., Ling, F., and Proakis, J. G. (1989). The LMS algorithm with delayed coefficient adaptation, , *IEEE Trans. Accost. Speech*,

Signal Processing, **37**, pp. 1397-1405, September 1989.

Miller, T. K., Alexander, S. T., and Faber L. J. (1986). An SIMD multiprocessor ring architecture for the LMS adaptive algorithm, *IEEE Trans. Commun.*, **34** (1), pp. 89-92, January 1986.

Meyer, M. D. and Agrawal, D. P. (1993). A high sampling rate delayed LMS filter architecture, *IEEE Trans. on Circuits and Systems-II: Analog and Digital Signal Processing*, **40** (11), pp. 727-729, November 1993.

The MathWorks Inc., (1992). SIMULINK User's Guide, pp.2.66-2.90.

Widrow, B., Glover, J.R., Mccool, J.M.,Kaunitz, J.,Williams, C.S., Hearn, R.H., Zeidler, J.R., Dong, E. and Goodlin, R.C. (1975). Adaptive noise cancelling: principles and applications. Proceedings of IEEE, 63, pp. 1692-1696.

PERFORMANCE EVALUATION OF HETEROGENEOUS ARCHITECTURES

D. N. Ramos-Hernandez and M. O. Tokhi

*The University of Sheffield, UK. Tel: +44 (0)114 222 5236.
email: ramos@acse.shef.ac.uk, o.tokhi@sheffield.ac.uk.*

Abstract: An investigation into the performance evaluation of heterogeneous architectures is presented in this paper. A task allocation strategy for maximum efficiency is proposed and evaluated. Compiler efficiency and code optimisation are also investigated in context of performance of an architecture. An adaptive filtering algorithm and a beam simulation algorithm are considered. These are implemented on a heterogeneous architecture incorporating the TMS320C40 digital signal processing devices, the Intel i860 vector processor and the T805 Inmos transputers. *Copyright © 1998 IFAC*

Keywords: Heterogeneous architectures, parallel processing, real-time signal processing and control, speedup, task allocation.

1. INTRODUCTION

The general guideline for the parallel real-time realisation of control-intensive algorithms is to increase the flexibility in hardware parallelism and to exploit software parallelism. Hardware and software design trade-offs also exist in terms of cost, complexity, expandability, compatibility, and performance. Compiling for multiprocessors is much more involved than for uni-processors (Hwang K., 1993).

Digital signal processing (DSP) devices are designed in hardware to perform concurrent add and multiply instructions and execute irregular algorithms efficiently. The Intel i860 vector processor, for instance, has been designed for high performance floating point computation and is able to efficiently process regular algorithms involving matrix manipulations. Purpose built PEs are not enough to bridge the ever-increasing software/hardware gap.

Alternative strategies where multi-processor based systems are employed, utilising high performance reduced instruction set computer (RISC) processors, DSP devices, transputers and parallel processing (PP) techniques, could provide suitable methodologies.

For PP with widely different architectures and different PEs, performance measurements such as MIPS, MOPS and MFLOPS of the PEs are meaningless. It is more important to rate the performance of each architecture with its PEs on the type of program likely to be encountered in a typical application. The different architectures and their different clock rates, inter-processor communication speed, etc. all confuse the issue of attempting to rate the architecture. This is an inherent difficulty in selecting a parallel architecture for better performance for algorithms in signal processing and control system development applications. The ideal performance of a parallel architecture demands a perfect match between the capability of the architecture and the program behaviour. Capability of the architecture can be enhanced with better hardware technology, innovative architectural features and efficient resources management. In contrast, program behaviour is difficult to predict due to its heavy dependence on application and run-time conditions. Moreover, there are many other factors that influence program behaviour. These include algorithm design, partitioning and mapping of the algorithm, inter-processor communication, data structures, language efficiency, programmer skill, and compiler technology.

Few investigations have been made to propose performance metrics for heterogeneous architectures (Yan et al., 1996; Zhang and Yan, 1995). However, these do not fully exploit the capability of all the PEs in the architecture. Tokhi et al. (1997) developed several performance metrics for sequential and parallel architectures, which ensure that the capabilities of the processors are exploited by maximising the efficiency of the architecture.

Compilers have a significant impact on the performance of the system (Hwang, 1993; Tokhi et al, 1996). The power of a parallelising compiler is derived from its ability to analyse loops and the precision of its dependence analysis. Several compiler optimisation techniques have been employed in parallelising compilers to eliminate some of the dependencies in the programs and make them parallelisable (Haghighat et al, 1992).

Performance is related to code optimisation facility of the compiler, which may be machine dependent. The goal of program optimisation is, in general, to maximise the speed of code execution. This involves several factors such as minimisation of code length and memory accesses, exploitation of parallelism, elimination of dead code, in-line function expansion, loop unrolling and maximum utilisation of registers. The optimisation often transforms a code into an equivalent but "better" form in the same representation language (Hwang, 1993). This paper continues with the description of the hardware architecture in Section 2. The application algorithms are presented in Section 3. The performance metrics used are described in Section 4. Finally, implementation results are given in Section 5 and the paper concluded in Section 6.

2. HARDWARE ARCHITECTURE

A heterogeneous architecture consists of various types of processors. The hardware used to carry out the experimental investigations is shown in Fig. 1. This comprises a Transtech TMB08 motherboard with two Inmos T805 (T8) transputers. The root T8

is connected to the Host computer and via its link 3 to the Transtech TMB16 motherboard that contains the Intel i860 RISC processor (i860). The i860 TRAM has 16 Mbytes shared memory and a T805 with 4 Mbytes of local memory. The second transputer on the TMB08 motherboard is connected to the root transputer via its link 1 and to a sub-network of three Texas Instruments TMS320C40 (C40) DSP devices.

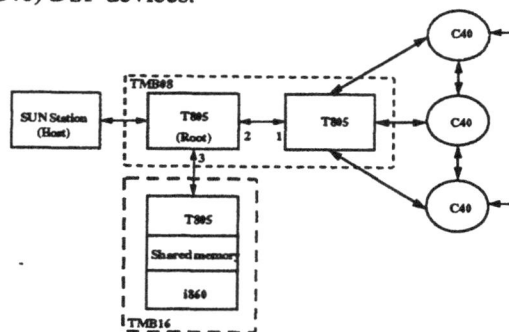

Fig. 1. The topology of the heterogeneous architecture.

3. APPLICATION ALGORITHMS

The first algorithm considered is the least mean square (LMS) adaptive filter. This is of irregular nature, with uneven inner and outer loops and many multiply and accumulate functions. The second algorithm is a cantilever beam simulation algorithm. This is of regular nature and is based on matrix multiplication and addition.

3.1 LMS algorithm

The LMS adaptive filter algorithm was developed by Windrow and his co-workers (Widrow et al., 1975). It is based on the steepest descent method where the weight vector is updated according to

$$W_{k+1} = W_k + 2e_k\mu X_k \qquad (1)$$

where W_k = weight vector, X_k = input signal vector and μ = constant that controls the stability, and the error is given by $e_k = y_k - W_k^T X_k$.

3.2 The Beam simulation algorithm

Consider a cantilever beam system with a force $U(x,t)$ applied at a distance x from its fixed end at time t . This force produces a deflection $y(x,t)$ of the beam from its stationary position at the point where the force has been applied. The dynamic equation that represent this system is

174

$$\mu^2 \frac{\partial^4 y(x,t)}{\partial x^4} + \frac{\partial^2 y(x,t)}{\partial t^2} = \frac{1}{m} F(x,t) \qquad (2)$$

where μ is a beam constant and m is the mass of the beam. Using a finite difference discretisation of the dynamic equation of the beam in time and distance yields (Tokhi et al., 1994)

$$Y_{k+1} = -Y_{k-1} - \lambda^2 S Y_k + \frac{(\Delta t)^2}{m} U(x,t) \qquad (3)$$

where S is a pentadiagonal matrix, entries of which depend on the physical properties and boundary conditions of the beam, Y_i $(i = k+1, k, k-1)$ represents the beam deflection at time step i, and Δt and Δx are increments along time and distance coordinates respectively, $\lambda^2 = (\Delta t)^2 (\Delta x)^{-4} \mu^2$.

4. PERFORMANCE METRICS

The performance of a processor, as execution time, in implementing an application algorithm generally evolves linearly with the task size (Tokhi et al., 1996a; Tokhi et al., 1996b). With some processors, however, anomalies in the form of change of gradient (slope) of execution time to task size are observed. These are mainly due to run-time memory management conditions of the processor where, up to a certain task size the processor may find the available cache sufficient, but beyond this it may require to access lower level memory. Despite this the variation in the slope is relatively small and the execution time to task size relationship can be considered as linear. This means that a quantitative measure of performance of a processor in an application can adequately be given by the average ratio of task size to execution time or the average speed. Alternatively, the performance of the processor can be measured as the average ratio of execution time per unit task size, or the average (execution time) gradient. In this manner, a generalised performance measure of a processor relative to another in an application can be obtained.

To obtain a comparative performance evaluation of a uni-processor and multi-processor for an application under various processing conditions, for example with and without code optimisation, the concept of generalised sequential speedup and task allocation theory can be applied. This concept is based on providing a maximum efficiency and maximum speedup in parallel architectures. A virtual processor is defined as a reference node that represent the characteristics of all the processing elements in a heterogeneous architecture. Let the generalised sequential speedup of processor i (in a parallel architecture) to the virtual processor be $S_{i/v}$;

$$S_{i/v} = \frac{V_i}{V_v} ; \quad i = 1, ..., N \qquad (4)$$

Equation (4) can alternatively be expressed in terms of fixed-load execution time increments as

$$S_{i/v} = \frac{\Delta T_v}{\Delta T_i} ; \quad i = 1, ..., N \qquad (5)$$

Thus, to allow the maximum utilisation of the processors in the architecture the task increments ΔW_i allocated to processors is given by

$$\Delta W_i = \frac{V_i}{V_v} \frac{\Delta W}{N} = S_{i/v} \frac{\Delta W}{N} ; \quad i = 1, ..., N \qquad (6)$$

Using equation (6), for the distribution of load among the processors, the speedup and efficiency achieved with a heterogeneous architecture with N processors are N and 100% respectively. However, real experimentation shows that due to communication overheads and run-time conditions the speedup and efficiency of the parallel architecture will be less than these values (Tokhi et. al, 1997).

It can be demonstrated that, in this manner, the architecture is conceptually transformed into an equivalent homogeneous architecture incorporating N identical virtual processors. This is achieved by the task allocation among the processors according to their computing capabilities to achieve maximum efficiency.

5. IMPLEMENTATION AND RESULTS

In implementing the LMS algorithm in these investigations the number of weights were varied from 10 to 200, in increments of 10. The operation at each step involved 1000 data points. The (convergence rate parameter) $\mu = 0.05$, was utilised throughout. For the beam simulation algorithm a cantilever beam of length L = 0.635 m and mass $m = 0.037$ kg was considered. The algorithm granularity was achieved by increasing the number of beam segments from 5 to 40, in increments of 5. The execution times of the processors were obtained over 20000 iterations at each step.

The beam simulation algorithm was only considered for the task allocation experiments. Investigations at compiler efficiency and code optimisation were carried out with both algorithms.

5.1 T8&C40 Implementation

Fig. 2 shows the execution times of the T8, C40 and the T8&C40 architectures in implementing the beam simulation algorithm. The characteristics of the virtual processor and of the corresponding theoretical T8&C40 heterogeneous architecture are also shown in Fig. 2. It is noted that among the uni-processors, the C40 is considerably faster that the

T8. The combination, thus, results in a virtual processor with characteristics closer to that of the C40. This implies that, for the T8&C40 to achieve maximum efficiency, a large proportion of the task is to be allocated to the C40. It is noted that the actual T8&C40 has performed slower than the corresponding theoretical model. This is due to communication between the processors.

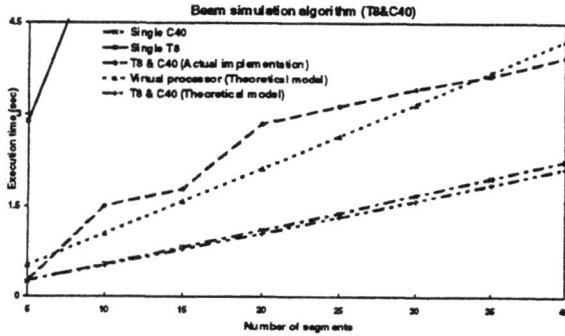

Fig. 2. Execution times of the architecture (T8&C40) implementing the beam simulation algorithm.

5.2 T8&i860 Implementation

Another heterogeneous configuration was used to implement the beam simulation algorithm. This consists of one T8 and one i860. Fig. 3 shows the execution times corresponding to this implementation. It is noted that the i860 is extremely faster than the T8. For this reason all the tasks were allocated to the i860, with the exception of the last two values (35 and 40 segments) where one segment was allocated to the T8. The performance of the actual T8&i860 is similar to the uni-processor experiment with the i860, except, for the last two values where inter-processor communication increases the execution time for the actual implementation.

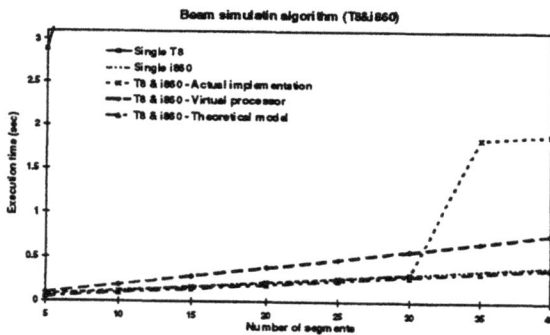

Fig. 3. Execution times of the architecture (T8&i860) in implementing the beam simulation algorithm.

5.3 C40&i860 Implementation

Fig. 4 shows the execution times of the C40, i860 and the actual, and theoretical execution times of the C40&i860 heterogeneous architecture in implementing the beam simulation algorithm. It is noted that the actual implementation is faster than the virtual processor and faster than the i860. It is also noted that the actual implementation is very close to the virtual parallel machine.

Comparing the three implementations, it is noted that with the T8&C40, the execution times of the actual implementation stay very close to the virtual processor. Task allocation was computed according to the theoretical model. However, communication overheads are evident with the T8&i860 implementation where no task is allocated to the T8 except for the last two numbers of segments. This resulted in the same execution times as the i860 processor, except for the 35 and 40 segment steps where the execution time has increased. The results of these two implementations show that due to the disparity in capabilities of the processors, communication overhead becomes a dominant factor in the implementation. The i860&C40 implementation shows that a better task allocation is obtained between both processors and communication overheads are minimum. In this manner the best performance for this algorithm is obtained with the combination of the C40&i860.

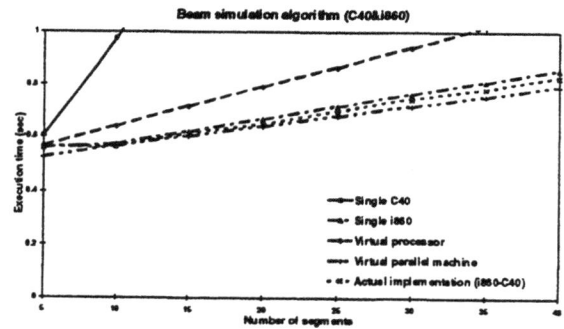

Fig. 4. Execution times of the architecture (C40&i860) implementing the beam simulation algorithm.

5.4 Compilers Efficiency

The beam simulation algorithm and the LMS algorithm were implemented for the performance evaluation of several compilers. The compilers involved are the 3L Parallel C version 2.1, Inmos ANSI C and Occam. All these compilers support PP and can be utilised with the heterogeneous architecture considered before. Occam is more hardware oriented and a straight-forward programming language for PP. It may not be as suitable as the Parallel C or ANSI C compilers for numerical computation (Tokhi et al., 1995). Both algorithms were implemented on one and two transputers to obtain a comparative performance evaluation of these compilers. Fig. 5a shows the execution times in implementing the beam simulation algorithm using the three compilers with one T8 and Fig. 5b shows the execution times with two T8s.

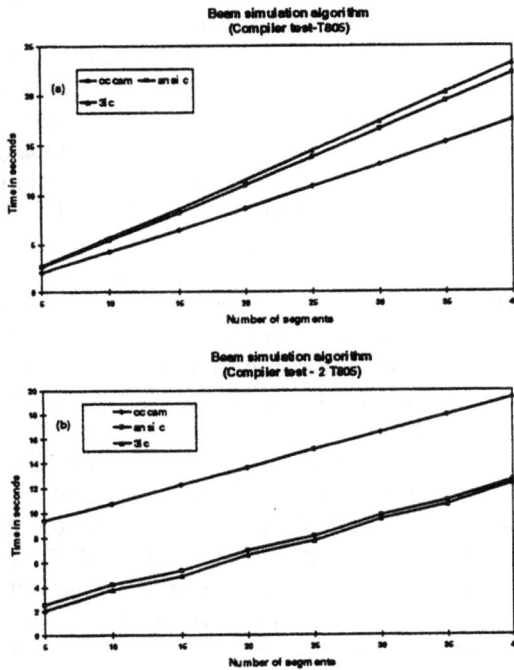

Fig. 5. Performance of the compilers in implementing the beam simulation algorithm (a) with one T8, (b) with two T8s.

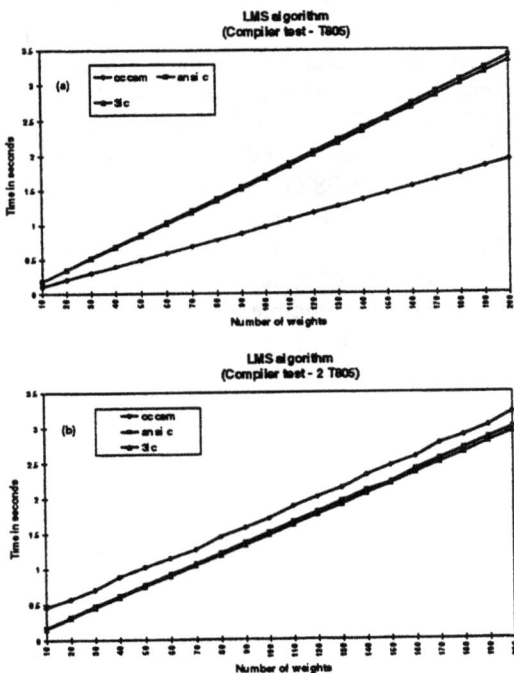

Fig. 6. Performance of the compilers in implementing the LMS algorithm (a) with one T8, (b) with two T8s.

Similarly, Fig. 6a and Fig. 6b show the execution times in implementing the LMS algorithm on single T8 and two T8s with the compilers. It is noted that better performances with Occam compiler have been achieved with single T8. However, with two T8s, Parallel C and ANSI C were more effective. This is due to large amounts of data handling, where run-time memory management problem can be solved with Parallel C and ANSI C more efficiently than with the Occam compiler (Tokhi et al, 1996a).

5.5 Optimisation

The i860 processor with its respective compiler is considered in investigating its optimisation facility. Using the optimisation levels of the i860, the beam simulation algorithm and the LMS algorithm were studied. The Portland Group (PG) C compiler is an optimising compiler for the i860. It incorporates four levels of optimisation. Level 0 causes no optimisation, Level 1 generates a fast code when the program is very irregular, i.e. it has few loops and little floating point arithmetic, Level 2, does good optimisation when the functions are short and there is a significant amount of floating point arithmetic, and levels 3 and 4 give the best performance when the code contains any conditional branches (Portland Group, 1991).

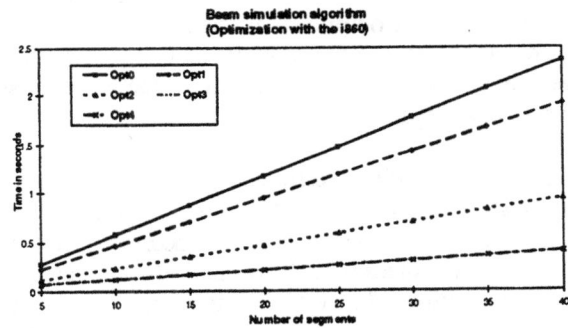

Fig. 7. Execution times of the i860 in implementing the beam simulation algorithm with the compiler optimiser.

Previous research has reported that in implementing the LMS algorithm the performance enhanced significantly with higher levels of optimisation and for the beam simulation algorithm, the enhancement was not significant beyond the first level (Tokhi et al, 1996a). The source code for both algorithms was optimised before carry out the experiments.

In the case of the beam simulation algorithm the implementation of the code used arrays with some loops, for this reason the compiler restructured the actual code. The results are show in Fig. 7. The best performance was obtained with optimisation level 3 and 4.

177

For the LMS algorithm the code has multiple nested loops so the compiler is able to recognise the structures involved and restructure the actual code. However for this algorithm few changes were done to the code. Fig. 8 shows levels 0 and four for the LMS. It can be noted that the performances with both optimisation levels are very similar.

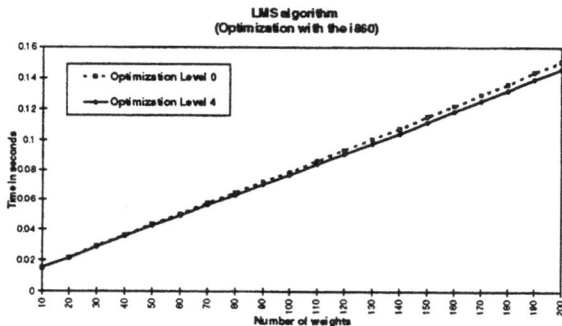

Fig. 8. Execution times of the i860 in implementing the LMS algorithm with the compiler optimiser.

6. CONCLUDING REMARKS

The effectiveness of task allocation theory developed in previous work has been demonstrated in new hardware configurations for heterogeneous architectures. Compiler efficiency and code optimisation have been investigated showing that these issues affect the performance of the processors in real-time applications.

Experiments with different compilers have demonstrated how these and different configurations modify the performance of the processors. In the optimisation experiments it was also shown how the regularity or irregularity of the algorithm, as well as the codification change the performance of the processors.

ACKNOWLEDGEMENTS

D.N. Ramos-Hernandez acknowledges the financial support of CONACYT-MEXICO.

REFERENCES

Haghighat, M. and Polychronopoulos C. (1992). Symbolic program analysis and optimization for parallelizing compilers Lecture Notes in Computer Science (757). Banerjee, U. Gelernter, D. Nicolau, A., Padua, D. (Eds).

Hwang, K. (1993). *Advanced Computer Architecture: Parallelism, Scalability, Programmability*. McGraw -Hill, Singapore.

Portland Group Inc. (1991). *PG tools user manual*, Portland Group Inc.

Tokhi, M. O. and Hossain, M. A. (1994). Self-tuning active vibration control in flexible beam structures. *Proceedings of IMechE-I: Journal of Systems and Control Engineering*, **208**, (14), pp. 263 277.

Tokhi, M. O. and Hossain, M. A. (1995). CISC, RISC and DSP processors in real-time signal processing and control. *Microprocessors and Microsystems*, **19**, (5), pp. 291-300.

Tokhi, M. O., Chambers, C. and Hossain, M. A. (1996a). Performance evolution with DSP and transputer based systems in real-time signal processing and control applications, *Proceedings of UKACC International Conference on Control-96*, Exeter, 02-05 September 1996, **1**, pp. 371-375.

Tokhi, M. O., Hossain, M. A. and Chambers, C. (1996b). *Performance evaluation in sequential real-time processing*. Research Report No. 645, Department of Automatic Control and Systems Engineering, The University of Sheffield, UK, October 1996.

Tokhi, M. O., Ramos-Hernandez D.N., Chambers, C. and Hossain, M. A. (1997). *Performance evaluation of DSP, RISC and transputer based systems in real-time implementation of signal processing and control algorithms*. The 4th IFAC Workshop on Algorithms and Architectures for real-time Control. Vilamoura, Portugal, 9-11 April 1997, pp. 287-292.

Yan, Y., Zhang, X. and Song, Y. (1996). An effective and practical performance prediction model for parallel computing on nondedicated heterogeneous NOW, *Journal of Parallel and Distributed Computing*, **38**, (1), pp. 63-80.

Zhang, X. and Yan, Y. (1995). Modeling and characterizing parallel computing performance on heterogeneous networks of workstations. *Proceedings of Seventh IEEE Symposium on Parallel and Distributed Processing*, San Antonio, Texas, 25-28 October 1995, pp. 25-34.

Widrow, B., Glover, J.R., Mccool, J.M., Kaunitz, J., Williams, C.S., Hearn, R.H., Zeidler, J.R., Dong, E. and Goodlin, R.C. (1975). Adaptive noise cancelling: principles and applications. Proceedings of IEEE, 63, pp. 1692-1696.

CONFIGURABLE PROCESSING FOR REAL-TIME SPECTRAL ESTIMATION

Margarida Madeira [*,1] **Stephen Bellis** [**] **Graça Ruano** [*,1]
William Marnane [***]

[*] *UCEH/SEC, Universidade do Algarve, Campus de Gambelas,
8000 FARO, Portugal*
[**] *National Microelectronics Research Centre, Prospect Row,
CORK, Ireland*
[***] *Department of Electrical Engineering and Microelectronics,
University College Cork, CORK, Ireland*

Abstract: This paper presents a system for real-time implementation of the fourth
order Modified Covariance spectral estimator, which, when used in conjunction with
pulsed Doppler blood flow detectors, has been shown to offer increased sensitivity
in atherosclerotic disease detection. The computational burden incurred with the
Modified Covariance method is considerably greater than that of the conventional
FFT method. This has led to separate studies to evaluate the cost and performance
of firstly, transputer/DSP based platforms and secondly, application specific custom
circuitry, for implementation of the algorithm in real-time. The advantages and
disadvantages of each of these different approaches are reviewed in this paper resulting
in the design of a combined custom/DSP based system. *Copyright © 1998 IFAC*

Keywords: Spectral estimation, DSP, FPGA, Heterogeneous architectures, Parallel
processing

1. INTRODUCTION

The growing need for computing resources has
been reflected by the emergence of parallel machines that are more powerful, faster and have
larger memories. However, increases in processing
capacity have not been fully transfered to the
effective computational throughput as parallelism
introduces the extra requirement of communication. The performance of the communication bus
restrains the overall system performance, which in
turn restricts the range of problems that can be
solved in real-time.

Advances in VLSI technology have resulted in two
developments of importance to parallel processing. The first is the availability of microprocessors
which have fast communication links to support
MIMD parallel processing. The other is in the area
of synthesis of algorithms into custom VLSI using systolic array processors. The implementation
route chosen, MIMD or VLSI, has its own advantages and disadvantages; ranging from flexibility
but with difficulties in programming in the MIMD
case, to high speed but with little re-use for the
VLSI solution. An alternative is available, Field
Programmable Gate Arrays (FPGAs) that possess the flexibility of the multiprocessor solution
and approach the speed of the custom solution.

Biomedical Engineering and in particular Doppler
ultrasound blood flow estimation require a real-
time implementation. Previous work in this field

[1] Financial support from Treaty of Windsor (B-115/96)
and JNICT (PBIC/2414/95).
[2] The authors gratefully acknowledge M. O. Tokhi and P.
J. Fish for their valuable contribution.

reported the implementation of a fourth order Auto Regressive Modified Covariance (AR/MC) spectral estimator on several different processing systems, such as homogeneous networks (Ruano et al., 1992) (Ruano et al., 1993), that is platforms containing either only transputers or DSPs, which enabled the possibility of consecutive spectral estimation of cardiac data segments. Heterogeneous parallel architectures, platforms combining both transputers and DSPs have also been considered (Madeira et al., 1997) (Nocetti et al., 1995). It has also been shown to be feasible to implement the AR/MC algorithm by synthesising systolic architectures onto application specific custom circuitry (Bellis et al., 1997a).

Through the analysis of the results achieved in the above mentioned studies, a strategy for the AR/MC spectral estimator using mixed technology has been studied. The complete system has been developed in a modular approach to enable the best partial solutions identified in each study, chosen by consideration of the suitability of the architectures to fulfill the computational requirements of functional blocks, to be merged. This paper reports the combined usage of custom pipelined processing, using field programmable gate arrays (FPGA's) and a software application implemented on DSPs.

2. MODIFIED COVARIANCE SPECTRAL ESTIMATION

The general description of the model order p AR/MC spectral estimator involves solving the following linear system of equations

$$
\begin{bmatrix}
c_{xx}[1,1] & c_{xx}[1,2] & \dots & c_{xx}[1,p] \\
c_{xx}[2,1] & c_{xx}[2,2] & \dots & c_{xx}[2,p] \\
\vdots & \vdots & \ddots & \vdots \\
c_{xx}[p,1] & c_{xx}[p,2] & \dots & c_{xx}[p,p]
\end{bmatrix}
\begin{bmatrix}
\hat{a}[1] \\
\hat{a}[2] \\
\vdots \\
\hat{a}[p]
\end{bmatrix}
= -
\begin{bmatrix}
c_{xx}[1,0] \\
c_{xx}[2,0] \\
\vdots \\
c_{xx}[p,0]
\end{bmatrix}
\quad (1)
$$

where each element $c_{xx}[i,j]$ of the covariance matrix and right-hand-side vector is obtained from

$$
c_{xx}[i,j] = \frac{1}{2(N-p)} \cdot (c1_{xx}[i,j] + c2_{xx}[i,j]) \quad (2)
$$

where

$$
c1_{xx}[i,j] = \sum_{n=p}^{N-1} x^*[n-i].x[n-j] \quad (3)
$$

and

$$
c2_{xx}[i,j] = \sum_{n=0}^{N-1-p} x[n+i].x^*[n+j] \quad (4)
$$

For the blood flow application, $p = 4$ is the optimal model order (Ruano and Fish, 1993) and the number of samples N, over the fixed time

duration window of 10ms, is either 64, 128, 256 or 512.

The $\hat{a}[k]$ ($k = 1, 2, ..., p$) filter parameter estimates are obtained by solution of the linear system (1), using the Cholesky, forward elimination and back substitution algorithms.

The signal white noise variance estimate, $\hat{\sigma}^2$, is calculated as

$$
\hat{\sigma}^2 = c_{xx}[0,0] + \sum_{k=1}^{p} \hat{a}[k] \cdot c_{xx}[0,k] \quad (5)
$$

and the power spectral density (PSD), $\hat{P}_{MC}(f_n)$, is obtained through

$$
\hat{P}_{MC}(f_n) = \frac{\hat{\sigma}^2}{|A(f_n)|^2} = \frac{\hat{\sigma}^2}{\left| 1 + \sum_{k=1}^{p} a[k] z^{-k} \big|_{z=e^{j2\pi f_n}} \right|^2} \quad (6)
$$

Hence, the AR/MC spectral estimator may be partitioned onto four different programming modules: calculation of the elements of the covariance matrix and right-hand side vector - module RMC; the solution of the linear system of equations - module Cholesky; the calculation of the white noise variance - module WNV; and the computation of the power spectral density - module PSD.

Implementing the algorithm in Matlab, using an Intel 80486 as a sequential processor reference, allows the computational requirements of the complete algorithm and its modular blocks to be evaluated in terms of floating point operations (FLOPS) (see Tables 1 and 2). Table 1 also presents the CPU time, in seconds, consumed by the processor while computing the complete algorithm for different data segment lengths.

Table 1. Matlab Implementation: number of floating point operations and CPU time (s) elapsed

N	Flops	CPU Time
64	16397	2.4
128	33357	4.74
256	67277	9.44
512	135177	18.9

Table 2. Number of floating point operations per module

N	RMC	Cholesky	WNV	PSD
64	9180	90	2674	4450
128	18908	90	5490	8866
256	38364	90	11122	17698
512	77276	90	22386	35362

From this study it is clear that a parallel processing solution is required to implement this system in real-time. As already stated there are two choices, a multiprocessor MIMD or a custom VLSI systolic array solution.

3. PARALLEL PROCESSING - THE MULTI-PROCESSOR SOLUTION

A previous study, comparing the performance of homogeneous and heterogeneous architectures (Madeira *et al.*, 1997) involving transputers, DSP's and a vector processor, identifies the DSP's homogeneous architecture as a cost-effective solution for the real-time implementation of the blood flow spectral estimator. For the multiprocessor solution, the topology, type of processor and number of processors necessary to get a real-time implementation, must be decided.

The multi-processor DSP architecture using Texas Instruments TMS320C40 DSP (C40) devices, is characterised by 40 MHz clock speed, 8 kbyte dedicated RAM, rated as 275 MOPS and 40 MFLOPS. Each processor has six high-speed parallel unidirectional links for communication with other processors and an asynchronous transfer rate of 20 Mbytes/sec. This architecture connects to a Sun-host using two T8s as routers and two topologies of one and three DSPs were considered. This allows one or three data segments to be estimated at the same time. Usage of the C language, with a 3L C compiler employed, enables portability among the different parallel processors used in the study.

The execution times obtained with these topologies, measured on the root T8, are summarised in Table 3. Thus the same time is required to estimate one segment on one C40 as it takes to estimate three data segments on three C40s. The results reveal a linear relationship between the execution times and the data segment length (N). It should also be noted that the execution times are substantially different when either measured on the module running on the T8 or on the C40.

Table 3. Execution times (ms) of the MC algorithm implementation on the multi-processor architecture

N	1xC40	3xC40
64	20.7	21.3
128	41.2	42.5
256	82.3	84.8
512	164.5	169.1

The usage of a root T8 limits the overall performance of the architecture. An attempt to eliminate this drawback leads to the study of a personal computer hosted DSP application. The main differences presented by the change of architecture

and the processor utilised are 50 MHz clock speed and 1 Mbyte SRAM plus 4 Mbyte DRAM. The correspondent execution times measured on the DSP are presented in Table 4 for each software module. The results show that for the higher data segment length, the most time consuming module is RMC.

The multiprocessor solution (results in Table 3), being scalable, presents the potential benefit of increasing the data throughput. A real time implementation of an algorithm can still be achievable using a scalable solution when the processing rate of an algorithm is lower than the input rate and a small latency is admissible. On the other hand, it can be noted from Table 4 that the C40 implementation without the usage of a T8 as root processor reduced the communication overhead drawback induced by the T8, reported in previous studies. The latter results motivate the use of a single PC hosted DSP topology on the configurable processing architecture proposed in this paper.

4. THE CUSTOM SOLUTION

The Modified Covariance spectral estimator can alternatively be implemented on an ASIC by designing systolic architectures for each section of the algorithm. Through pipelining and parallel processing the systolic model of computation can be used to easily achieve the throughput necessary for real-time operation. Systolic architectures, which consist of a regular, locally interconnected array of processing elements (PE's), where, upon a global clock signal, data is transferred to a neighbouring PE, are well suited to VLSI implementation. Each PE computes a simple task such as multiply-accumulate or division and the systolic model can be applied at both the word-level, to obtain hardware solutions to the algorithms under consideration, and the bit-level to replace word-parallel arithmetic units with bit-serial alternatives thus reducing IO connectivity, communication and hardware.

The data dependence graph mapping methodology can be used to design systolic arrays to perform the RMC computation (2) (Bellis *et al.*, 1994). The symmetries present within the

Table 4. Modular execution times (ms) of a PC hosted DSP system

N	RMC	Cholesky	WNV	PSD
64	1	0	3	5
128	2	0	4	6
256	4	0	5	6
512	9	1	6	8

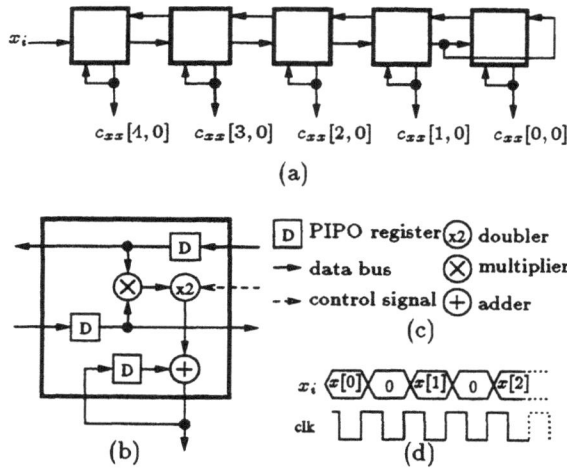

Fig. 1. (a) Linear systolic array for covariance matrix element computation, (b) PE configuration, (c) key to symbols and (d) input waveforms.

covariance matrix allow the number of elements that need to computed to be reduced to nine. Elements $c_{xx}[i, 0]$ $(0 \leq i \leq 4)$ may be computed on the linear systolic array of Fig. 1(a) where each PE performs multiply accumulate function (Fig. 1(b)). Similar arrays of length 3 and 1 are used to compute $c_{xx}[i, 1]$ $(1 \leq i \leq 3)$ and $c_{xx}[2, 2]$ but with slightly different input sequence requirements. The operation of the systolic arrays is very simple. Doppler data from the A/D converter is piped in word-parallel from left to right across the array and once this data reaches the rightmost PE's it gets piped back along each array from right to left. The product of the two data words from these streams stored in any particular PE on a certain clock cycle is then accumulated with the previous total via each PE feed-back bus. A control signal doubles the accumulated product to allow for the condition when the two multiplications in (3) and (4) are the same.

The computation of the filter parameters is far less demanding than the RMC computation for a fourth order system (Table 2). Systolic arrays for the Cholesky algorithm (Bellis et al., 1997b), require 10 PE's for decomposing a dimension 4 by 4 matrix with the PE's required to perform division and inner product step functions (Fig. 2). After decomposition the filter parameters could then be derived in conjunction with arrays for forward elimination and back substitution (Kung and Leiserson, 1979).

The WNV computation is the most simple section requiring just five multiply accumulate operations, as the custom RMC section has the capability for computation of the $c_{xx}[0, k]$ elements. The five remaining operations could be carried out on a PE similar to that in Fig. 1(b) but there is

no requirement for the multiply by two function in this case.

The bulk of the PSD computation is the Fourier Transform computation. In hardware this section could be implemented using systolic arrays for either the Discrete Fourier Transform (DFT) or the Fast Fourier Transform (FFT). The DFT implementations have the advantage of low control complexity and localised communication. However, the DFT array's computational burden is higher than that of the FFT whose implementations are less amenable to VLSI due to global communications. For the Modified Covariance application zero padding can be used to substantially reduce the hardware cost in the DFT method leading to its preference.

5. THE CONFIGURABLE SOLUTION

Table 2 shows that the RMC section presents the greatest processing burden in the AR/MC spectral estimator. It is also apparent from Table 4 that the execution time of this operation occupies the bulk of the allowable 10 ms time frame in the PC hosted DSP system for the $N = 512$ samples case. Despite this, the RMC calculation is a relatively simple series of multiply-accumulate operations and could be easily implemented in custom hardware. Also the dynamic range of such a computation is narrow making fixed point arithmetic

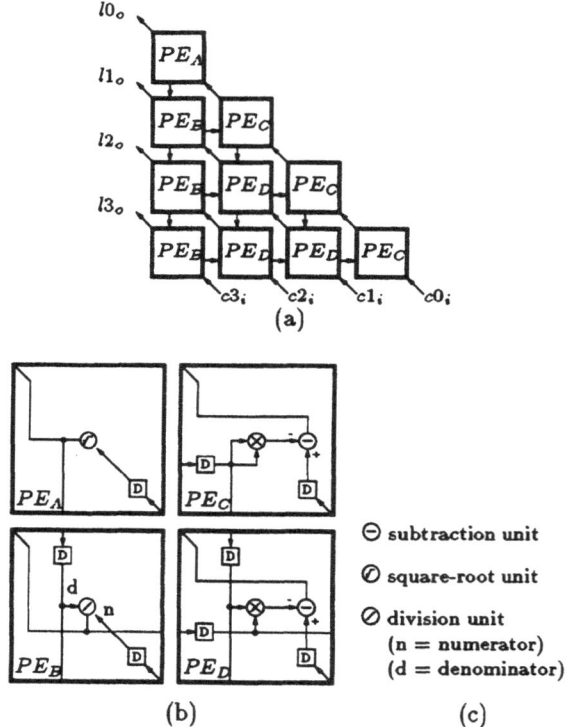

Fig. 2. (a) Brent & Luk's Cholesky decomposition systolic array (b) PE configuration, (c) symbols.

182

perfectly adequate. Therefore the high precision floating point capacity of the DSP microprocessor seems somewhat wasted, especially since analysis shows that a 12 bit fixed point data output provides enough resolution (Bellis *et al.*, 1997*a*). An important issue is the time consumed loading and storing data to and from the DSP's (Madeira *et al.*, 1997). The input of data varies from 321% to 228% and the output of the spectral estimates varies from 226% to 76% of the global computational time. In the custom solution of Fig. 1(a) the data may be input as soon as it becomes available from the A/D converter eliminating the data loading problem.

The Cholesky decomposition module presents very different requirements as opposed to the RMC section with a much reduced computational requirement. Despite this it is much harder to design a custom circuit to perform this calculation since division is required and data intercommunication is more involved. In particular there is a bottleneck between bit-serial LSB first multiplication and MSB first division (Marnane *et al.*, 1997). Although the regularity of the decomposition arrays makes them amenable to VLSI implementation, the hardware cost incurred cannot be justified for the required throughput. Speed can be traded off for area by mapping the array into a single programmable function PE. The control however would be very complex suggesting that the software implementation is more suitable for use in this application, especially since the execution times are low (Table 4).

A considerable portion of the WNV computation is taken up with computing the elements $c_{xx}[0, k]$ (5). Due to the matrix symmetries the computation of these elements can be performed on the RMC custom section leaving just a series of $p + 1 = 5$ multiply accumulate operations. These could easily be performed on an application specific circuit or in software.

One of the problems with the PSD computation is that of dynamic range resolution. Representing this computation using fixed point arithmetic would require scale factoring and bit selection procedures to maintain a reasonable word-length throughout the computation. Introduction of such procedures could lead to complex control and a detailed analysis would be required in order to determine the most appropriate scale factors and bit selections. For these reasons a floating point processor such as that used by the software implementation would be preferred for this section of the problem.

Thus the optimum implementation of each of the modules is: RMC in custom hardware, Cholesky in software, WNV in either and the PSD in software. In order to maintain the overall goal of flexibility

an FPGA is proposed to implement the custom hardware, thus giving a completely configurable solution.

An overview of the proposed configurable system is shown in Fig. 3. It consists of an A/D converter which outputs the sampled Doppler signal in bit-serial to a Xilinx 4028 which is used to perform the RMC computation (2-4) and formation of the elements $c_{xx}[0, k]$ for the WNV computation (5). The A/D is synchronised with the RMC processing by dedicating a segment of the 4028 FPGA to control. This control section also sets clock frequencies dependent upon N and enables the storage of the matrix element computations in the on board FPGA RAM. In addition the FPGA relays the system parameters and controls the loading of the computed data to the DSP card which computes the remainder of the AR/MC modules for output display on the host computer.

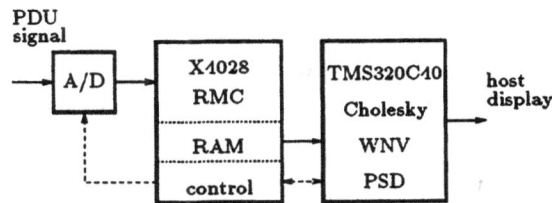

Fig. 3. Configurable system.

The longest data segment length $N = 512$ over the fixed time window of 10ms gives the maximum sampling frequency f_s=51.2kHz. To avoid having to store the data, it must be processed immediately in the 4028 FPGA as it becomes available. Due to zero interleaving (Fig. 1(d)) the maximum systolic clock frequency would therefore need to be twice f_s allowing up to $9.7\mu s$ per multiply accumulate. This is quite a long period for such an operation suggesting that the algorithm should be pipelined onto fewer PE's. Taking the linear array of Fig. 1(a), a single PE performing these tasks would have to run at 256kHz, since zero interleaving is not required for a single PE working by itself, and this leaves about $3.9\mu s$ per operation. Using multi-projection the linear array operations can be mapped onto a single multiplier as shown in Fig. 4(a) and the data input waveforms relative to the clock for this are shown in Fig. 4(b). The multiplication operations performed in the two other arrays are also computed in the PE of Fig. 4(a) , the only difference being in the way these products are accumulated. Therefore two extra controlled accumulator sections on the multiplier output would enable computation of all the required matrix elements.

Also many A/D converters output data in bit-serial and the real-time requirement of this custom PE is such that bit-serial arithmetic units can provide the necessary throughput at a 12 bits

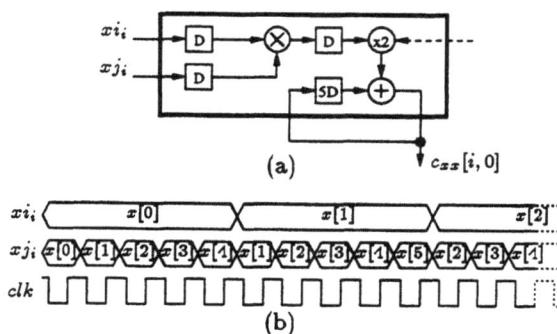

Fig. 4. (a) PE for computation of Modified Covariance matrix elements, (b) input waveforms.

word-length even when using the slower FPGA technology. For example, synthesis of a 12 bit LSB first multiplier onto a Xilinx 4028 FPGA shows double precision multiplications can be mapped onto 81 configurable logic blocks (CLB's) and throughput at over 1.5 MHz, compared to the required working rate for the RMC PE of only 256 kHz. A bit-serial approach has the added advantage of hardware cost saving over an equivalent word-parallel approach and output connectivity is much reduced to ease storage of the computed elements. The majority of the CLB usage in the RMC computation is from the SISO registers used for the input data scheduling and in the accumulator sections, making the overall CLB usage for this section approximately 450 CLB's.

The FPGA RAM in Fig. 3 allows pipelining of the the overall calculation. Thus while the FPGA is processing the RMC module for one data segment, the DSP processor can calculate the Cholesky, WNV and PSD modules of the previous data segment. Given that the major part of the WNV calculaton is implemented in the FPGA and examining Table 4 it is clear that only one 50Mhz TMS320C40 processor is necessary to implement the Cholesky and PSD calculations in the 10 ms time frame.

6. CONCLUSION

In this paper a real-time implementation of the Modified Covariance spectral estimator has been proposed. This implementation has adopted a hardware/software co-design approach that implements the individual modules in the most advantageous technology. The use of FPGA's means that the flexibility and re-use of the software approach is maintained. The proposed system can also be used to implement other spectral estimation techniques, such as the Choi-Williams Distribution, (Madeira *et al.*, 1998), which also has a module that performs many computations on the inputs, similar to the RMC module.

7. REFERENCES

Bellis, S. J., P. J. Fish and W. P. Marnane (1997*a*). Optimal systolic arrays for real-time implementation of the Modified Covariance spectral estimator. *Parallel Algorithms and Applications* **11**(1-2), 71–96.

Bellis, S. J., W. P. Marnane and P. J. Fish (1997*b*). Alternative systolic array for non-square-root Cholesky decomposition. *IEE Proceedings: Computers and Digital Techniques* **144**(2), 57–64.

Bellis, S. J., W. P. Marnane, D. Wilde and P. J. Fish (1994). Systolic arrays for Modified Covariance spectral estimation used with ultrasonic Doppler blood flow detectors. In: *Signal Processing VII - Theories and Applications, Proceedings of EUSIPCO-94*. Vol. 3. Edinburgh, Scotland. pp. 1361–1364.

Kung, H. T. and C. E. Leiserson (1979). Systolic arrays (for VLSI). In: *Sparse Matrix Proceedings*. Society for Industrial and Applied Mathematics. pp. 256–282.

Madeira, M. M., M. O. Tokhi and M. G. Ruano (1997). Parallel processing architectures for real-time doppler signal blood flow estimation: a comparative study. In: *Preprints of AARTC'97*. pp. 293–298.

Madeira, M. M., M. O. Tokhi and M. G. Ruano (1998). A time-frequency spectral implementation for a real time biomedical application. In: *Euromicro Workshop on Real Time Systems'98*. (In review).

Marnane, W. P., S. J. Bellis and P. Larsson-Edefors (1997). Bit-serial interleaved high-speed division. *Electronics Letters* **33**(13), 1124–1125.

Nocetti, D. F. G., J. M. Flores and J. S. González (1995). Improving signal processing performance using a transputer-dsp parallel architecture. In: *Preprints of AARTC95*. pp. 569–573.

Ruano, M. G. and P. J. Fish (1993). Cost/benefit criterion for selection of pulsed Doppler ultrasound spectral mean frequency and bandwidth estimation. *IEEE Transactions on Biomedical Engineering* **40**(12), 1338–1341.

Ruano, M. G., D. F. G. Nocetti, P. J. Fish and P. J. Fleming (1992). A spectral estimator using parallel processing for use in a Doppler blood flow instrument. In: *Parallel Computing: from Theory to Sound Practice* (W. Joosen and E. Milgrom, Eds.). ISO Press. pp. 397–400.

Ruano, M. G., D. F. G. Nocetti, P. J. Fish and P. J. Fleming (1993). Alternative parallel implementations of an AR-modified covariance spectral estimator for diagnostic ultrasound blood-flow studies. *Parallel Computing* **19**(4), 463–476.

HIGH-PERFORMANCE REAL-TIME IMPLEMENTATION
OF A SPECTRAL ESTIMATOR

**M. M. Madeira *, L. A. Aguilar Beltran+ , J. Solano Gonzalez+
F. Garcia Nocetti+, M.O. Tokhi •, M. G. Ruano ***

* UCEH/SEC, Universidade do Algarve, Campus de Gambelas, 8000 FARO Portugal
+ Depto. de Ingenieria de Sistemas Computacionales y Automatización, IIMAS, UNAM,
Apdo. Postal 20-726, Del. A. Obregón, México D.F., 01000, MEXICO
• Depart. of Automatic Control & Systems Eng., University of Sheffield,
Mappin Street, Sheffield, S1 3JD United Kingdom

Abstract: Doppler blood flow spectral estimation is a common technique of non-invasive cardiovascular disease detection. Blood flow velocity and disturbance may be evaluated by measuring spectral mean frequency and bandwidth respectively. Aiming at minor stenosis diagnosis, parametric spectral estimators may be employed. These models present better spectral resolution than the FFT based ones, at the expense of higher computational burden. Seeking for an efficient real-time implementation of a blood flow spectral estimation system, high performance techniques are being investigated. This paper compares the implementation of the Modified Covariance (MC) spectral estimator on two different DSP architectures: the TMS320C40 and the ADSP2016x (SHARC). Implementations are described and their performance assessed. Considerations about portability of algorithms, compiler optimisation levels and system dependence features are addressed. *Copyright © 1998 IFAC*

Keywords: Spectral estimation, performance evaluation, DSP architectures.

1. INTRODUCTION

Doppler blood flow spectral estimation is a common technique of non-invasive cardiovascular disease detection. the evaluation of blood flow velocity and disturbance through accurate estimation of spectral mean frequency and bandwidth respectively enables an early diagnosis of cardiovascular diseases (Ruano *et al.*, 1992). some parametric spectral estimators have been evaluated as possible blood flow estimators (Ruano *et al.*, 1993), revealing better spectral performance than the traditional estimators but at the expense of higher computational processing times.

This lead to the investigation of alternative techniques of real-time implementation of the parametric spectral estimators, in particular the investigation of parallel processing architectures.

The auto regressive Modified Covariance (MC) spectral estimator with model order of 4 was elected on behalf of its accuracy inspite of its algorithmic complexity (Ruano, Fish, 1993).

Previous work (Madeira *et al.*, 1997) investigated the feasibility of a real-time implementation of the MC spectral estimator on several homogeneous and heterogeneous parallel processing architectures using up to three transputers (T8), up to three TMS320C40 (C40) and an i860. The results obtained indicated that a real time implementation was achievable with the C40s. It was concluded that the usage of T8's as routers was degrading the overall performance of the architecture.

Aiming a low-cost PC-hosted system and taking into account that the results obtained with the C40 implementation did not allow the real time

185

estimation of the signal over the complete range of data segment lengths to be considered, led to further investigation.

Separate studies were conducted to evaluate the performance of firstly, some of the fastest digital signal processor currently available and secondly, custom parallel processing using an heterogeneous architecture comprising field programmable gate arrays and a C40 (Madeira *et al.*, 1998). Two digital signal processors, C40 and ADSP21062 (SHARC), were selected based upon their availability and cost. Special attention was given to the add-on characteristics of the modules enclosing the processors, ensuring the compatibility of a final solution.

The performance evaluation of the architectures is typically done in terms of computational execution time and speedup. In this case study, execution time is particularly relevant since real-time implementation is envisage. For clinical diagnosis several cardiac cycles should be spectral estimated, each cardiac cycle being windowed approximately a hundred times, leading to signal data segments (lengths varying from 64 to 512 points) to be spectral estimated within 10ms. Execution time was therefore adopted as a performance measurement evaluator. In order to compare different architectures, the gradient performance metric was considered. Gradient measures the mean time per data element..

2. MODIFIED COVARIANCE SPECTRAL ESTIMATION

The general description of the autoregressive MC spectral estimation (Kay, 1988) involves solving the following linear system of equations,

$$
\begin{bmatrix}
c_{xx}[1,1] & c_{xx}[1,2] & \ldots & c_{xx}[1,p] \\
c_{xx}[2,1] & c_{xx}[2,2] & \ldots & c_{xx}[2,p] \\
\ldots & \ldots & \ldots & \ldots \\
c_{xx}[p,1] & c_{xx}[p,2] & \ldots & c_{xx}[p,p]
\end{bmatrix}
\times
\begin{bmatrix}
a[1] \\
a[2] \\
\ldots \\
a[p]
\end{bmatrix}
$$

$$
= -
\begin{bmatrix}
c_{xx}[1,0] \\
c_{xx}[2,0] \\
\ldots \\
c_{xx}[p,0]
\end{bmatrix}
\tag{1}
$$

where each element $c_{xx}[i,j]$ of the covariance matrix and right-hand-side vector is obtained by

$$
c_{xx}[i,j] = \frac{1}{2 \times (N-p)} \times
$$

$$
\left(
\begin{array}{c}
\sum_{n=p}^{N-1} x^*[n-i] \times x[n-j] \\
+ \\
\sum_{n=0}^{N-1-p} x[n+i] \times x^*[n+j]
\end{array}
\right)
\tag{2}
$$

The $a[k]$ $(k=1,...,4)$ estimates, solution of the above linear system, are achieved using the Cholesky algorithm.

The signal white noise variance estimate is calculated as

$$
\sigma^2 = c_{xx}[0,0] + \sum_{k=1}^{p} a[k] \times c_{xx}[0,k]
\tag{3}
$$

and the power spectral density , $P_{MC}(f_n)$, is obtained through

$$
P_{MC}(f_n) = \frac{s^2}{\left| A(f_n) \right|^2}
$$

$$
= \frac{s^2}{\left| 1 + \sum_{k=1}^{p} a[k] z^{-k} \right|^2_{z=e^{j2\pi f_n}}}
\tag{4}
$$

On this particular application, signal time windowing is performed each 10 ms producing data segments of varying lengths: 64, 128, 256 and 512 elements.

3. PROCESSING ARCHITECTURES

The block diagrams of the processing architectures considered are presented in Fig.1.

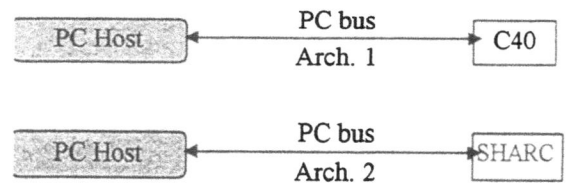

Fig.1. Block diagram of the architectures considered.

The C40 architecture is a PC-host system using a 50MHz floating point C40 device, 1 MB SRAM plus 4 MB DRAM, rated as 50 MFLOPS, 25 MIPS, 275 MOPS. This architecture has six high-speed parallel

Model regularity

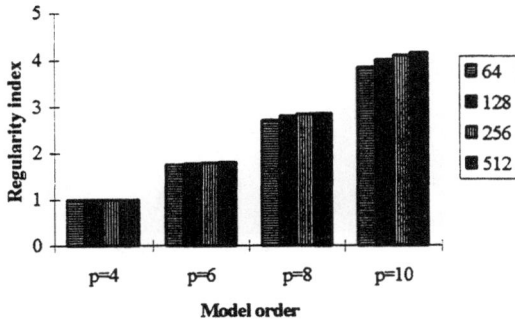

Fig. 4. MC algorithm growth rate with model order.

Regularity of MC implementation Arch. 1

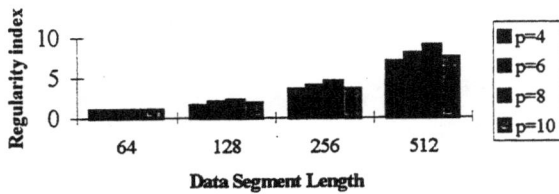

Fig. 5. Regularity of MC implementation on Arch. 1 with model order varying from 4 to 10.

Regularity of MC implemention Arch. 2

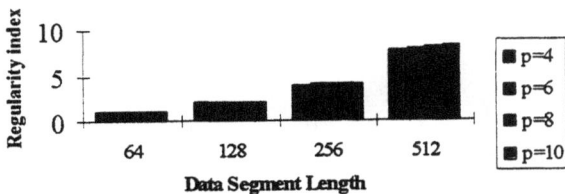

Fig. 6. Regularity of MC implementation on Arch. 2 with model order varying from 4 to 10.

6. CONCLUSION

The real-time implementation of an autoregressive Doppler signal spectral estimator has been addressed. Two high performance processors were utilised: a TMS320C40 and a ADSP21062. The MC algorithm presents a regular nature and therefore regular execution times were expected.

The execution times obtained proved that real-time implementation requirements were not achieved with architecture 1 for the higher data segment length. The other architecture including the SHARC as processing element achieved the real-time requirements for all the data segment lengths to be considered.

Performance

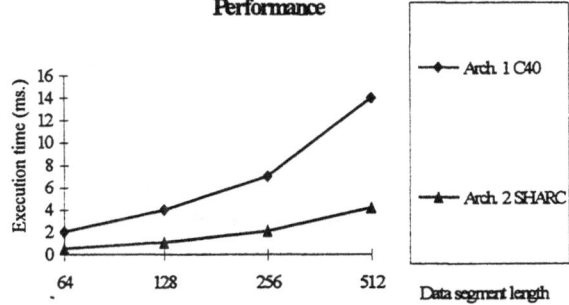

Fig. 7. Execution time performance on C40 and SHARC architectures.

Architecture 1 (C40) performed best upon model order variation, presenting a smaller growth index with the higher data segment lengths. Fig. 7 allows the comparison between the initial and final execution times, in milliseconds, obtained in implementing the algorithm on the architectures.

The performance enhancements achieved by adapting the algorithm to the processing architecture were due to the usage of compiler optimisation and the reduction of function calls, leading to significant improves on the overall performance with Arch.1 (reducing 20 to 60% of the execution times).

In the particular case of architecture 2 (SHARC), in addition to the usage of compiler optimisation and reduction of function calls, the utilisation of the architecture file (to distribute the data through the internal memory blocks) enabled usage of the dual internal bus and the intrinsic parallelism of the Sharc processor architecture leading to execution times 93% less than the original ones.

7. ACKNOWLEDGEMENTS

The authors gratefully acknowledge the Treaty of Windsor(B-115/96), JNICT(PBIC/2414/95), UNAM (PAPIIT-IN106796) and CONACYT(2146P-A9507)

REFERENCES

Alex Parallel Computers Inc. (1996). *SHARC1000 User's Manual*, Alex Parallel Computers I Inc. Ithaca, NY14850, USA.

3L Ltd. (1995). *C4x Parallel C V2.0.2*, 3L Limited, Scotland.

Kay S.M. (1988). *Modern Spectral Estimation - Theory & Application*, Prentice Hall.

Madeira, M. M., Tokhi, M.O., Ruano, M. Graça (1997). *Parallel Processing Architectures for Real-Time Doppler Signal Blood Flow Estimation: a Comparative Study*, Preprints of AARTC'97, pp 293-298.

Madeira, M.M., Bellis, S.J., Marnane, W.P., Ruano, M.G. (1998) *Configurable processing for real-time spectral estimation*, submitted to AARTC'98.

Ruano M. G, Garcia Nocetti D. F., Fish P., Fleming P.J. (1992). *A Spectral Estimator using Parallel Processing for use in a Doppler Blood Flow Instrument*, Parallel Computing: from Theory to Sound Practice, ed.: W. Joosen and E. Milgrom, IOS Press, pp 397-400.

Ruano M.G., García Nocetti D.F., Fish P., Fleming P.J. (1993). *Alternative Parallel Implemetations of an AR-Modified Covariance Spectral Estimator for Diagnostic Ultrasonic Blood Flow Studies*. Parallel Computing, Vol. 19, pp. 463-476.

Ruano M.G., Fish P.J. (1993). Cost/benefit Criterion for Selection of Pulsed Doppler Ultrasound Spectral Mean Frequency and Bandwidth Estimators, IEEE Trans. Biomedical Engineering, Vol. 40, No. 12, pp 1338-1341.

Texas Instruments (1991). TMS320C40 User's Guide, Texas Instruments, USA.

The Math Works Inc. (1991). *Matlab users' guide*, The Math Works Inc., South Natick, MA01760, USA.

189

ALGORITHM OF CALCULATING THE MAXIMUM PERMISSIBLE
SPEED OF TRAFFIC

E. GAVRILOV *, N. ZHEMCHUGOV **

* professor, doctor of technical science, vice- rector of Kharkiv State Automobile &
Highway Technical University, head of the Highway Design and Pioneering Chair,
e- mail: gavr@khadi.kharkov.ua
** graduate engineer of the Highway Design and Pioneering Chair

Abstract: The criterion and algorithm of calculating the maximum permissible speed of
traffic on the automobile road in use are examined in the paper. The evaluation of the
maximum permissible speed proceeds from technical, ergonomic and ecological
requirements specified for the automobile road and traffic. The fulfillment of these
requirements ensures the compatibility of technosphere and biosphere.
Copyright © 1998 IFAC

Keywords: Speed limit, performance criterion, the evaluation of the transportation
system status, the algorithm of calculating the maximum permissible speed.

1. INTRODUCTION

The expediency of limiting speed of traffic on
automobile roads in use is based upon the necessity
for increasing the road safety level, reducing the
harmful effect produced by traffic on the environment
and saving fuel. The maximum permissible speeds
are different for each of the listed requirements since
they are calculated on the basis of the operational
condition analysis of different components of the
'man-automobile-traffic environment' system. The
absolute prohibition of automobile traffic proves to
be the best solution for the system on the whole. On
the other hand, the very idea of limiting traffic speed
is in conflict with the requirement of the most
effective use of automobile roads for their designated
purposes, which causes the increase of speed.

The article suggests one of the possible ways of
solving the conflict being considered.

2. THE CRITERION FOR LIMITING
TRAFFIC SPEED

We see the way out of contradictoriness of local
requirements in the transition to more important goals
of traffic organization, i.e. to the requirements set by
systems of the higher level. To our mind, such a goal
can be ensuring the compatibility of technosphere and
biosphere in the broad sense and compatibility of
biosphere and the 'man -automobile-traffic
environment' (MATE) system of human activity in
the narrow sense. The structure of traffic environment
includes a road and the part of biosphere the effect of
which upon the system is not equal to zero.

The result of the compatibility of technosphere and
biosphere is viability of the MATE system. Therefore
either the probability of the given system's
destruction or the time of its existence (the system's
life span) can serve as indexes of such compatibility.
The task of automobile roads design and use and
traffic organization is the creation of conditions
ensuring the maximum time of the MATE system's
existence. One of such conditions is a selection of the
optimum behavior of the system's components.

The simplest mathematical model of behavior optimization for the system's components can be formulated as:

Suppose we have the 'man-automobile-traffic environment' system the status of which is determined by the n-dimensional vector $\bar{F}$. The coordinates of the system's components status f_1, $f_2,....f_n$ increase with constant speeds $\bar{U}(u_1, u_2,...u_n), u_i \geq 0$. Biosphere is described by a set of conditions $C=(c_1, c_2,...c_n)$ in which the system can be. In the i-th condition the effect is produced on the corresponding i-th coordinate of the system f_i which reduces it at a speed v_i, i.e. in the condition C the 'man-automobile-traffic environment' system moves at a speed $\bar{V}_i(u_1, u_2,...u_i - v_i,...v_n)$. At any moment the system can pass from one condition to another. No constraints are imposed.

Let's also consider the system existing if the vector $\bar{F}$ does not fall outside the limits of accepted values domain bounds of which are specified by the equation $L(\bar{F}) = 0$. At the bound of accepted values domain $\bar{F} = \bar{F}_g$, where $\bar{F}_g$ is an accepted value of the vector $\bar{F}$. Values of controlled variables can lie only in the following range: $f_i \geq 0$, i=1,n.

It is necessary to find the optimum behavior tactics (the rule of conditions change) for the 'man-automobile-traffic environment' system that would maximize the time of the system's existence T, i.e.

$$T \rightarrow \max. \quad (1)$$

If accepted values domain of the vector F is bounded by a n-dimensional ellipsoid and $L(\bar{F}) = \sum_i f_i^2 k_i - 1 = 0$ then according to the research conducted by I. Linnik (1996) the local rule of conditions change takes the following form:

$$\max_i K_i f_i v_i. \quad (2)$$

In line with this rule, the control action maximizing the time T must be directed to such a component of the 'man-automobile-traffic environment' system for which the product k f v is a minimum. Therefore the solution of the problem of limiting speed of traffic on automobile roads can be executed on the basis of status analysis only of that component of the MATE system for which the product $k_i f_i v_i$ is also a minimum. If those products are equal then limiting speed of traffic should be executed proceeding from the requirements of all components of the system.

3. THE EVALUATION OF THE SYSTEM STATUS

The description of the system status is very significant in formulating the defined problem. For the quantitative evaluation of the status vector it is offered to use the complex index of adequacy of technosphere to biosphere in the following form (E.Gavrilov, 1997):

$$\bar{F} = F_1 + b + \gamma, \quad (3)$$

where F_1 is an index of adequacy provided that factors of interaction between technosphere and biosphere do not exceed the accepted values,
b is an index which takes account of the action of factors exceeding the accept values,
γ is index of gain or attenuation of interacting factors.

The index of adequacy F_1 is calculated on the basis of the formula:

$$F_1 = \frac{\sum_{i-1}^{n} f_i \alpha_i}{\sum_{i-1}^{n} \alpha_i}, \quad (4)$$

where α_i is a weight coefficient of the i-th coordinate of the status.
The index 'b' is calculated on the basis formula:

$$b = 0, \quad \text{at } \Phi_i < \Phi_{gi},$$
$$b = \sum_{i=1}^{m} [1 + (f_i - \frac{1}{\alpha_i})], \quad \text{at } \Phi_i \geq \Phi_{gi}, \quad (5)$$

where Φ_i, Φ_{gi} is an interaction factor and it's accepted value;
m is a number of such components of the system for which $\Phi_i \geq \Phi_{gi}$.

The index 'γ' is calculated on the basis of the formula:

$$\gamma = \frac{1}{S}\bigcup_{j=1}^{S}\gamma_j, \text{ at } S > 1,$$
$$\gamma = 0, \quad \text{at } S = 0, \quad (6)$$

where S is a number of components which enhance or reduce load on biosphere at the expense of their joint operation;

$$\gamma_j = \frac{\Psi_{1,2,i\ldots n}\sum\limits_{i=1}^{n}f_i\alpha_i}{\sum\limits_{i=1}^{n}\alpha_i} \; ; \qquad (7)$$

Ψ_i is function of f_i which adopt numerical values in view of the direction of mutual effect of components (the value Ψ_i can be signed plus as well as minus). Functions Ψ_i are calculated by experiment.

Coordinates of the status of the system's components and their weight coefficients are calculated on the basis of factor of interaction:

$$f_i = \frac{\Phi_i}{\Phi_{hi}},$$
$$\alpha_i = \frac{\Phi_{hi}}{\Phi_{gi}}, \qquad (8)$$

where Φ_{hi} is a normal value of a factor of interaction (the functional norm).

By a normal value of a factor of interaction we mean it's functional optimum value, i.e. the value of a factor complying to the highest degree with tasks and conditions of the functioning of the i- th component of the system.

The evaluation of coordinates of the status can also be executed by the results of the functioning of the system's components. In the latter case

$$f_i = \frac{V}{V_{hi}},$$
$$\alpha_i = \frac{V_{hi}}{V_{gi}}, \qquad (9)$$

where V, V_{hi} and V_{gi} are actual, normal and accepted (permissible) speeds of traffic for the i- th component of the system.

Let's suppose that at the bound of accepted values domain of the vector $\bar{F}$ the actual value of the factor of interaction is equal to the accepted value, i. e.

$$\Phi_i = \Phi_{gi}, \qquad (10)$$

Then in view of (8), (9) and (10) the accepted value of the vector of the MATE system status is equal to:

$$\bar{F}_g = \frac{3}{\sum\limits_{i=1}^{3}\alpha_i} . \qquad (11)$$

4. THE CALCULATION OF THE MAXIMUM PERMISSIBLE SPEEDS OF TRAFFIC

In view of everything stated above the maximum permissible speed of traffic on a road in use can be calculated on the basis of the condition

$$\bar{F} = \bar{F}_g \qquad (12)$$

or

$$\sum\limits_{i=1}^{n}f_i\alpha_i = 3 . \qquad (13)$$

If coordinates of the status of the system's components are evaluated on the basis of traffic speeds then in view of (13) the maximum permissible speed will be calculated on the basis of the formula:

$$V_{g\Sigma} = \frac{3V_{gr}V_{ga}V_{gc}}{V_{ga}V_{gc} + V_{gr}V_{gc} + V_{gr}V_{ga}} , \qquad (14)$$

where $V_{g\Sigma}$ is the maximum permissible speed of traffic by indexes of the MATE system status; V_{gr}, V_{ga}, V_{gc} are the maximum permissible speeds of traffic by indexes of the status of the main, automobile and traffic environment accordingly.

The maximum permissible speed of traffic based upon the status of the man is calculated on the basis of the formula (E. Gavrilov 1988):

$$V_{gr} = \frac{2}{3}V_{mt}, \text{at } H_m > (H_m + r),$$
$$V_{gr} = \frac{2}{3}V_{mt} + \frac{V_{mt}}{3}(1 - \frac{H_m - H_{mo}}{r}), \qquad (15)$$
$$\text{at } H_{mo} < H_m < (H_m + r),$$

where V_{mt} is the maximum possible speed of the automobile under standard conditions (the automobile's values from the registration certificate);
$r = H_{mk} - H_{mo}$;
H_{mk} is the maximum entropy of the man's perception field at the traffic rate equal to the road's traffic capacity;
H_{mo} is the maximum entropy of the man's perception field at the traffic rate equal to zerro;
H_m is the maximum entropy of the man's perception field at the calculated traffic rate,

$$H_m = m^2 ; \qquad (16)$$

m is a number of data carriers within the perception field of the man driving the automobile.

The maximum permissible speed of traffic based upon the automobile status is calculated proceeding from the theory of interaction between the automobile and the road. Thus, at driving along narrow curves in the plan:

$$V_{ga} = \sqrt{127R(\mu \pm i_n)} , \qquad (17)$$

where R is the radius of the curve;
i_n is the cross slope, fraction of one;
μ is a coefficient of the lateral force ($\mu = 0.15$).

At curves in the plan at the limited visual range

$$V_{ga} = \sqrt{\frac{127(\varphi^2 - i^2)}{K_\varepsilon \varphi}(L_B - 5)} , \qquad (18)$$

where φ is a coefficient of longitudinal adhesion;
i is the longitudinal slope of the road, fraction of one;
K_ε is a coefficient of braking operational conditions (K_ε is ≈ 1.45 for a passenger car, $K_\varepsilon \approx 1.8$ for a truck);
L_B the visual range, $L_B = 2\sqrt{RB}$;
B is the width of the roadbed.

At concave vertical curves the maximum permissible speed is:

$$V_{ga} = \sqrt{13aR} , \qquad (19)$$

where 'a' is the centripetal acceleration ($a \approx 0.5 - 0.7$ m/s^2).
At grades 'i' (up to 2%) ending with horizontal section

$$V_{ga} = \sqrt{\frac{127(\varphi + i)}{K_\varepsilon}(L_B - 5)} . \qquad (20)$$

At the convex change of slope with mating slopes 'i_1' and 'i_2'

$$V_{ga} = \sqrt{\frac{127(\varphi + i_1)(\varphi + i_2)}{K_\varepsilon \varphi}(L_B - 5)} . \qquad (21)$$

The maximum permissible speed based upon the traffic environment status is defined as the speed at which the running exhaust emission of pollutants produced by the automobile reaches the maximum permissible value. In it's turn, the latter one is calculated on the basis of the maximum allowable concentration of pollutants at the border of a sanitary zone or at the level of a residential construction.
The running exhaust emission Q (g/km) of pollutants is calculated on the basis of the formula (N. Govorushenko 1990):

$$Q = 0.0548 \cdot M_x \cdot \rho_T \cdot x \cdot \alpha \cdot P_T , \qquad (22)$$

where M_x is the molecular weight of a pollutant, g/mole;
x is the content of pollutants, %;
α is a coefficient of excess of air;
ρ_T is the fuel density, g/sm^3;
P_T is the fuel consumption, l/100 km.

The fuel consumption is calculated on the bases of the formula:

$$P_t = \frac{1}{\eta}[A_{i_k} + B_{i_k}^2 V + D(G_a \varphi_g + 0.077k\,s\,V^2 \pm 0.1\beta\,G_a \frac{dV}{dt})], \qquad (23)$$

where η is the indicator coefficient of efficiency;
i_k is the gear ratio;
G_a is the calculated weight of the automobile;
φ_g is the total road resistance to traffic;
k is a coefficient of air resistance;
s is the frontal area of the automobile;

$$A = \frac{7.95\,\alpha_1\,V_h\,i_o}{H_h\,\rho_t\,r_k} ;$$

$$B = \frac{0.69\,b_1\,V_h\,l_n\,i_o}{H_h\,\rho_t\,r_k^2} ;$$

$$D = \frac{100}{H_h\,\rho_t\,\eta_{tr}} ;$$

V_h is the working volume of engine cylinders;
H_h is the lowest heat of fuel combustion;
l_h is the piston stroke;
i_o is the axle ratio;
η_{tp} is a coefficient of efficiency of transmission;
r_k is the radius of the automobile's wheel rolling;
a_1 and b_1 are experimental coefficients,

$a_1 \approx 45$ kPa for Otto engines,
$a_1 \approx 48$ kPa for diesel engines,
$b_1 \approx 13$ kPa·s·m^{-1} for Otto engines,
$b_1 \approx 16$ kPa·s·m^{-1} for diesel engines.

The maximum permissible emission Q_g at the level of a residential construction is calculated on the basis of the solution of Bozanke- Pirson differential equation in the form:

$$Q_g = \frac{V_B\,P\,X}{1,000} \cdot \frac{(PDK - C_\Phi)T_1 \cdot l}{\exp[-h/(P \cdot X)]} , \text{at } x < (l_3 - l_T),$$

$$\qquad (24)$$

$$Q_g = \frac{V_B\,P \cdot X}{\lambda\,1,000} \cdot \frac{(PDK - C_\Phi)T_1 \cdot l}{\exp[-h/(P \cdot X)]} , \text{at } (l_3 - l_T) \le x \le l_3.$$

where PDK is the maximum allowable concentration of the pollutant in the atmosphere;

C_Φ is the background concentration of the pollutant;

V_B is the speed of the wind perpendicular to the road direction;

X is the distance from the pollution source to the calculated point;

h is the altitude of the automobile's exhaust pipe above the traffic road;

P is a coefficient of the effect of pollutant's diffusion angle in the vertical plane;

l_3 is the distance from the pollution source to the construction;

l_T is the width of the winded aerodynamic shadow of the construction;

λ is a coefficient of the effect of the construction,

$\lambda = 1 + 0.044(X - l_3 + l_t) + 0.0013(X - l_3 + l_3)^2$;

l is the length of the calculated section of the road;

T_1 is the time of covering 1 km of the way, $T = 3,600 / V$.

Thus, in view of (22),(23) and (24) under the set traffic conditions the condition of calculating the maximum permissible speed of traffic on the basis of traffic environment takes the following form:

$$D \cdot 0.077 \cdot k \cdot s \cdot V^3 + B \cdot i_k^2 \cdot V^2 +$$
$$+ (A_{i_k} + D \cdot G_a \cdot \varphi_g)V - \frac{R \cdot 3,600 \cdot \eta}{z} = 0 \qquad (25)$$

where $R = \frac{V_6 \cdot P \cdot X}{\lambda \cdot 1,000} \cdot \frac{(PDK - C \cdot \Phi) \cdot L}{\exp[-h /(P \cdot X)]}$;

$z = 0.0548 \cdot M_X \cdot \rho_t \cdot x \cdot \alpha$.

The solution of the equation (25) for V makes it pissible to calculate the maximum permissible speed of traffic V_{gc} on the basis of the traffic environment status.

5. CONCLUDING REMARKS

The given algorithm of calculation the maximum permissible speed of traffic on the automobile road i use is practically implemented as a computer program for Windows 95 written in the C++ programming language. The results of calculating by the given program make it possible to calculate the maximum permissible speed of traffic for automobiles of different types and transport on the whole. In the data base traffic environment is represented by 36 data carriers including: geometrical characteristics of the road, opposing automobiles and automobiles gang in the same direction, road signs, indexes, marking- out, planting of trees and gardens, equipping the road, etc. Speeds of traffic are calculated with due account of patterns of driver's behavior on a road.

The given paper extends the ideas suggested at the 8th IFAC Symposium 'Transportation System'.

REFERENCES

Gavrilov E., Turenko A. (1997). *Traffic efficiency and human factor*. In: Preprints of the 8th IFAC/IFIP/IFORS Symposium 'Transportation System', Chania, Greece. (Papageorgion M, Pouliezos A.(Ed.)), pp. 1268- 1270.

Gavrilov E., Alekseev O., Tumanov B. (1988). *Computer in designing automobile roads*. EMK HE Publishing House Kiev, Ukraine..

Govorushenko N. (1990). *Fuel saving and toxicity reducing for automobile transport*. 'Transport' Publishing House, Moscow, Russia.

Linnik I. (1996). *The optimization model for volumes of work at reducing fuel consumption and amount of pollutants*. In: The Bulletin of Kharkiv State Automobile & Highway Technical University. The 4th issue, RIO KhSAHTU, Kharkiv, Ukraine, pp. 28- 31.

SPECIAL SYSTEMS FOR COMPLICATED OBJECTS
OF EXTENDED TYPE MONITORING

Anatoly Turenko*, Oleg Alekseyev **

**Professor, academician, rector of Kharkiv State Automobile and Highway Technical
University, e-mail: admin@ khadi.kharkov.ua*
***Professor, doctor of technical science, head of the Electrical Engineering Depart-
ment of Kharkiv State Automobile and Highway Technical University,
e-mail: alex@khadi.kharkov.ua*

Abstract: Problems of analysis and synthesis of special systems for complicated objects
of extended type monitoring are investigated. The results of creation and practical realization of such systems on example of monitoring system of transport communications
are given. *Copyright © 1998 IFAC*

Keywords: Monitoring , information technology, algorithm, real-time, transport communication

1. INTRODUCTION

Let's define the process of observation and evaluation
of condition of complicated long objects (or objects
of an extended type) as monitoring. It is not an ordinary monitoring of a complicated system. Monitoring
is not a process of evaluation. It is a process of data
origination for management. The results of researches
are development into a theory of evaluation of conditions of complicated systems earlier expressed in
other researches by Alekseev et. al. (1997), Gavrilov
and Turenko et. al. (1997). Let's consider how to
organize monitoring of such rather complicated object or transport communication as a highway. Usually, for this purpose a special built-in computer system is created equipped on an automobile. This mobile item moving along a highway registers information about it and forms a data base to evaluate appropriate transport communication.

So such system can be defined as a automobile microsystem. It consists of a measuring complex, a
controller and a computer. All components of this
complex form a hardware-software subsystem of
information technology of evaluation transport communication condition.

2. TRANSPORT COMMUNICATIONS
EVALUATION

The system under consideration, the equipment of an
the automobile saloon are shown on fig.1, where the
computer is a main subsystem.

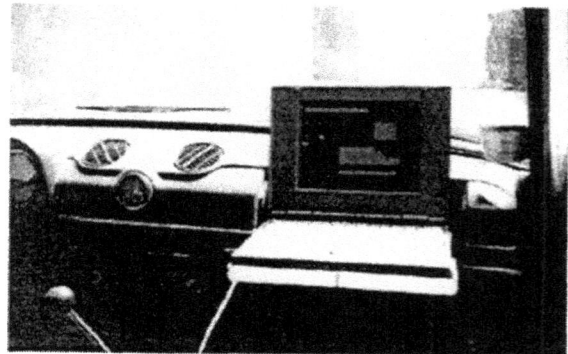

Fig.1. The equipment of a saloon of an automobile

Special information technology (IT) with automobile
microsystem to evaluate condition of transport communications is developed in Kharkiv State Automobile and Highway Technical University. Such system
represents an information computer complex of the
fifth updating (ICC-5). It allows to measure parame-

ters of a road, to register and to store data about parameters of a highway as an object of an extended type and to form an electronic data bank for preparation management decisions about repair, reconstruction and organization of traffic.

The main technical subsystem of IT is a mobile information complex (ICC). It consists of a measuring system, road controller and computer of notebook type. ICC can be installed practically on any means of transport (car, microbus). It allows to measure and register data with velocity from 25 up to 70 km/hour.

ICC is a universal system. The road controller allows analog signal to enter computer by 16 channels and discrete signals by 12 channels. Therefore this system can be used for ecological monitoring, for evaluation driver's state of health and as a movable office. Such movable data bank allows to decide problems of traffic optimization on any road an any transport communications. ICC is developed according individual orders taking into account features of using hardware and the list of parameters of a formed movable data bank. In table 1 information about various updating of this system are given.

Table 1 Updating of ICC

Characters	ICC-1	ICC-2	ICC-3	ICC-4	ICC-5
Number of information channels	2	4	5	6	8
Base vehicle	micro-bus	micro-bus	micro-bus	motor-car	motor-car
Productivity km/hour	25	25	40	60	70
Staff	3	3	3	1	2
Relative cost (value)	1	1.5	2	3	3

The road controller and measuring system of ICC are indicated in fig. 2.

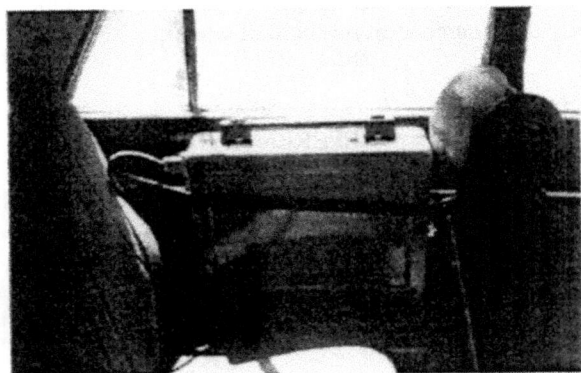

Fig.2. Controller

In table 2 the information about technical parameters of the fifth updating of this system of highway monitoring are given (ICC-5).

Table 2 Technical parameters of ICC-5

Measuring parameters	Range	Relative precision
1. Angle of a turn of a road	± 180 °	3.5 %
2. declivity of a road	± 60 % .	4 %
3. Distance	Up to 50 km/h	1 %
4. Radius of a curve in a plain	Up to 5 km/h	6 %
5. Radius of a curve in the longitudinal profile	Up to 30 km/h	6 %
6. Plain of a road	0-400 sm	Qualitative parameter is commensurate with to an expert evaluation
7. Velocity MIC	25 - 55 km/h	1.0 %
8. Factor of tripping of a road	0.1 – 1	4.0 %

3. ALGORITHM OF DATA LOGGING

Let's consider a problem of monitoring of such transport communication as a highway. Let's limit to evaluation of geometric parameters only. First we define transport communication as a geometric object. Its main parameter is expansion or length.

Let's deliver a set A as a researched object which consists of geometric elements of two types: "straight" and "turns" accordingly $\{p_i\}$ and $\{k_i\}$. As the extended objects are continuous they are such pairs, that they are possible to be noted $A = \{p_i \times k_i\}$. A case is possible when the objects consist only of $\{p_i\}$ or of $\{k_i\}$.

A special case of monitoring research process is an evaluation of one parameter $\{Z_i\}$, i = 1,2... N. The principle of deviations monitoring is used for minimization the volume of computer used memory. The parameters are registered if there is a difference of current significance of a parameter from the previous significance of a parameter. Residual of significances of parameters should be more than significance of a threshold of a distinguishability ξ_z. The appropriate algorithm of identification consists of following procedures:

1. Measurements Z_i;
2. Comparisons Z_i and Z_{i-1};
3. If $(Z_i - Z_{i-1}) > \xi_z$, the pitch 6 is fulfilled, pitch 4 is fulfilled differently;
4. $Z_i - Z_{i-1}$;
5. Pitch 8 is fulfilled;
6. Registration value Z_i and value t_j appropriate to

the moment of measurement;

7. $i = i + 1$;

8. If $i \leq N$ a pitch 1 is fulfilled , differently- end.

This algorithm controls measurement, evaluation and dataorigination for identification of researched transport communications.

It is difficult to describe analytically algorithm of identification as usual arithmetical and logical relations. Therefore let's present this problem as a solution of a problem of identification of elementary section of a road j and the result of evaluation of medium traffic to be defined by a sequence $\{q_1; q_2\}$, if q_1 and q_2 are evaluations of the previous and following sections. Significances q_1 and q_2 - (-1); (+1); (0) if the section is on the "left turn", "right turn" or "straight" accordingly. The offered algorithm of identification of the researched transport communications is based on replacement of logical expressions by simple arithmetical formulas:

$$N = Q1 + Q2, \qquad (1)$$

where

$$Q_1 = \begin{cases} 1, \text{if } g_1 = 1; \\ 2, \text{if } g_1 = -1; \\ 3, \text{if } g_1 = 0. \end{cases} \qquad Q_2 = \begin{cases} 0, \text{if } g_1 = 1; \\ 3, \text{if } g_1 = -1; \\ 6, \text{if } g_1 = 0. \end{cases}$$

They corresponds to data of table 3.

Table 3 Parameters of sections

Section	q_1	q_2	N
"turn"	±1	±1	1
"straight"	0	0	5 9
begin of "straight"	±1	0	8 7
begin of "turn"	0	±1	6 3
end and begin of another "turn"	±1	±1	2 4

4. DYNAMIC ANALYSIS

Let's consider a mathematical simulation and dynamic analysis of the process of monitoring of transport communications. Obviously the application of not traditional methods of analog signals research for this simulation is needed. Modern digital methods of measurement and datalogging of condition of the transport communications is used for it.

It is necessary to notice that the problems of registration and data processing of monitoring are connected with analog-digital conversion of signals. In the considered system the generators of impulses are applied. So, source of digital signals of sequence of generated

impulses are automobile wheels and engine. Their mathematical description is a continuous sequence of numbers, time or number of revolutions of an automobile wheel.

Such system works in a mode which can be defined as "asynchronous". The entries are formed in instants t_i to be determined by signals of wheels for each revolution. The computer controls the process of registration. The "synchronous" mode differs by forming entries on signals of the timer at regular intervals. The "asynchronous" mode precisely determines parameters of digital sequences and "synchronous" precisely determines parameters of analog signals. The "synchronous and asynchronous" operational modes determine overlapping in time of operations of datalogging. The computing procedures are executed in parallel in real- time.

Thus input ICC are signals which arrive from the automobile. Therefore automobile is the gauge of this system. Evaluating condition of an automobile it is possible to make the conclusion about the state of transport communications.

5. CONCLUSIONS

A. Special system of monitoring of transport communications consists of computer, road controller and special measuring system;

B. In a mobile system of monitoring of a highway an automobile (microbus or motor-car) is used;

C. Measurements, registration, processing and submission of data of monitoring are executed in parallel. Digital process of datalogging causes their processing in real- time;

D. Modern digital methods of measurement and datalogging of condition of the transport communications is used for monitoring transport communication. Input and output are described by mathematical variable of a sequence of numbers, time, number of rotation of wheels.

REFERENCES

Alekseev O. P.(1997) The integrated approach to valuation condition of transport systems: transportion systems, Preprints of the 8-th IFAC/IFIP/IFOPS Symposium, Chania, Greece, 16-18 June 1997, 175-177.

Gavrilov E.V., Turenko A.N. (1997) Traffic efficiency and human factor: Preprints of the 8-th IFAC/IFIP/IFOPS Symposium, Chania, Greece, 16-18 June 1997, 1268-1270.

PARALLEL ALGORITHMS FOR THE CHOLESKY FACTOR OF GENERALIZED LYAPUNOV EQUATIONS *

David Guerrero * Vicente Hernández * José E. Román *
Antonio M. Vidal *

* Universidad Politécnica de Valencia.
Dept. Sistemas Informáticos y Computación.
Camino de Vera, s/n, 46071 Valencia, Spain. Tel. 34-6-3877356
{dguerrer, vhernand, jroman, avidal}@dsic.upv.es

Abstract: In this work the solution of generalized Lyapunov equations using an extension of Hammarling's method is described. Lyapunov equations are used in multiple applications, e.g. model reduction in linear control problems. The sequential algorithms have been implemented making use of standard linear algebra libraries BLAS and LAPACK, and they include block versions. In the parallel algorithms, libraries PBLAS and ScaLAPACK have been used. Experimental results show good performances. A cluster of PC's connected via Fast Ethernet under Linux operating system has been used to obtain the experimental results. *Copyright ©1998 IFAC*

Keywords: Control Algorithms, Model Reduction, Lyapunov Equation, Parallel Computation

1. INTRODUCTION

One of the most frequent problems in control theory is model reduction. This problem consists in obtaining, starting from a linear control system, a lower-order model which maintains a great part of the properties from the original system.

There exist many methods for model reduction, but most of them are only applicable to concrete cases. Among the most general methods, we focus on methods of truncation of components, because of their greater algebraic nature (Fortuna *et al.*, 1992). These methods work with the state-space description of a system and consist in obtaining an equivalent equal-order system with the state variables ordered in accordance with their weight with respect to some interesting properties. Then a truncation of components is done in

this new system: state variables with less weight with respect to the properties which want to be conserved are rejected, following some criterion.

One of these methods is that based on system balancing. In this truncation method, the properties of the system to be conserved are controllability and observability. For this purpose, a transformation in the state-space description is computed, which leads the system to a description in which the state variables are ordered in accordance with their weight with respect to these controllability and observability properties. Then, the truncation of components is done in this system.

An important step in this model reduction method is the computation of Cholesky factors of the controllability and observability Grammians (for a complete description see (Petkov *et al.*, 1991, sec. 4.5)). Both controllability and observability Grammians verify Lyapunov equations. So, the necessity of solving these equations appears.

---

* This work has been developed under the support of the Spanish government, project TIC96–1062–C03–01.

In many real applications, the balancing algorithms have to deal with dense matrices of the order of several hundreds (Fortuna *et al.*, 1992, sec. 5.4). In these cases, the traditional methods are not applicable in sequential computers because of their physical limitations, thus making it necessary to apply parallel techniques.

Lyapunov equations also play an important role in other control problems, for example the linear-quadratic optimal control problem.

In this work the generalized versions of these equations are considered, the continuous time version

$$A^T X E + E^T X A = -Y \qquad (1)$$

and the discrete time one (generalized Stein equation)

$$A^T X A - E^T X E = -Y, \qquad (2)$$

where A, E and Y are real square matrices of size n. Matrix Y is symmetric, as well as the solution matrix X when unique.

There are various methods for solving these equations. The conceptually simplest method is based on the solution of a system of linear equations, which is obtained by means of Kronecker products. However, this method produces systems of n^2 equations, which cannot be faced in practice except in cases where the size of the problem, n, is a very small value. Another possibility is to use the Bartels-Stewart method (Bartels and Stewart, 1972), on which the work described in (Blanquer *et al.*, 1998) is based.

The next sections focus on the continuous case of the equation. In section 2, an extension of Hammarling's method (Hammarling, 1982) for solving the generalized Lyapunov equation efficiently is presented. In section 3, a sequential implementation for this method is described, which uses calls to the BLAS and LAPACK linear algebra standard libraries. In section 4, the parallel implementation is presented, which uses the parallel versions of those libraries: PBLAS and ScaLAPACK. In the last section, some performance results obtained with a cluster of PC's are shown.

2. AN EXTENSION OF HAMMARLING'S METHOD

Hammarling's method (Hammarling, 1982) is an alternative to the Bartels-Stewart method (Bartels and Stewart, 1972) for solving the standard Lyapunov equation, when the equation is stable and its right hand side is positive semidefinite. Both methods can be extended for solving the generalized Lyapunov equation. Here it is presented the extension of Hammarling's method.

The condition of stability is given by the location in the open left half plane, $\text{Re}(z) < 0$, of the eigenvalues of the pencil $A - \lambda E$, for the case of the generalized continuous equation

$$A^T X E + E^T X A = -B^T B, \qquad (3)$$

in which B is a real matrix of size $m \times n$ whereas A, E and X are $n \times n$ matrices.

Conditions of stability and right hand side positive semidefiniteness guarantee that the solution X is positive semidefinite. Under these assumptions, Hammarling's method allows us to obtain the Cholesky factor of X without having to calculate the product $B^T B$ explicitly.

Hammarling's method approaches the problem in three stages: transformation of the equation to a reduced form, solution of the reduced equation, and back transformation.

2.1 *Transformation to reduced form*

The transformation to reduced form is carried out in order to obtain the equation

$$A_s^T X_s E_s + E_s^T X_s A_s = -B_s^T B_s, \qquad (4)$$

where E_s and B_s are upper triangular matrices of size $n \times n$, and A_s is an upper quasi-triangular matrix of the same dimensions.

This form of the equation is obtained by reducing the pair of matrices (A, E) to the generalized Schur form (A_s, E_s) using the QZ algorithm. This algorithm computes two orthogonal matrices Q and Z which verify

$$A_s = Q^T A Z, \qquad (5)$$
$$E_s = Q^T E Z, \qquad (6)$$

being A_s and E_s an upper quasi-triangular matrix and an upper triangular matrix, respectively. The term upper quasi-triangular refers to a matrix which is upper triangular with the exception of some nonzero elements in the main subdiagonal. A_s can be seen as an upper block triangular matrix. The diagonal of this matrix is formed by 1×1 or 2×2 blocks corresponding to real or complex eigenvalues of the pencil $A - \lambda E$, respectively.

For this transformation there is an available shared memory implementation in the LAPACK library (Anderson *et al.*, 1992), namely the subroutine DGEGS. For distributed memory platforms, some work is done for the case of the

$$A_{11}^T U_{11}^T U_{11} E_{11} + E_{11}^T U_{11}^T U_{11} A_{11} = -B_{11}^T B_{11} \tag{7}$$

$$A_{22}^T U_{12}^T + E_{22}^T U_{12}^T M_1 = -B_{12}^T M_2 - A_{12}^T U_{11}^T - E_{12}^T U_{11}^T M_1 \tag{8}$$

$$A_{22}^T U_{22}^T U_{22} E_{22} + E_{22}^T U_{22}^T U_{22} A_{22} = -B_{22}^T B_{22} - B_{12}^T B_{12}$$
$$-(A_{12}^T U_{11}^T + A_{22}^T U_{12}^T)(U_{11} E_{12} + U_{12} E_{22})$$
$$-(E_{12}^T U_{11}^T + E_{22}^T U_{12}^T)(U_{11} A_{12} + U_{12} A_{22})$$
$$= -B_{22}^T B_{22} - yy^T. \tag{9}$$

Fig. 1. Equations that appear in the solution of the reduced equation.

standard Schur form (Henry et al., 1997) and the next version of the ScaLAPACK library (Choi et al., 1992) will hopefully include routines for the generalized case.

In this transformation to reduced form (4), the matrix B_s is obtained as the upper triangular matrix R from the QR factorization of the product BZ.

2.2 Solution of the reduced equation

To solve equation (4) in U_s, Cholesky factor of the solution ($X_s = U_s^T U_s$), the involved matrices A_s, E_s, B_s and U_s are partitioned in the same way:

$$A_s = \begin{pmatrix} A_{11} & A_{12} \\ 0 & A_{22} \end{pmatrix}, \; E_s = \begin{pmatrix} E_{11} & E_{12} \\ 0 & E_{22} \end{pmatrix}, \tag{10}$$

$$B_s = \begin{pmatrix} B_{11} & B_{12} \\ 0 & B_{22} \end{pmatrix}, \; U_s = \begin{pmatrix} U_{11} & U_{12} \\ 0 & U_{22} \end{pmatrix}. \tag{11}$$

The upper left blocks are 1×1 or 2×2 blocks depending on whether the first diagonal block of the upper quasi-triangular matrix A_s is of size 1×1 or 2×2, which represents a real eigenvalue or a pair of complex conjugate eigenvalues of the pencil $A - \lambda E$, respectively.

Applying this partition to the equation

$$A_s^T U_s^T U_s E_s + E_s^T U_s^T U_s A_s = -B_s^T B_s, \tag{12}$$

the equations shown in figure 1 are obtained, where U_{11} and E_{11} are assumed to be nonsingular. Matrices M_1, M_2 and y are given by

$$M_1 = U_{11} A_{11} E_{11}^{-1} U_{11}^{-1}, \tag{13}$$

$$M_2 = B_{11} E_{11}^{-1} U_{11}^{-1}, \tag{14}$$

$$y = B_{12}^T - (E_{12}^T U_{11}^T + E_{22}^T U_{12}^T) M_2^T. \tag{15}$$

Under the condition of stability of the equation, it can be proved that the matrix E is nonsingular and hence the inverse of E_{11} exists (Penzl, 1996). Moreover, it can be proved that if $B_{11} \neq 0$ then U_{11} is nonsingular as well. As a consequence, it is possible to compute the matrices M_1 and M_2.

Equation (7) is a version of order 1 or 2 of the reduced equation (12). The small order of this equation makes it possible to be solved by means of a system of linear equations obtained by Kronecker products.

When (7) is solved, matrices M_1 and M_2 can be computed. Then, equation (8) is solved as a generalized Sylvester equation.

If the right hand side of equation (9) had the form of an upper triangular matrix premultiplied by its transpose, it would be possible to solve this new Lyapunov equation, in which the problem size has been reduced, applying the same method described for equation (12).

For this purpose it is needed an upper triangular matrix $\tilde{B}_{22}$ that verifies

$$B_{22}^T B_{22} + yy^T = \tilde{B}_{22}^T \tilde{B}_{22}. \tag{16}$$

This upper triangular matrix $\tilde{B}_{22}$ can be obtained with the following QR factorization

$$\begin{pmatrix} B_{22} \\ y^T \end{pmatrix} = Q_{\tilde{B}} \begin{pmatrix} \tilde{B}_{22} \\ 0 \end{pmatrix}. \tag{17}$$

Therefore, equation (9) can be expressed as

$$A_{22}^T U_{22}^T U_{22} E_{22} + E_{22}^T U_{22}^T U_{22} A_{22} = -\tilde{B}_{22}^T \tilde{B}_{22} \tag{18}$$

and it can be solved using the same procedure used in the solution of the reduced equation, whose size has now decreased in 1 or 2.

The reduced equation (12) can be solved in a recursive way by successively applying this method to the Lyapunov equations that appear. The finalization of this procedure is guaranteed because the size of these equations decrease in every step.

A special care will have to be taken in the step where matrices M_1 and M_2 are computed, since it is needed to invert the sub-matrices U_{11} and E_{11} which may be ill-conditioned. In (Hammarling, 1982, sec. 6) a method for computing matrices M_1 and M_2 in a reliable way is shown.

2.3 *Back transformation*

Once the solution of the reduced equation has been computed, another transformation is required to obtain the solution of the original equation.

In Hammarling's method, this transformation consists in obtaining the QR factorization of the product of matrix U_s, solution of the reduced equation, and the transpose of matrix Q, coming from the transformation to generalized Schur form:

$$U_s Q^T = Q_U U. \tag{19}$$

With this transformation, it is obtained the upper triangular matrix U, Cholesky factor of the solution X of equation (3).

3. SEQUENTIAL IMPLEMENTATION

In the sequential implementation, the libraries BLAS and LAPACK (Anderson *et al.*, 1992) have been used to achieve portability and good performances. Portability is guaranteed by the availability of versions of these libraries for numerous platforms. Performance is enhanced by the utilization of these libraries, since their routines are optimized for every architecture. If algorithms are implemented using basic computational kernels in standard libraries, good performances can be obtained in very different machines without having to optimize the program for all of them.

We have started from a sequential version coded in FORTRAN 77 (Penzl, 1996). In this version, the different steps of the algorithm are performed by calling routines of BLAS and LAPACK. In the transformation to reduced form, the LAPACK routine DGEGS is used, which computes the Schur factorization by means of the QZ algorithm. The routine DGEQRF is also called to compute the QR factorization in both the transformation to reduced form and the back transformation of the solution. For the QR factorization needed in each iteration of solution of the reduced equation a factorization based on the application of Givens rotations has been used. This results more efficient because of the structure of the matrix (see equation (17)).

Transformation to reduced form and back transformation steps are carried out easily with a few calls to BLAS (products of matrices) and LAPACK (Schur and QR factorizations). However, the solution of the reduced equation is somewhat more complex and requires many calls to routines of BLAS with submatrices. It has been necessary to make a block version of this part of the algorithm in order to obtain good performances in the solution of the reduced equation.

Hammarling's method is also used in the implementation of the block version that solves the reduced equation. But in this case, operations that involve matrices of greater order (M_1, M_2, y) appear. This does not require any substantial modification of the algorithm in the case of updating and multiplication of matrices, but it becomes important in the case of Sylvester equation (8). This equation is now formed by matrices of greater dimension, and a new routine for solving it has been implemented. In the unblocked version, the solution matrix of this Sylvester equation does not have more than two rows, because of the partitioning induced by the 1×1 or 2×2 diagonal blocks corresponding to real or complex conjugate generalized eigenvalues.

4. PARALLEL IMPLEMENTATION

In the parallel implementation, standard libraries such as BLACS (Dongarra and Whaley, 1995), PBLAS and ScaLAPACK (Choi *et al.*, 1992) have been used.

BLACS is a communications library for message passing in distributed memory environments. It uses matrices as the unit for communication. This is why it has been used as communications support in the libraries PBLAS and ScaLAPACK.

PBLAS is a library for parallel computation of linear algebra operations. It is equivalent to BLAS, but oriented to distributed matrices. It allows to perform basic linear algebra operations such as products of matrices and solution of triangular systems.

ScaLAPACK is a version of LAPACK adapted to distributed memory platforms. As well as LAPACK, ScaLAPACK solves common linear algebra problems such as general systems of equations. Its routines are implemented by means of calls to routines of BLAS for local operations, calls to PBLAS for matrix operations in parallel, and calls to BLACS when communication is required.

In the parallel implementation of the generalized Hammarling's method, the methodology suggested by ScaLAPACK has been used. Calls to BLAS in the sequential algorithm become calls to PBLAS in the parallel algorithm, and calls to LAPACK in the sequential one are calls to ScaLAPACK in the parallel one. For the steps of transformation to reduced form and back transformation of the solution, the parallel algorithm is obtained directly from the sequential one by exchanging the calls to the sequential libraries with calls to the parallel ones. However, in the solution of the reduced equation, this methodology does not lead to a parallel algorithm with good performance, since calls to BLAS and LAPACK deal with too small submatrices and therefore the parallel routines of PBLAS and ScaLAPACK cannot benefit from these operations. A parallel algorithm has been implemented for this part of the process, starting from the sequential block algorithm and using calls to BLAS for local operations and to BLACS for communications.

When observing the algorithm for solving the reduced equation, one can find mainly two points to optimize, namely the preparation and solving of the Sylvester equation and the QR factorization.

Due to the data distribution scheme of ScaLAPACK (2D block cyclic distribution) only the row of processors that own the blocks A_{12}, E_{12}, B_{12} work in the preparation and solving of the Sylvester equation (8). The remaining processors will communicate in order to send them the matrices A_{22} and E_{22}. However, there exists a high degree of parallelism among processors in the same row, which can compute updating operations simultaneously during the solving of Sylvester equation. A new routine has been implemented, which solves this equation in parallel for this special case inside the algorithm that solves the reduced equation.

The QR factorization, that is computed in the sequential version by means of Givens rotations, has a high degree of parallelism among processors of the same row, which can apply the same rotation to different blocks of the matrices simultaneously. Among processors of the same column there exists a vertical dependence: before applying a rotation, they have to wait for the processors in the same column to finish with it. However, this dependence can be used to apply the rotations in a segmented fashion if the size of the distribution block has more rows than rows of processors. Vertical parallelism is achieved by making each row of processors apply a different rotation at the same time.

5. EXPERIMENTAL RESULTS

In this section, execution times are presented for the different algorithms that solve the reduced equation, which has been the center of attention in the orientation to blocks as well as in the parallelization of the algorithm.

All the times have been obtained taking measures on a cluster of personal computers, with up to four nodes, with Pentium processor at 100 MHz and Linux operating system, connected by Fast Ethernet.

All the tests have been carried out for different block sizes: 8, 16, 32, 64, 128, in the cases where this parameter can be varied (block algorithm and parallel algorithm). In each case the best result obtained for these block sizes have been considered.

In figure 2, execution times of the different algorithms are shown. A 40% reduction in the execution time is observed in the blocked version versus the unblocked version for matrices of order 1000×1000. This is a consequence of a better exploitation of the data locality and their reutilization when blocks fit in cache memory, closer to the processor.

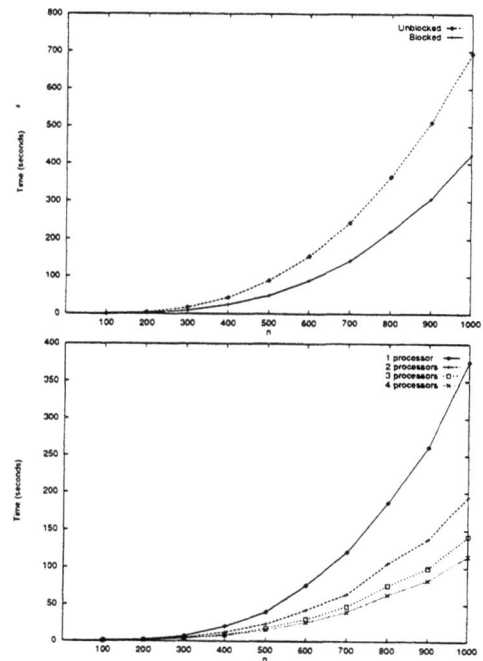

Fig. 2. Execution times of the sequential and parallel algorithms.

The times shown for the parallel algorithm are the best ones obtained among the different block sizes tested and the different configurations of the 2D mesh. As an example, with 4 processors the mesh configurations 1×4, 2×2, and 4×1 have been tested. In all the cases, horizontal meshes $(1 \times 2, 1 \times 3, 1 \times 4)$ have always given better results. This happens because in the solution of the Sylvester equations only processors in the first row work, whereas the rest communicate. In meshes with only one row, all processors are active in this step of the algorithm. However, as it has been mentioned in the previous section, vertical parallelism is also used in the application of Givens rotations.

Speed-ups and efficiencies obtained with the parallel algorithm are shown in table 1. These relative metrics have been obtained with respect to the parallel algorithm executed with only one processor, that has resulted faster than the sequential blocked algorithm. This has turned out to be the best algorithm running on a single processor, even slightly better than the blocked sequential algorithm, due to a better cache memory exploitation.

Table 1. Speed-ups and efficiencies in various processors.

S_p	100	200	500	1000
p=1	1.00	1.00	1.00	1.00
p=2	0.86	1.24	1.67	1.93
p=3	0.62	1.28	2.22	2.66
p=4	0.70	1.31	2.59	3.26
E	100	200	500	1000
p=1	1.00	1.00	1.00	1.00
p=2	0.43	0.62	0.84	0.97
p=3	0.21	0.43	0.74	0.89
p=4	0.18	0.33	0.65	0.82

These results were taken from test cases generated randomly. It must be said that the performance of the algorithms is expected to maintain with other input matrices since the influence of the characteristics of the matrices in the evolution of the algorithms is negligible.

6. CONCLUSIONS

Distinct algorithms have been developed for solving the generalized Lyapunov equation using an extension of Hammarling's method. This equation has many fields of application, for example system balancing for model reduction in linear control systems.

In the developed implementations, standard linear algebra libraries have been used trying to obtain high portability and good performance.

It has been proved in the experimental results how the block version obtains better performance than the basic sequential version. This is due to the better data reutilization when they are in cache memory.

In the parallel version, it has been shown the utility of a library of high performance computing such as ScaLAPACK for facilitating the implementation tasks. However, in order to achieve good performance in certain parts of the algorithm, it has been necessary to have recourse to libraries of lower level such as BLAS and BLACS.

As a conclusion, the figures show how execution time of solving the reduced equation has decreased. In the case of matrices of order 1000 × 1000, execution time has passed from 694.3 seconds for the original algorithm in one processor, to a time of 115.5 seconds for the execution of the parallel version of the algorithm in four processors.

7. REFERENCES

Anderson, E., Z. Bai, C. Bischof, J. Demmel, J. Dongarra, J. Du Croz, A. Greenbaum, S. Hammarling, A. McKenney, S. Ostrouchov and D. Sorenson (1992). *LAPACK Users' Guide*. Society for Industrial and Applied Mathematics. Philadelphia, PA, USA.

Bartels, R. H. and G. W. Stewart (1972). Solution of the equation $AX + XB = C$. *Comm. ACM* **15**, 820–826.

Blanquer, I., H. Claramunt, V. Hernández and A.M. Vidal (1998). Solving the Generalized Lyapunov Equation by the Bartels-Stewart Method using Standard Software Libraries for Linear Algebra Computations. In: *to be published in Proceedings of the IFAC Conference on System Structure and Control, Nantes, 8-10 July, 1998*.

Choi, Jaeyoung, Jack J. Dongarra, Roldan Pozo and David W. Walker (1992). ScaLAPACK: a scalable linear algebra library for distributed memory concurrent computers. Mathematical Sciences Section, Oak Ridge National Laboratory.

Dongarra, Jack J. and R. Clint Whaley (1995). LAPACK working note 94: A user's guide to the BLACS v1.0. Technical Report UT-CS-95-281. Department of Computer Science, University of Tennessee.

Fortuna, L., G. Nunnary and A. Gallo (1992). *Model Order Reduction Techniques with Applications in Electrical Engineering*. Springer-Verlag.

Hammarling, S. J. (1982). Numerical solution of the stable, non-negative definite Lyapunov equation. *IMA J. of Numerical Analysis* **2**, 303–323.

Henry, G., D. Watkins and J. Dongarra (1997). LAPACK working note 121: A parallel implementation of the nonsymmetric QR algorithm for distributed memory architectures. Technical Report UT-CS-97-352. Department of Computer Science, University of Tennessee.

Penzl, Thilo (1996). Numerical solution of generalized Lyapunov equations. Technical Report SFB393/96-02. Numerische Simulation auf massiv parallelen Rechnern. Technische Universität Chemnitz-Zwickau.

Petkov, P. Hr., N. D. Christov and M. M. Konstantinov (1991). *Computational Methods for Linear Control Systems*. Prentice Hall Int. Series in Systems and Control Engineering. Prentice Hall.

ON-LINE GENETIC AUTO-TUNING OF PID CONTROLLERS
FOR AN ACTIVE MAGNETIC BEARING APPLICATION

P. Schroder*, B. Green+, N. Grum+, P. J. Fleming*

* Dept. Automatic Control and Systems Engineering, The University of Sheffield,
Mappin Street, Sheffield, S1 3JD, UK. Tel: +44 114 222 5250,
Fax: +44 114 273 1729, Email: P.Schroder@Sheffield.ac.uk.
+ Rolls-Royce and Associates Limited, PO Box 31, Derby, DE24 8BJ, UK.
Tel: +44 1332 661 461 x5959, Fax: +44 1332 622 948.

Abstract: A prototype large electrical machine running on active magnetic bearings is described. This rig is controlled by a digital signal processor connected by a custom interface to MATLAB/Simulink hosted by a PC. The on-line tuning of a PID controller is set up as an optimisation problem from MATLAB and a multiobjective genetic algorithm is used to drive the optimisation. The results of an optimisation are presented and analysed. *Copyright © 1998 IFAC*

Keywords: magnetic bearings, hardware-in-the-loop design, on-line controller tuning, genetic algorithms.

1. INTRODUCTION

The high reliability and minimal maintenance of magnetic bearings makes them ideal for use in completely sealed or *canned* applications. Canned machines have a number of advantages, including the elimination of potentially unreliable seals, prevention of contamination and a reduction in through-life maintenance costs. Rolls-Royce and Associates Limited (RRA) have constructed a prototype large canned pump levitated on magnetic bearings in order to achieve a very long maintenance free operating life.

This paper demonstrates a novel, but practical approach to the design of active magnetic bearing (AMB) control systems. Modern advanced control techniques have been shown to control AMB systems effectively (Nonami *et al.,* 1994 and Fittro *et al.,* 1996). However, these techniques can be hampered by an involved design process and require

the development of accurate models. Consequently, most industrial AMB systems are controlled by PID controllers which are tuned manually on a prototype plant. This work demonstrates a convenient method for automating this PID tuning process to produce an optimal design.

2. THE APPLICATION

The application consists of a large electric motor driven pump levitated on active magnetic bearings. The rotor is mounted vertically and weighs approximately 200 kg. The AMB system fitted to the machine provides rotor control in two orthogonal directions radially at the drive end of the pump, two orthogonal directions radially at the non-drive end and one direction vertically (the thrust/axial bearing). The pump's impeller is mounted at the bottom of the rotor and the entire rotor is sealed in a stainless steel can.

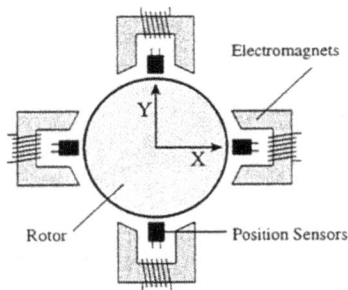

Figure 1: Configuration of the radial bearings

Each of the four radial bearing systems consists of a pair of electromagnets mounted one either side of the rotor journal as shown in Figure 1. A position sensor is also mounted on either side of the journal. The analogue voltages from each sensor pair are conditioned and the differential voltage routed to a digital control system. The control system generates an analogue output demand signal which is fed to a power amplifier and this in turn, drives current into the appropriate pair of electromagnets situated on each side of the rotor. Each electromagnet produces an attractive force acting on the rotor. The net radial force generated provides levitation at the bearing. The thrust bearing operates as a single electromagnet mounted vertically above the rotor. This levitates the rotor and is countered by the rotor weight acting downwards against gravity.

3. THE MULTI-OBJECTIVE OPTIMISATION

The rig is controlled by a digital controller running on a digital signal processor (DSP) card mounted inside a PC. Controllers are specified in MATLAB/ Simulink, a dynamic simulation environment. An auto-code generator and a custom interface to Simulink allow the adjustment of controller parameters on the DSP from MATLAB without interrupting control. Sensor data from the rig can be logged by MATLAB in real-time. It is thus possible to formulate a hardware-in-the-loop optimisation problem in MATLAB with the controller parameters as design variables and measures of the rotor's response as optimisation objectives.

This type of problem would prove difficult for a conventional optimiser as the optimisation is highly non-linear and subject to random noise. Genetic algorithms, however, are comparatively robust to these problems as they use a population of potential solutions and are stochastic in nature. A multiobjective genetic algorithm (MOGA) described in Fonseca and Fleming (1993), has been successfully applied to AMB control design problems in simulation, Schroder et al. (1997 a,b) and is therefore chosen as the optimisation engine for this on-line PID design. The advantage of using a multiobjective optimiser is that different measures of

performance can be optimised simultaneously without defining their relative importance a priori. Different measures of tracking performance are used as optimisation objectives, with their design specifications as targets for the optimiser.

The optimisation objective functions operate the rig in two modes. First, a known stabilising controller is used to levitate the rotor. Second, the parameters under evaluation are switched in and their performance monitored. If the controller is unstable, the stabilising controller is switched back in before the rotor moves too far from its desired location. A 5Hz square wave is constantly applied as a demand signal, (see Figure 2). This excites the system and allows various performance measures to be taken. The metrics used here are the peak overshoot (rising and falling) and the mean absolute error when the demand signal is high and when it is low. After a controller has been evaluated the rig is reset with the stabilising controller installed, to ensure that every controller is evaluated against the same metrics.

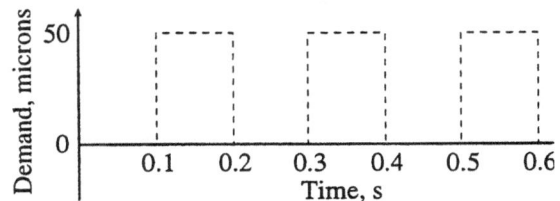

Figure 2: Demand signal for controller evaluation

4. RESULTS

An optimisation was performed for a PID controller on the non-drive end x-axis, with the other axes levitated using a stabilising controller. Figure 3 shows a typical trade-off graph for the AMB PID control system. The x-axis shows the design objectives and the y-axis shows the objective domain performance of the controllers. Each line represents a single solution's performance against each objective. The 'x' marks in the figure represent the optimisation targets, or system specifications. Trade-offs between adjacent objectives result in the crossing of lines whereas concurrent lines represent non-competing objectives. Table 1 shows the objectives and their displayed ranges.

The most striking feature of the trade-off graph is the difference in performance between rising and falling step demands, with much better performance being achievable with zero demand. This anisotropic behaviour is due to the off-centre location of the high demand position causing the plant to operate in a more non-linear region. This trade-off surface encompasses the achievable performance with this controller configuration and specification. It is possible to improve on any of the objectives beyond what is apparent here, but doing this will violate the

No.	Objective Name	Range
1	Percentage overshoot (rising)	0% - 250%
2	Mean absolute error (high)	0 - 50 microns
3	Percentage overshoot (falling)	0% - 250%
4	Mean absolute error (low)	0 - 50 microns

Table 1: The objectives

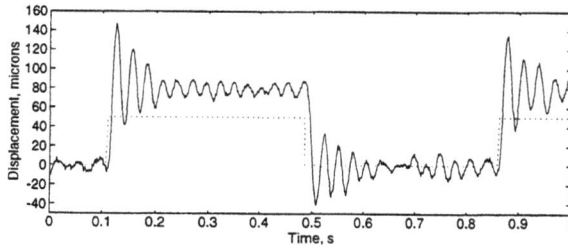

Figure 3: Sample trade-off graph

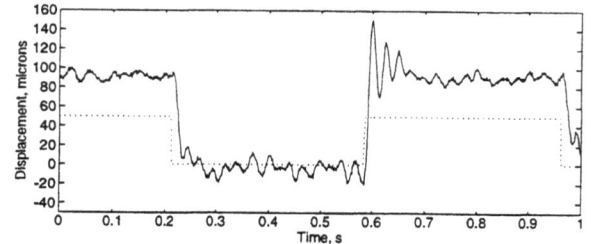

Figure 4: Rotor's response under controller 1

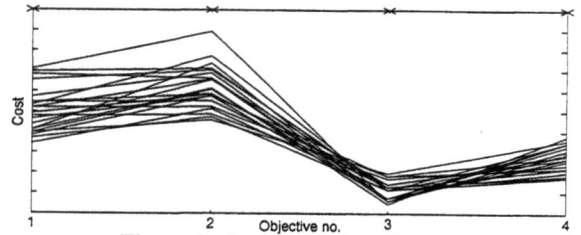

Figure 5: Rotor's response under controller 2

goal boundary for some or all the other objectives. Study of Figure 3 reveals that there is some trade-off between all the objectives. Objectives 1 and 2 only exhibit a little conflict, minimising one will tend to minimise the other, although it is not possible to completely minimise each one simultaneously. There is a greater conflict between objectives 2 and 3 and between 3 and 4, minimising one will tend to maximise the other.

To further illustrate the inherent trade-offs present in the system, response graphs of two controllers from Figure 3 are shown in Figures 4 and 5. The solid line represents the rotor's response and the dotted line represents the demand signal. Controller 1 exhibits the smallest mean absolute error when the demand is high and has correspondingly poor performance in the other metrics. Controller 2 achieves a slightly better zero demand error, but shows much worse performance in all the other objectives, as would be expected from the trade-off graph.

5. CONCLUDING REMARKS

A multiobjective genetic algorithm is used as a design tool for generating optimal active magnetic bearing controllers for a Rolls-Royce large electric machine application. The MOGA is used to search the PID parameter selection problem for the non-linear AMB system. The optimisation is performed directly onto the AMB rig, and several measures of AMB performance are used as objectives for the optimisation. From these a great deal of information about the limiting characteristics of the many possible controllers can be inferred. A collection of satisfactory controllers is generated, from which a controller with a good performance for the application can be selected. This powerful design technique not only gives insight into the behaviour of the system, but allows the designer to select the

most appropriate compromise control solution for the particular system under development. In the end, it is the designer who makes the decision about what controller parameters are to be used. The MOGA is simply used as an efficient way to explore the possibilities offered by each alternative.

6. ACKNOWLEDGEMENTS

The authors gratefully acknowledge the support of this research by Rolls - Royce & Associates Limited and UK EPSRC grants on "Evolutionary Algorithms in Systems Integration and Performance Optimisation" (GR/K 36591) and "Multiobjective Genetic Algorithms" (GR/J 70857).

REFERENCES

Fittro R. L., C. R. Knospe, L. S. Stephens, (1996). Experimental results of μ-synthesis applied to point compliance minimisation, *Proc. 5th International Symposium on Magnetic Bearings*, Kanazawa, Japan, pp. 203-208.

Fonseca C. M., P. J. Fleming, (1993). An overview of evolutionary algorithms in multiobjective optimisation. *Evolutionary Computation*, Vol. 1, No. 1, pp 25-49.

Nonami K., W. He, H. Nishimura, (1994). Robust control of magnetic levitation systems by means of H_∞ control / μ-synthesis. *JSME International Journal, Series C, Vol. 37*, No. 3.

Schroder P., A. J. Chipperfield, P. J. Fleming, N. Grum (1997a). Robust multivariable control of active magnetic bearings. *Proc. European Control Conference*, Brussels.

Schroder P., A. J. Chipperfield, P. J. Fleming, N. Grum (1997b). Multiobjective optimisation of distributed active magnetic bearing controllers. *Proc. GALESIA 97*, Glasgow.

ADAPTIVE INVERSE CONTROL BASED ON NONLINEAR ADAPTIVE FILTERING

Bernard Widrow[1], Gregory Plett[2], Edson Ferreira[3] and Marcelo Lamego[4]

Information Systems Lab., EE Dep., Stanford University

Abstract: Many problems in adaptive control can be divided into two parts; the first part is the control of plant dynamics, and the second is the control of plant disturbance. Very often, a single system is utilized to achieve both of these control objectives. The approach of this paper treats each problem separately. Control of plant dynamics can be achieved by preceding the plant with an adaptive controller whose transfer function is the inverse of that of the plant. Control plant disturbance can be achieved by an adaptive feedback process that minimizes plant output disturbance without altering plant dynamics. The adaptive controller is implemented using adaptive filters. *Copyright © 1998 IFAC.*

Keywords: Adaptive Control, Inverse Control, Adaptive Filters, Neural Networks, Nonlinear Systems.

1. INTRODUCTION

At present, the control of a dynamic system (the "plant") is generally done by means of feedback. This paper proposes an alternative approach that uses adaptive filtering to achieve feedforward control. Precision is attained because of the feedback incorporated in the adaptive filtering. The control of plant dynamic response is treated separately, without compromise, from the optimal control of the disturbance. All of the required operations are based on adaptive filtering techniques (Widrow and Walach, 1996). Following the proposed methodology, knowledge of adaptive signal processing allows one to go deeply into the field of adaptive control.

In order for adaptive inverse control to work, the plant must be stable. If the plant is not stable, then conventional feedback methods should be used to stabilize it. Generally, the form of this feedback is not critical and would not need to be optimized. If the plant is stable to begin with, no feedback would be required.
If the plant is linear, a linear control system would generally be used. The transfer function of the

controller converges to the reciprocal of that of the plant. If the plant is minimum phase, an inverse is easily obtained. If the plant is non-minimum phase, a delayed inverse can be obtained. The delay in the inverse results in a delay in overall system response, but this is inevitable with a non-minimum-phase plant. The basic idea can be used to implement "model-reference control" by adapting the cascaded filter to cause the overall system response to match a pre-selected model response.

Disturbance in a linear plant, whether minimum phase or non-minimum phase, can be optimally controlled by a special circuit that obtains the disturbance at the plant output, filters it, and feeds it back into the plant input. The circuit works in such a way that the feedback does not alter the plant dynamic response. So disturbance control and control of dynamic response can be accomplished separately. The same ideas work for MIMO systems as well as SISO systems.

Control of nonlinear plants is an important subject that raises significant issues. Since a nonlinear plant does not have a transfer function, how could it have an inverse? By using a cascade of the nonlinear adaptive filter with the nonlinear plant, the filter can learn to drive the plant as if it were the plant's inverse. This works surprisingly well for a range of training and operation signals. Control of dynamic response and plant disturbance can be done.

---

[1] Professor.
[2] Ph.D. student.
[3] Visiting Professor sponsored by CAPES/UFES, Brazil.
[4] Ph.D. student sponsored by CNPq/UFES, Brazil.

This paper introduces adaptive inverse control by first discussing adaptive filters. Then, inverse plant modeling for linear plants is described. The ideas are extended to nonlinear control, examples are presented and conclusions made.

2. ADAPTIVE FILTERS

An adaptive digital filter, shown in fig. 1 has an input, an output, and another special input called the "desired response". The desired response input is sometimes called the "training signal".

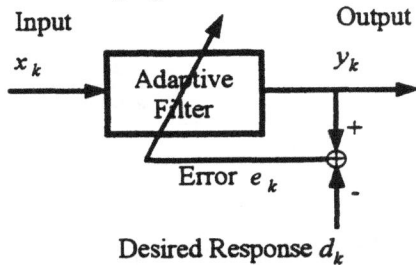

Fig. 1. *Symbolic representation of an adaptive transversal filter adapted by the LMS algorithm*

The adaptive filter contains adjustable parameters that control its impulse response. These parameters could, for example, be variable weights connected to the taps of a tapped delay line. The filter would thus be FIR, finite impulse response.

The adaptive filter also incorporates an "adaptive algorithm" whose purpose is to automatically adjust the parameters to minimize some function of the error (usually mean square error). The error is defined as the difference between the desired response and the actual filter response. Many such algorithms exist, a number of which are described in the text-books by Widrow and Stearns (1985) and by Haykin (1996).

3. INVERSE PLANT MODELING

The plant's controller will be an inverse of the plant. Inverse plant modeling of a linear SISO plant is illustrated in Fig 2. The plant input is its control signal. The plant output, shown in the figure, is the input of an adaptive filter. The desired response for the adaptive filter is the plant input (sometimes delayed by a modeling delay, Δ). Minimizing mean square error causes the adaptive filter $\hat{P}^{-1}$ to be the best least squares inverse to the plant P for the given input spectrum. The adaptive algorithm attempts to make the cascade of plant and adaptive inverse behave like a unit gain. This process is often called deconvolution. With the delay Δ incorporated as shown, the inverse will be a delayed inverse.

For sake of argument, the plant can be assumed to have poles and zeros. An inverse, if it also had poles and zeros, would need to have zeros where the plant had poles and

poles where the plant had zeros. Making an inverse would be no problem except for the case of a non-minimum phase plant. It would seem that such an inverse would need to have unstable poles, and this would be true if the inverse were causal. If the inverse could be non-causal as well as causal, however, then a two-sided stable inverse would exist for all linear time-invariant plants in accord with the theory of two-sided z-transforms. For useful realization, the two-sided inverse response would need to be delayed by Δ. A causal FIR filter can approximate the delayed version of the two-sided plant inverse. The time span of the adaptive filter (the number of weights multiplied by the sampling period) should be made adequately long, and the delay Δ needs to be chosen appropriately. The choice is generally not critical.

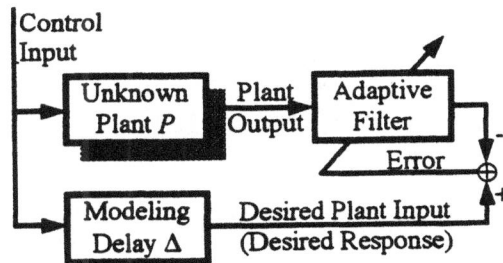

Fig.2. *Delayed inverse modeling of an unknown plant*

The inverse filter is used as a controller in the present scheme, so that Δ becomes the response delay of the controlled plant. Making Δ small is generally desirable, but the quality of control depends on the accuracy of the inversion process, which sometimes requires Δ to be of the order of half the length of the adaptive filter.

Fig. 3. *Adaptive inverse model control system*

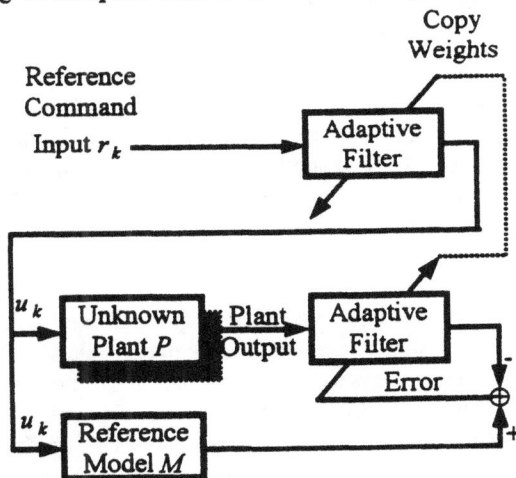

A model-reference inversion process is incorporated in the feedforward control system of Fig. 3. A reference model is used in place of the delay of Fig. 2. Minimizing mean square error with the system of Fig. 3 causes the cascade of the plant and its "model-reference inverse" to closely approximate the response of the reference-model M. Much is known about the design of model-reference systems (Landau, 1979). The model is chosen to give a desirable response for the overall system.

Thus far, the plant has been treated as disturbance free. But, if there is disturbance, the scheme of Fig. 4 can be used. A direct plant modeling process, not shown, yields $\hat{P}$, a close fitting FIR model of the plant. The difference between the plant output and the output of $\hat{P}$ is essentially the plant disturbance.

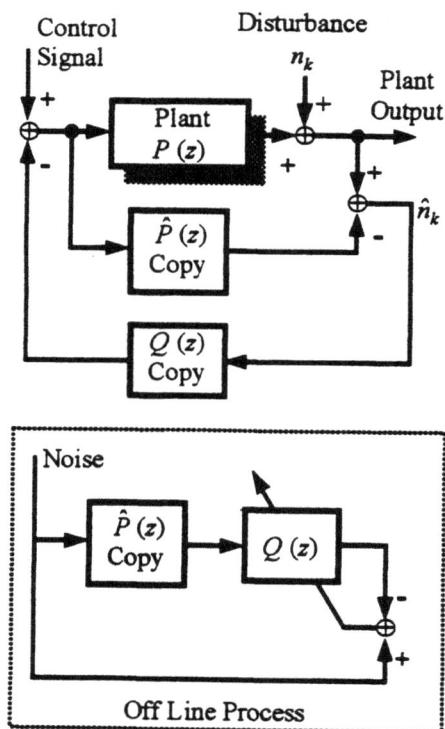

Fig. 4. *Optimal adaptive plant disturbance canceler*

Now, using a digital copy of $\hat{P}$ in place of P, an off line process, shown in Fig. 4, calculates the best least-squares plant inverse Q. The off line process can run much faster than real time, so that as $\hat{P}$ is calculated, the inverse Q is immediately obtained. The disturbance is filtered by digital copy of Q and subtracted from the plant input. For linear systems, the scheme of Fig. 4 has been shown to be optimal in the least-squares sense (Widrow and Walach, 1996).

To illustrate the effectiveness of adaptive inverse control, a non-minimum phase plant has been simulated, and its impulse response is shown in Fig. 5(a). the output of this plant and the output of its reference model are plotted in Fig. 5(b), showing dynamics tracking when the command input signal is a random first-order Markov process. The gray line is the desired output and the black line is the actual plant output. Tracking is quite good. With disturbance added to the plant output, Fig. 5(c) shows the effect of disturbance cancelation. Both the desired and actual plant outputs are plotted in the figure, and they become close when the canceler is turned on, at 300 samplings.

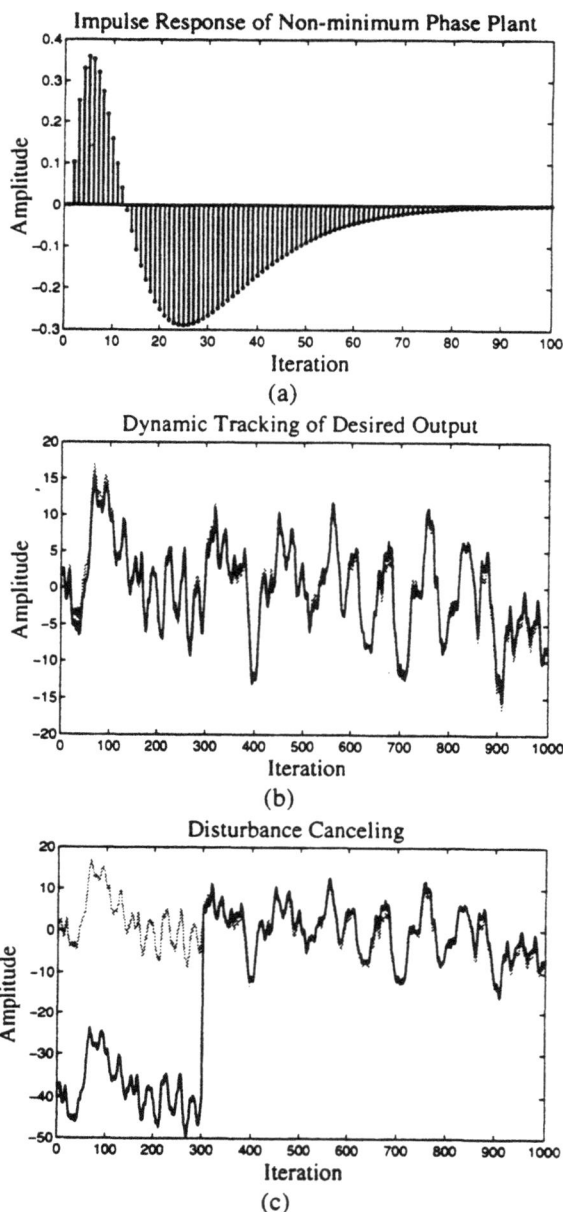

Fig. 5. *(a) Impulse response of the non-minimum phase plant used in simulation; (b) Dynamics tracking of desired output by actual plant output when the plant was not disturbed. (c) Cancelation of plant disturbance.*

4. NONLINEAR ADAPTIVE INVERSE CONTROL WITH NEURAL NETWORKS

Nonlinear inverse controllers can be used to control nonlinear plants. Although the theory is in its infancy, experiments can be done to demonstrate this. A nonlinear adaptive filter is shown in Fig. 6. It is composed of a neural network whose input is a tapped delay line connected to the exogenous input signal. In addition, the input to the network might include a tapped delay line connected to its own output signal. This type of nonlinear filter is called a Nonlinear Auto-Regressive with eXogeneous Input (NARX) filter, and has recently been shown to be a universal dynamical system (Siegelmann, et al., 1997). Algorithms such as real-time-recurrent-learning *(RTRL)* (Williams and Zipser, 1989) and the backpropagation-through-time *(BPTT)* (Werbos, 1990) may be used to adapt the weights of the neural network to minimize the mean squared error. If the feedback connections are omitted, the familiar backpropagation may be used (Werbos, 1974), (Rumelhart, et al., 1986). In the nonlinear adaptive inverse control scheme of Fig.7, such filters are used as the plant emulator and controller.

Nonlinear systems do not commute. Therefore, the simple and intuitive block-diagram method of Figs. 2 and 3, for adapting a controller to be the inverse of the plant, will not work if the plant is nonlinear. Instead, a lower-level mathematical approach is taken. We use an extension of the RTRL learning algorithm to train the controller. This method can be briefly summarized using the notation of ordered derivatives, proposed by Werbos

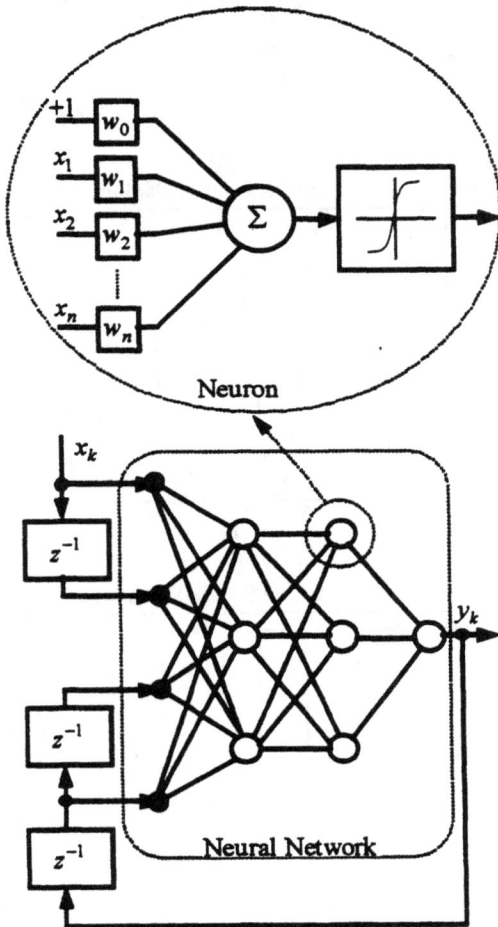

Fig. 6. *An adaptive nonlinear filter composed of a tapped delay line and a three-layer neural network.*

(1974). The goal is to adapt the weights of the controller to minimize the mean squared error of the output of the system. We use the fact that the controller computes a function of the form

$$u_k = g(u_{k-1}, u_{k-2}, ..., u_{k-m}, r_k, r_{k-1}, ..., r_{k-q}, W)$$

where W are the weights of the controller's neural network. We also use the fact that the plant model

computes a function of the form

$$y_k = f(y_{k-1}, y_{k-2}, ..., y_{k-n}, u_k, u_{k-1}, ..., u_{k-p}).$$

The weights of the controller are adapted using steepest descent. The change in the weights at each time step is in the negative direction to the gradient of the system error with respect to the weights of the controller. To find the gradient, we use the chain-rule expansion for ordered derivatives

$$\frac{\partial^+ \|e_k\|^2}{\partial W} = -2e_k \frac{\partial^+ y_k}{\partial W}$$

$$\frac{\partial^+ u_k}{\partial W} = \frac{\partial u_k}{\partial W} + \sum_{j=1}^{m} \left(\frac{\partial u_k}{\partial u_{k-j}} \right) \left(\frac{\partial^+ u_{k-j}}{\partial W} \right) \qquad (1)$$

$$\frac{\partial^+ y_k}{\partial W} = \sum_{j=0}^{p} \left(\frac{\partial y_k}{\partial u_{k-j}} \right) \left(\frac{\partial^+ u_{k-j}}{\partial W} \right) + \\ + \sum_{j=1}^{n} \left(\frac{\partial y_k}{\partial y_{k-1}} \right) \left(\frac{\partial^+ y_{k-j}}{\partial W} \right) \qquad (2)$$

Each of the terms in Eqs. (1) and (2) is either a Jacobian matrix, which may be calculated using the *dual-subroutine* (Werbos, 1992) of the backpropagation algorithm, or is previously calculated value of $\partial^+ u_k / \partial W$ or $\partial^+ y_k / \partial W$.

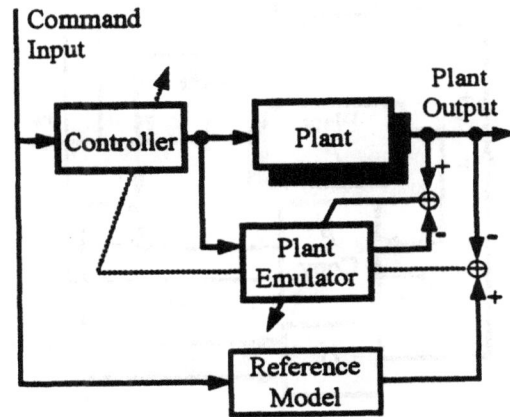

Fig. 7. *A method for adapting a nonlinear controller*

To be more specific, the first term in Eq. (1) is the partial derivative of the controller's output with respect to its weights. This term is one of the Jacobian matrices of the controller and may be calculated with the dual subroutine of the backpropagation algorithm.

The second part of Eq. (1) is a summation. The first term of the summation is the partial derivative of the controller's current output with respect to a previous output. However, since the controller is externally

214

recurrent, this previous output is also a current input. Therefore the first term of the summation is really just a partial derivative of the output of the controller with respect to one of its inputs. By definition, this is a submatrix of the Jacobian matrix for the network, and may be computed using the dual-subroutine of the backpropagation algorithm.

The second term of the summation in Eq. (1) is the ordered partial derivative of a previous output with respect to the weights of the controller. This term has already been computed in a previous evaluation of Eq. (1), and need not be re-computed.

A similar analysis may be performed to determine all of the terms required to evaluate Eq. (2). After calculating these terms, the weights of the controller may be adapted using the weight-update equation

$$\Delta W_k = 2\mu e_k \frac{\partial^+ y_k}{\partial W}$$

Continual adaptation will minimize the mean squared error at the system output.

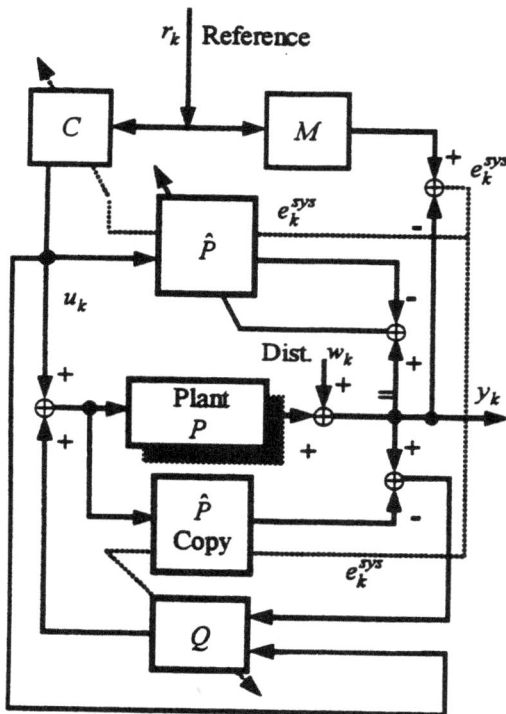

Fig. 8. *A fully integrated nonlinear adaptive inverse control scheme.*

Disturbance canceling for a nonlinear system is performed by filtering an estimate of the disturbance with the nonlinear filter Q and adding the filter's output to the control signal. An additional input to Q is the control signal to the plant u_k, to allow the disturbance canceler knowledge of the plant state. The same algorithm which was used to adapt the controller can be used to adapt the disturbance canceling filter. The entire control system is shown in Fig. 8.

An interesting discrete-time nonlinear plant has been studied by Narendra and Parthasarathy (1990)

$$y_k = \frac{y_{k-1}}{1 + y_{k-1}^2} + u_{k-1}^3 .$$

The method just described for adapting a controller and disturbance canceler were simulated for this plant, and the results are presented here.

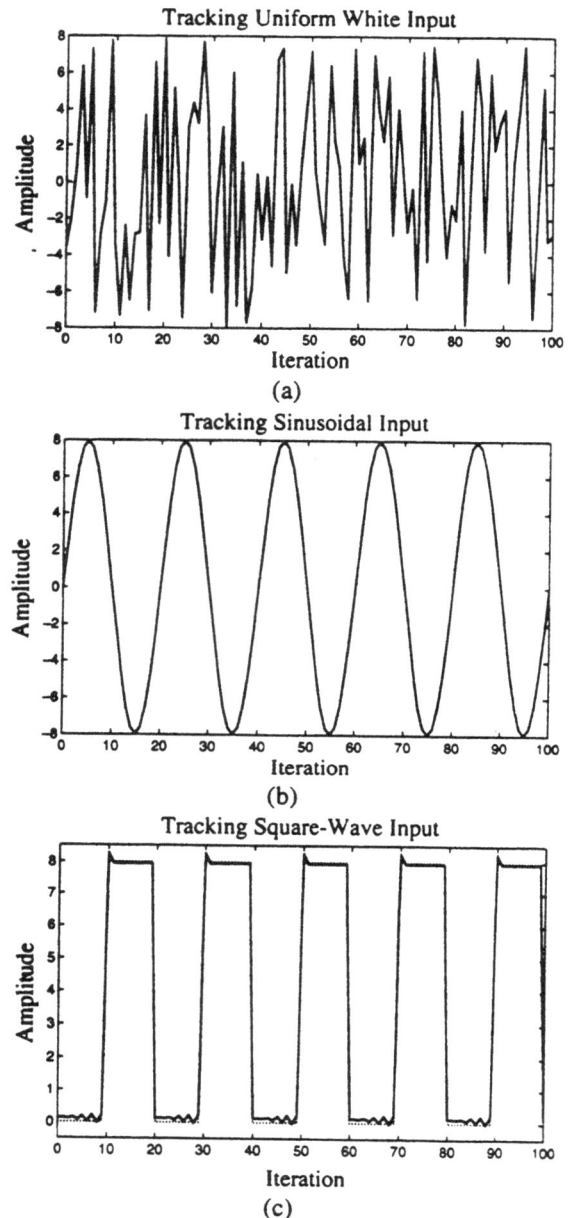

(a)

(b)

(c)

Fig. 9. *Feedforward control of a nonlinear system. The controller was trained with uniform distributed (white) random input (a). Plots (b) and (c) show the plant tracking sinusoidal and square waves, having been previously trained with the random input.*

With the reference model being a simple delay, and the command input being an i.i.d (independent and identically distributed) uniform process, the system adapted and learned to track the model output. The result

215

is shown in Fig. 9(a). The desired plant output (gray line) and the true plant output (solid line) are shown, at the end of training, when the training signal was used to drive the controller. The gray line is completely covered by the black line, indicating near-perfect control. With the weights fixed at their trained values, the next two plots show the generalization ability of the controller. After training with the random input, the adaptive process was halted. With no further training, the system was tested with inputs of different character in order to demonstrate the generalization ability of the controller. The first test was a sine-wave command input. Tracking was surprisingly good, as shown in Fig. 9(b). Again, without further training, the system was tested with a square-wave command input, and the results, shown in Fig. 9(c), are excellent.

A disturbance canceler was also trained for this plant, were the disturbance was a first-order Markov signal added to the plant output. Fig.10 shows the results of disturbance cancelation. The power of the system error is plotted versus time. The disturbance canceler was turned on at iteration 500. Dramatic improvement may be seen.

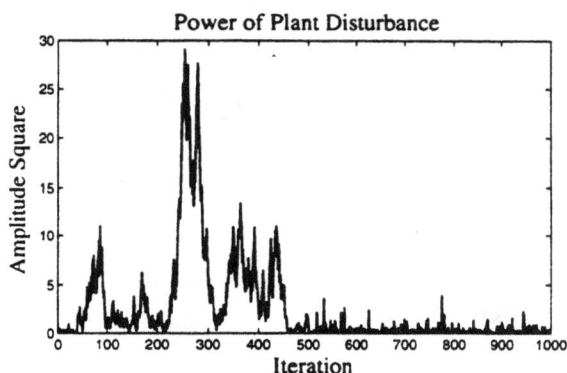

Fig. 10. *Cancelation of plant disturbance for a nonlinear plant. The disturbance canceler was turned on at iteration 500.*

CONCLUSIONS

Adaptive control is seem as a two part problem, (a) a control of plant dynamics, and (b) control of plant disturbance. Conventionally, one uses feedback control to treat both problems simultaneously. Tradeoffs and compromises are necessary to achieve good solutions, however.

The method proposed here, based on inverse control, treats the two problems separately without compromise. The method applies to SISO and MIMO linear plants, and to nonlinear plants.

An unknown linear plant will track an input command signal if the plant is driven by a controller whose transfer function approximates the inverse of the plant transfer function. An adaptive inverse identification process can be used to obtain a stable controller, even if the plant is non-minimum phase. A model-reference version of this idea allows system dynamics to closely approximate

desired reference-model dynamics. No direct feedback is used, except that the plant output is monitored and utilized by an adaptive algorithm to adjust the parameters of the controller. Although nonlinear plants do not have transfer functions, the same idea works well for nonlinear plants.

Control of internal plant disturbance is accomplished with an adaptive disturbance canceler. The canceler does not affect plant dynamics, but feeds back plant disturbance in a way that minimizes plant output disturbance power. This approach is optimal for linear plants and works surprisingly well with nonlinear plants.

A great deal of work will be needed to gain greater understanding of this kind of behavior, but the prospects for useful and unusual performance and for development of this new approach seem very promising.

REFERENCES

Haykin, S. (1996). *Adaptive Filter Theory*. Prentice Hall, third edition, Upper Saddle River, NJ.

Landau, I. D. (1979). *Adaptive Control. The Model Reference Approach*, volume VIII of Control and Systems Theory Series. Marcel Dekker, New York.

Narendra, K. S. and K. Parthasarathy (1990). Identification and control of dynamical systems using neural networks. *IEEE Transactions on Neural Networks*, **Vol. 1(1)**, March.

Rumelhart, D. E., G. E. Hinton and R. J. Williams (1986). Learning internal representations by error propagation. *Parallel Distributed Processing*. (D. E. Rumelhart and J. L McClelland editors), volume 1, chapter 8. MIT Press, Cambridge, MA.

Siegelmann, H. T., B. B. Horne and C. L. Giles (1997). Computational capabilities of recurrent NARX neural networks. *IEEE Transactions on Systems, Man and Cybernetics — Part B: Cybernetics*, **Vol. 27(2)**, April, pp. 208–215.

Werbos, P. (1974). *Beyond Regression: New Tools for Prediction and Analysis in the Behavioral Sciences*. PhD thesis, Harvard University, Cambridge, MA, August.

Werbos, P. (1990). Backpropagation through time: What it does and how to do it. *Proceedings of the IEEE*, **Vol. 78(10)**, October, pp. 1545–1680.

Werbos, P. (1992). Neurocontrol and supervised learning: An overview and evaluation. *Handbook of Intelligent Control: Neural, Fuzzy and Adaptive Approaches*. (D. White and D. Sofge editors), chapter 3. Van Nostrand Reinhold, New York.

Widrow, B. and S. D. Stearns (1985). *Adaptive Signal Processing*. Prentice Hall, Englewood Cliffs, NJ.

Widrow, B. and E. Walach (1996). *Adaptive Inverse Control*. Prentice Hall PTR, Upper Saddle River, NJ.

Williams, R. J. and D. Zipser (1989). Experimental analysis of the real-time recurrent learning algorithm. *Connection science*, **Vol. 1(1)**, pp.87–111.

BIOLOGICALLY INSPIRED REAL-TIME RECONFIGURATION
TECHNIQUE FOR PROCESSOR ARRAYS

Cesar Ortega and Andy Tyrrell

Department of Electronics
University of York
York, YO1 5DD, UK
E-mail: (cesar, amt)@ohm.york.ac.uk

Abstract: Real-time control applications involve the interaction of multiple components. As systems become more complex their reliability tends to decrease, hence, fault tolerance must be incorporated to keep reliability within specified levels. A reconfiguration mechanism for processor arrays inspired by mechanisms that take place during the embryonic development of living beings is proposed in this paper. It is illustrated using an example that the rapid fault-recovery characteristic of the embryonic system makes it a promising approach for real-time control applications. *Copyright © 1998 IFAC*

Keywords: Fault-tolerant systems, Parallel processors, Binary decision systems, Bio control, Embryonics.

1. INTRODUCTION

Practical experience has demonstrated that the goal of building fault-free real-time control systems, although attractive, is impossible to achieve. Hardware deteriorates with time, and software has become so complex that design faults are difficult, if not impossible, to avoid. Hence, the more viable alternative for applications requiring high levels of dependability is to implement systems capable of tolerating faults, *i.e.* Fault-Tolerant Systems.

The complexity of real-time control problems solved by computers and the complexity of computers themselves is increasing with time, therefore, the risk of failure is greater now than when computers and the problems they solved were simple. Complexity implies unreliability. Hence, it is necessary to look for new methodologies and strategies to deal with complex systems. One approach is the refinement of traditional design techniques, but the techniques themselves are

becoming too complex to be considered error-free. Evidently, it is necessary to look somewhere else for the answers.

Nature offers some remarkable examples of how to deal with complexity and its associated unreliability. The human body is one of the most complex systems ever known. Failures are not rare, but the overall function is highly reliable because of self-diagnosis and self-healing mechanisms that work ceaselessly throughout our bodies. To borrow the main principles that sustain this mechanisms and applying them to the design of electronic systems, could result in a new approach for the design of fault-tolerant systems (Avizienis, 1997).

Of particular interest are the fault tolerance attributes of massively parallel processing networks or processor arrays, similar in structure to cellular automata, capable of self-diagnosis and self-repair. In this approach the "knowledge" is distributed throughout multiple processing elements, therefore, if one or a small subset of processors fail, the overall functionality can be maintained (Chean and Fortes, 1990a; Dyer, 1995). For the purposes of this paper processor arrays will be considered as a special case of cellular automata, therefore the terms processor and cell will be used interchangeably.

Self-healing mechanisms found in nature and the implicit redundancy in processor arrays constitute the inspiration for the presented work. Embryonics' proposal is to construct field programmable processor arrays with self-diagnosis and self-reconfiguration abilities able to tolerate the presence of failing cells.

The embryonics concept was initially presented in (Mange, *et al.*, 1996 and Marchal, *et al.*, 1996). Embryonics (Embryology + Electronics) is inspired by the embryonic development of multicellular organisms, *i.e.* a new individual is evolved from a single cell (the fertilised egg) through a process of multiple divisions and specialisation. Just after conception there is only one copy of the organism's DNA, a replica of which is passed to every cell during embryo's development. The function of a particular cell of group of cells is determined by that of their neighbours (Nüsslen-Volhard, 1996), in other words, a cell's function would be different if located in a different position inside the embryo. This is possible because every cell possess a complete copy of the DNA describing the organism.

2. THE PRESENT APPROACH TO FAULT TOLERANCE IN PROCESSOR ARRAYS

Fault tolerance in processor arrays implies the mapping of a logical array onto a physical non-faulty array, *i.e.* every logical cell must have a correspondent physical cell. When faults arise, a mechanism must be provided for reconfiguring the physical array such that the logical array can still be represented by the remaining non-faulty cells. All reconfiguring mechanisms can be considered to be based on one of two types of redundancy: Time redundancy or hardware redundancy (Chean and Fortes, 1990b).

In time redundancy the tasks performed by faulty cells are distributed among its neighbours. When reconfiguration occurs, processors share their time between performing their own tasks and the faulty cells functions, resulting in some degradation of system's performance. In addition, the algorithm being executed must be flexible enough to allow a simple and flexible division of tasks in real time.

In hardware redundancy physical spare cells and links are used to replace the faulty ones. Therefore, reconfiguring algorithms must optimise the use of

spares. In the ideal case a processor array with N spares must be able to tolerate N faulty cells; but, in practice, limitations on the interconnection capabilities of each cell prevents this goal from being achieved. The work presented in this paper is focused on this type of redundancy giving a new slant to the structure and reconfiguration process.

Most of hardware redundancy reconfiguration techniques rely on complex algorithms to re-assign physical resources to the elements of the logical array. In most cases these algorithms are executed by a central processor which also performs diagnosis functions and co-ordinates the reconfiguration of the physical array (Dutt and Mahapatra, 1997; Fortes and Raghavendra, 1985). This approach has demonstrated to be effective, but its centralised nature makes it prone to collapse if the processor in charge of the fault tolerance functions fails. Furthermore, the timing of signals involved in the global control from a host computer is often prohibitively long, therefore, unsuitable for real-time fault tolerance. Only distributed processing is feasible (Kung, *et al.*, 1989).

An alternative approach is to distribute the diagnosis and reconfiguration algorithms among all the cells in the array. In this way no central agent is necessary and consequently the reliability and time-response of the system should improve. However, this decentralised approach does tend to increase the complexity of the reconfiguration algorithm and the amount of communications within the network.

Embryonics' proposal is to embed some distinctive characteristics of biological cellular systems into the design of programmable processor arrays in order to achieve rapid reconfiguration suitable for real-time control applications.

3. THE BIOLOGICAL APPROACH TO FAULT TOLERANCE

A human being consists of approximately 60 trillion (60×10^{12}) cells. At each instant, in each of these 60 trillion cells, the genome, a ribbon of 2 billion characters, is decoded to produce the proteins needed for the survival of the organism. This genome contains the ensemble of the genetic inheritance of the individual and, at the same time, the instructions for both the construction and the operation of the organism. The parallel execution of 60 trillion genomes in as many cells occurs ceaselessly from conception to death of an individual. Faults are rare and, in the majority of cases, successfully detected and repaired (Mange, *et al.*, 1996). Which part of the DNA is interpreted will depend on the physical location of the cell with respect to its neighbours.

Embryonics is inspired by the basic processes of molecular biology. By adopting the features of biological cellular organisation, and by transposing them to the two-dimensional world of cellular arrays, it can be shown that properties unique to the living world, such as self-reproduction and self-repair, can also be applied to integrated circuits. Figure 1 shows the basic architecture of an embryonic system.

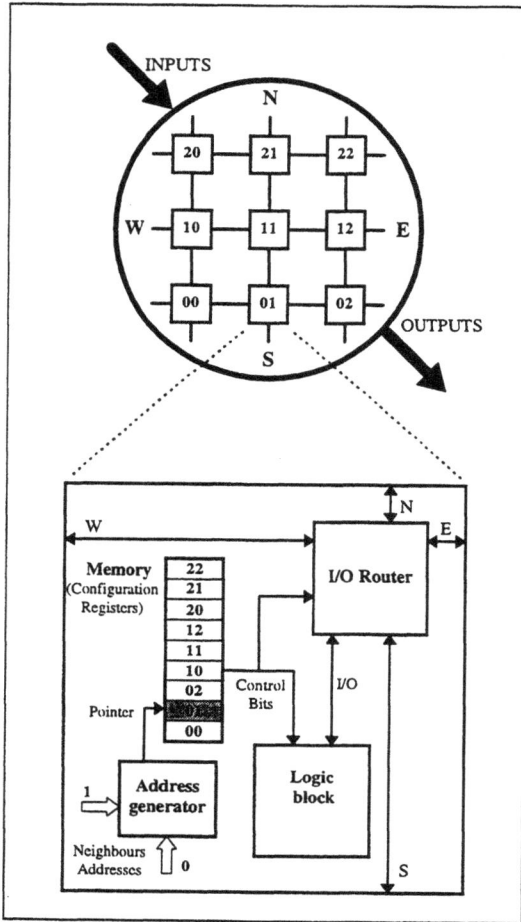

Fig. 1 Basic Components of an Embryonic Cell.

This architecture presents the following advantages:
- It is highly regular, which simplifies its implementation on silicon.
- The actual functionality of the logic block is independent from the function of the remaining blocks. This modularity could be exploited to produce a family of Embryonic FPGAs, each member offering a particular logic function, *e.g.* ALU, MUX, DEC, etc.
- The simplicity of blocks' architecture allows the implementation of built-in self test (BIST) logic to provide self-diagnosis without excessively incrementing the silicon area.

A full description of the self-diagnosis system can be found in (Tempesti, 1997). If one of the cells fails, it becomes transparent allowing another cell to take its co-ordinates and therefore its function. The block corresponding to error-detection and error-handling is not shown in figure 1. This cell differs from those proposed to improve the yield of VLSI or WSI cellular architectures in that it is autonomous and the reconfigurability can be used both during manufacturing and normal operation.

Every cell has communication links with its north east west and south neighbours. The programmable I/O router block allows the spread of information all over the array. This block is controlled by the configuration register selected by the cell.

The reconfiguration logic is designed so that when a cell auto-diagnoses faulty, the row for the corresponding cell is eliminated and replaced by a spare row. Co-ordinates for the cells above the row being eliminated are recalculated and new configuration registers are selected accordingly. No "code" is being communicated around the array when reconfiguration takes place, only Boolean signals are passed between cells. The reconfiguration is implemented by simply using a different local memory location. This strategy is far from being optimal with respect to the use of spare resources, but the short time needed to recover from a failure makes it attractive to implement real-time control systems (Allworth, 1990). A detailed description of the cell can be found in (Ortega and Tyrrell, 1998).

4. EXAMPLE

A simple application is presented next in order to highlight the reconfiguration properties of the embryonics strategy. A multiplexer is selected to be the functional part of the basic cell. The multiplexer has the particular advantage of being able to implement any node from a binary decision diagram (BDD), which in turn can represent any combinatorial or sequential logic function (Akers, 1978). Therefore, the architecture presented is ideal for implementation on a field programmable gate array (FPGA) (Tempesti *et al.*, 1997).

The goal was to design a programmable frequency divider. Figure 2 shows the block diagram of the circuit. It is composed by a 3-bit selector which latches either the division factor **n** or the next state of a 3-bit down-counter according to the output of a zero detector. In this way, a 1-cycle wide pulse will be generated every **n** cycles of **F**.

The output of the circuit is taken from the output of the zero detector. It will be 1 during one cycle of **F** when the down-counter reaches the 000 state.

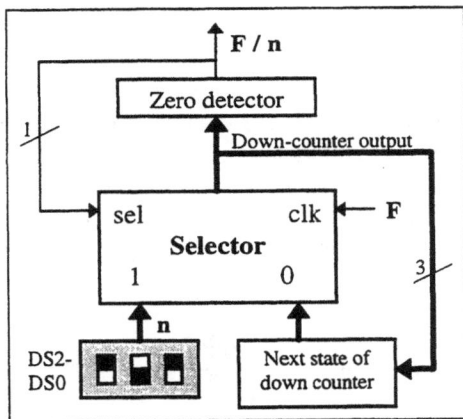

Fig. 2 Programmable frequency divider.

Figure 3 shows the binary decision diagrams for the combinatorial down-counter and the zero detector. A,B,C are the outputs of the 3-bit down counter, C being the most significant bit. The diagrams must be evaluated from top to bottom. If the variable on each node takes the value 0, then the branch with a broken line is selected. Otherwise, if the variable takes the value 1, then the branch represented with a solid line is followed. The final nodes on a BDD represent the output of the function for a particular set of input values. A good tutorial on BDDs can be found in (Bryant, 1986; Rauzy, 1996).

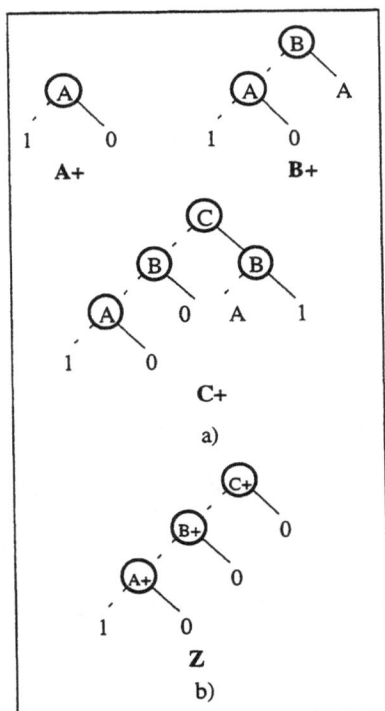

Fig. 3. BDDs for a) 3-bit down counter; and b) 3-bit zero detector.

Figure 4 shows the hardware implementation of the BDDs shown in figure 3.

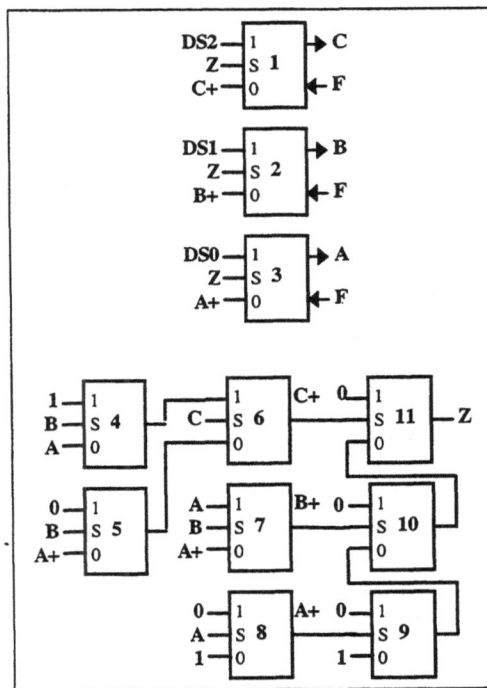

Fig. 4. Hardware implementation of BDDs in figure 3 using multiplexers.

From figure 3 note that the BDD for A+ is repeated in the diagrams for B+ and C+, therefore the corresponding network of multiplexers is simpler. Multiplexers 1, 2 and 3 implement the selector block in figure 2. These multiplexers operate in synchronous mode, *i.e.* their outputs will be updated on the rising edge of F. DS2, DS1 and DS0 are used to set the value of n.

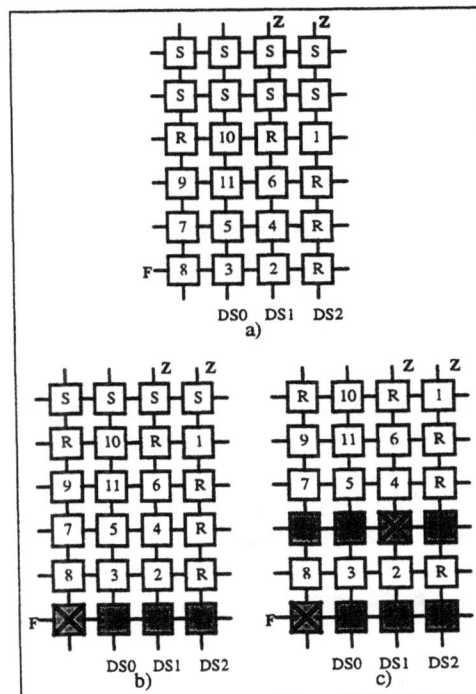

Fig. 5. Frequency divider: a) Without fails; b) with one faulty cell; c) with two faulty cells.

Figure 5a shows the final distribution of multiplexers needed to implement this particular application. The numbers on each cell correspond with the numbers assigned to the multiplexers in figure 4. Cells labelled S are spare cells, two rows in this example. Cells labelled R are routing cells. Routing cells are needed because every cell has direct connections only to its cardinal neighbours. Figures 5b and 5c show the reconfiguration process when one and two cells become faulty, respectively. Remember that during the reconfiguration process the cells' co-ordinates are re-calculated so that every cell selects the configuration register of the cell it is substituting.

Figure 6 shows the simulation results obtained for the frequency divider. Labels correspond with those of figure 2. **OK4** and **OK8** simulate fails in cells 4 and 8 respectively. Reconfiguration takes from 3 to 5 clock cycles.

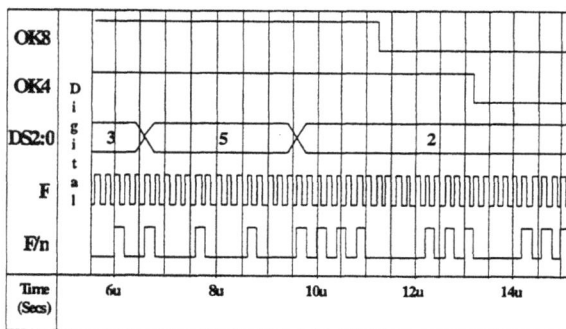

Fig. 6. Simulation results for frequency divider.

5. CONCLUSIONS

A novel paradigm for designing fault-tolerant processor arrays inspired by biological processes has been presented. The regularity of the architecture makes it suitable to be implemented using state of the art FPGAs or WSI circuits. The array's rapid reconfiguration characteristics should be ideal for the use of embryonic systems in real-time control applications where response time is a critical constraint. Embryonics is a nascent discipline, therefore much research must be done to investigate in depth the real-time fault-tolerant properties of these systems. The embryonics' approach is coherent with a recent uproar about the application of biological concepts to the solution of engineering problems, *e.g.* evolutionary computing, evolvable hardware and genetic algorithms.

ACKNOWLEDGEMENTS

This work was partially supported by the Mexican Government under grant CONACYT-111183.

REFERENCES

Akers S. (1978). Binary Decision Diagrams. *IEEE Trans. on Computers*, **Vol.27-6**, pp.509-516

Allworth S. and Zobel R. (1990). *Introduction to Real-time Software Design*, Macmillan, Hong Kong

Avizienis A. (1997). Toward Systematic Design of Fault-Tolerant Systems. *IEEE Computer*, **April**, pp.51-58

Bryant R. (1986). Graph-based Algorithms for Boolean Function Manipulation. *IEEE Trans. on Computers*, **Vol.35-8**, pp.677-691

Chean M. and Fortes J. (1990a), The Full Use of Suitable Spares (FUSS) Approach to Hardware Reconfiguration for Fault-Tolerant Processor Arrays, *IEEE Trans. on Computers*, **Vol.39-4**

Chean M. and Fortes J. (1990b). A Taxonomy of Reconfiguration Techniques for Fault-Tolerant Processor Arrays. *Computer*, **January**, pp. 55-69

Dutt S. and Mahapatra N. (1997). Node-covering, Error-correcting Codes and Multiprocessors with Very High Average Fault Tolerance. *IEEE Trans. on Computers*, **Vol.46-9**, pp.997-1014

Dyer M. (1995). Toward Synthesising Artificial Neural Networks that Exhibit Cooperative Intelligent Behaviour: Some open issues in Artificial Life. In: *Artificial Life: an Overview* (Langton C. (Ed)), pp.111-134. MIT Press, USA.

Fortes J. and Raghavendra C. (1985). Gracefully Degradable Processor Arrays. *Trans. on Computers*, **Vol.34-11**, pp.1033-1043

Mange D. *et al.* (1996). Embryonics: A new family of coarse-grained FPGA with self-repair and self-reproduction properties. In: *Towards Evolvable Hardware* (Sanchez E. *et al.* (Ed)), pp.197-220. LNCS 1062, Springer-Verlag, Berlin.

Marchal P. *et al.* (1996). Embryonics: The Birth of Synthetic Life. In: *Towards Evolvable Hardware* (Sanchez E. *et al.* (Ed)), pp.166-196. LNCS 1062, Springer-Verlag, Berlin.

Nüsslein-Volhard C. (1996). Gradients That Organize Embyo Development. *Scientific American*, **August**, pp.38-43

Ortega C. and Tyrrell A. (1998). Design of a Basic Cell to Construct Embryonic Arrays. IEE Procs. on Computers and Digital Techniques, in print.

Rauzy A. (1996). A Brief Introduction to Binary Decision Diagrams. *Journal Européen des Systèmes Automatisés*, **Vol. 30-8**, pp.1033-1050

Kung S. *et al.* (1989). Fault-Tolerant Array Processors Using Single-Track Switches, *Trans. on Computers*, **Vol.38-4**, pp. 501-513

Tempesti G. *et al.* (1997). A robust multiplexer-based FPGA inspired by biological systems. *Journal of Systems Architecture*, **Vol.43**, pp.719-733

COMPLETE DISAGREEMENT IN REDUNDANT REAL-TIME CONTROL APPLICATIONS

G. Latif-Shabgahi, J. M. Bass ([1]) and S. Bennett

Department of Automatic Control and System Engineering, The University of Sheffield,

Mappin Street, Sheffield, S1 3JD, UK

Abstract: Redundant application algorithm "variants" are used to minimise errors in fault-tolerant real-time control systems. Voting algorithms arbitrate between variant results to select a single, hopefully correct, output. Both variants and voting algorithms can be implemented in either hardware or software. The behaviour of voting algorithms in multiple error scenarios is considered in this paper. Complete disagreement is defined as those cases where no two variant results are the same. A novel algorithm for real-time control applications, the smoothing voter, is introduced and its behaviour compared with previously published algorithms. Software implemented error-injection tests, reported here, show that the smoothing voter achieves a compromise between the result selection capabilities of the median voter and the safety features of the majority voter. *Copyright © 1998 IFAC*

Keywords: Fault-tolerance, Embedded Systems, Redundancy, Safety-Critical Systems.

1. INTRODUCTION

In real-time control redundant subsystems are used to achieve stringent safety and / or reliability constraints. Voting algorithms arbitrate between the outputs of these redundant subsystems. The problem of voting on inexact results arises when correctly functioning variants produce slightly different results due to, for example, rounding or truncation errors. Inexact voters may use thresholding techniques to define boundaries within which agreement can be said to exist. Results that lie outside the *threshold*, or lie a large distance away from some average value, are not considered to be in agreement. Complete

---

[1] Now at, Department of Electronics, University of Hertfordshire

disagreement is defined as those cases where no two variant results are the same.

Related work on voting algorithms is presented in section 2. A novel voting technique, the smoothing voter, is introduced in section 3. The software implemented error-injection test harness is discussed in section 4 and results presented in section 5. Conclusions of the work are discussed in section 6.

2. RELATED WORK

2.1 Voting Algorithms

Voting algorithms are required to identify those variant results in agreement and select one of these as the voter output. The majority voter outputs a result from among n variant results where at least $[(n + 1)/2]$ variant results agree. The formalised plurality voter implements m-out-of-n voting, where m is less than a strict majority (e.g. 2-out-of-5 voting). A weighted average voter calculates the mean of the variant results. This mean value is output as the voter result and need not be identical to any of the voter inputs. The median voter is a mid-value selection algorithm in which those pairs of variant results with the largest distance from the mid-value are repeatedly selected and discarded. These algorithms have been extended from triple-modular redundant to n-modular redundant configurations in (Lorczak et al., 1989). An optimal voter can be proposed, where the probability of error arrivals is known (or can be assumed) (Blough and Sullivan, 1990). Thus, an ideal voter can be described mathematically. However, in real systems exact knowledge of error arrival probabilities is not known. Thus, the optimal voter output for any given set of variant results is also unknown. Gersting, (1991) considered and proposed the problem of voting on vectors of results, rather than individual results; algorithms that arbitrate between vectors with different elements in disagreement. While (Parhami, 1994) considered the performance, in terms of execution time, of voters, and proposed efficient implementations of a variety of algorithms.

Embedded control applications are typically cyclic systems in which there exists some relationship between the result in one cycle and the result in the next. Knowledge of this relationship between successive results is used in predictive voters (Bass, 1995) to produce results in cases of disagreement. A history of previous voter results is used to generate an expected result value where disagreement is detected. The expected result is compared with each of the variant results in order to make a selection.

2.2 Comparison of Voters

The cases in which each algorithm is capable of generating incorrect results is tabulated in (Lorczak et al., 1989). The authors show that the majority and plurality algorithms can produce exceptions (i.e. benign as opposed to catastrophic errors) where disagreement is detected. The weighted average and median voters, in contrast, can produce catastrophic errors in such cases of disagreement. Algorithms are categorised in (Parhami, 1994) depending upon the type of voting algorithm used (exact, inexact and approval), the rule for output selection (plurality or threshold) and properties of the input space.

There is relatively little published work contrasting voter behaviour in multiple error scenarios. Many authors take the view that multiple simultaneous errors are sufficiently unlikely that they can be ignored. However, such multiple error scenarios can produce catastrophic failures and should therefore be analysed. In safety-critical systems any catastrophic failure is unacceptable and potential causes should explored. The error detection capabilities of the majority and plurality voters are obtained at a price. It is easy to imagine double simultaneous errors (in a triple modular redundant configuration) which can be masked by selecting the mid-value variant result. The median voter, in particular, is shown to be successful at masking such double errors. The smoothing voter has been designed to achieve a compromise between the error detection capabilities of the majority voter and the results selection capabilities of the median voter.

The empirical study, reported in Sections 4 and 5 below, contrasts the behaviour of algorithms in these multiple error cases. Firstly a new algorithm is introduced in Section 3, the smoothing voter, whose behaviour has been specifically designed considering complete disagreement cases.

3. SMOOTHING VOTER

3.1. Definition

This algorithm extends the majority voter by adding a special kind of acceptance test. The acceptance test is based on the assumption that a discontinuity between consecutive variant results is indication of an error. This assumption is valid in many real-time embedded control applications where there is feedback control and periodic computation. In the smoothing voter, when there is no agreement between variant results, the closest result to the previous voter output is selected as the probable output for this cycle. If the measured distance is smaller than a pre-defined value named as *'smoothing threshold'*, then that result is taken as the voter output, otherwise, the voter fails to produce answer. It is assumed that the first cycle of algorithm is successful and its output is available for next cycle. The selection of a value for the *'smoothing threshold'* parameter of this algorithm is critical.

Although arbitrary values can be used, improved performance will be obtained if information about the probable discontinuity size of consecutive results of the system during its mission time is available. This algorithm can be defined more formally as follows:

S1. Let A = { d_1 d_2 d_3 ... d_n } denote the set of N variant results

S2. Sort the set A in ascending order to construct the new set AS = { x_1 x_2 ... x_n }

S3. Construct the following partitions from AS , for all j = 1: m , in which m = (N+1) / 2 :

$$V_j = \{ x_j \ x_{j+1} \ ... \ x_{j+m-1} \}$$

S4. If at least one of the partitions V_j (j = 1, 2, ...m) satisfies the property $d(x_j, x_{j+m-1}) \leq \varepsilon$, then the majority for original set A is satisfied. Here ε is the voter threshold and d is a distance metric between elements.

S5. If none of the partitions V_j satisfies the property $d(x_j, x_{j+m-1}) \leq \varepsilon$, then determine the output x_k in A such that $d(x_k, X) = \min\{ d(x_1, X), d(x_2, X) ... d(x_n, X) \}$, where X is the previous successful voter result.

S6. If x_k satisfies the condition $d(x_k, X) \leq \beta$, where β is the smoothing threshold, then select x_k as the voter output, otherwise, the voter has not found agreement.

3.2. Basic Smoothing voter

The algorithm defined in section 3.1 is called *Basic smoothing voting algorithm* in which the smoothing threshold β is fixed. In table 1 the results of this voter for a stream of inputs including seven samples have been illustrated. It is assumed that voter threshold ε is equal to 0.1 and smoothing threshold β is 1. In the first sample, agreement has been found and the result of variant one has been selected as voters' output. There is no majority for the second sample, therefore according to S5 the closest variant output of this cycle to the previous successful output of voter, 2.1, is chosen. Note that the difference between this selected output and previous output of voter is smaller than β. The output of third cycle is 3.1 due to the existing majority between variant results. Again, in cycle 4, there is no agreement, the candidate output is 4.2 which does not satisfy the necessary condition $d(4.2 , 3.1)$ <β. Thus a *no vote exception* is raised in sample 4. The following two cycles have a similar situations while for 7th sample a majority is achieved. There is   a problem which occurs with this fixed smoothing threshold. As stated, at sample 4 there is disagreement between var1, var2 and var3 and now var1 differs by 1.1 from previous value hence no output is produced. Since no output is produced the reference for future computations remains 3.1 produced at sample 3. Hence the voter

cannot produce an agreed output until two of the variants satisfy the threshold β. This is shown at sample 7.

Table 1. Basic smoothing alg. ($\varepsilon = 0.1, \beta = 1$)

sample	var1	var2	var3	result
1	1	1.1	1.4	1
2	2.1	2.3	3	2.1
3	3.1	3.2	7	3.1
4	4.2	4.9	5.5	no result
5	5	5.3	5.8	no result
6	6	6.8	6.5	no result
7	7	7.1	7.5	7

3.3. Modified Smoothing voter

A more rapid recovery from a transient error is achieved, by the introduction of a cumulative smoothing threshold which is dynamically adjusted after each cycle where *no result* is found. Where each successive variant result is more positive than its predecessor then the initial smoothing threshold β is added to the cumulative smoothing threshold. In contrast, where each successive variant result is more negative than its predecessor then the initial smoothing threshold is subtracted from the cumulative smoothing threshold. The behaviour in this case is shown in table 2.

Table 2. Modified smoothing alg. ($\varepsilon = 0.1, \beta = 1$)

sample	var1	var2	var3	result
1	1	1.1	1.4	1
2	2.1	2.3	3	2.1
3	3.1	3.2	7	3.1
4	4.2	4.9	5.5	no result
5	5	5.3	5.8	5
6	6	6.8	6.5	6
7	7	7.1	7.5	7

At sample 4 *no result* has been produced, then the value of β has been changed to a new threshold, 2β, to be used in the following cycle. At sample 5 there is again disagreement between variants output but the candidate output, var1 = 5, has satisfied the condition $d(5, 3.1)$ <2β. Hence the result 5 has been selected for this cycle, also the value of 2β has been changed to its initial value. By using this strategy there is no problem for sample 6 to produce an output. If β is selected very large then the smoothing algorithm approaches the performance of the median voter.

4. EXPERIMENTAL METHOD

4.1. Assumptions

The experimental work reported here is based on the following assumptions and terminology:

- the voter is used in a cyclic system where there exists some relationship between correct results from one cycle to the next;

- that faults cause errors whose symptoms appear to the voter as numerical input values perturbed by varying amounts;

- that perturbations (below some predefined accuracy threshold) in voter input values are considered as acceptable inaccuracies;

- for the purposes of the results reported here, the issues associated with ensuring synchronisation of the inputs to the voter (or indeed, synchronisation of the inputs to variants) is ignored;

- there exists some "notional correct result", which can be calculated from the current inputs and system state, based on the history of previous system states;

- the "notional correct result" is the desired voter result even where voter inputs are erroneous;

- a comparator is used to check for agreement between the notional correct result and the voter output; and

- an accuracy threshold is used, in the comparator, to determine if the distance between the notional variant result and the voter output is within acceptable limits.

These assumptions are representative of a class of voters common in embedded real-time control systems. The assumptions exclude those applications where inputs are randomly distributed through the input space in successive cycles.

4.2. Error Model

Each voter input is defined as a member of a tuple; $\{c, \ \varepsilon \pm, \ A_T \pm \}$; in application specific value units. Where,

c, notional correct value,

$A_T +/ A_T -$, upper/lower accuracy threshold.

$\varepsilon +/ \varepsilon -$, upper/lower voter threshold (in those voting algorithms that depend on thresholds to perform inexact voting),

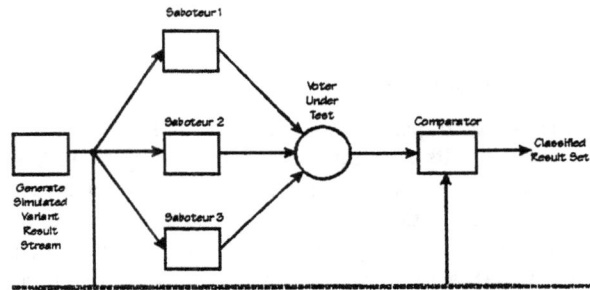

Figure 1. Experimental harness

This represents a simplified version of the error model presented in (Bass, 1995). The voter is tested in a triple modular redundant configuration, as shown in Figure 1. One notional correct result is produced by the input generator in each test cycle. This sequence of numbers simulate identical correct results generated by redundant variants. Copies of the notional correct result are presented to each saboteur, in every cycle. The saboteurs can be programmed to introduce selected variant error amplitudes, according to selected random distributions. In a given set of tests one, two or three saboteurs may be activated to simulate variant result errors on the voter inputs.

4.3. Experimental scenario

The voters tested are: formalised traditional voters (majority voter, plurality voter, median voter and weighted average voter) and smoothing voter. The performance of voters is dependent on many parameters such as: the relationship between consecutive inputs (linear, sinusoidal, saw-tooth, random walk, etc.), injected error amplitude (in comparison with the voter threshold), number of perturbed inputs, error distribution (normal or uniform), the value of voter threshold, accuracy-threshold (small, medium or large), and smoothing-threshold for the smoothing voter. Various scenarios can be constructed by different combination of these parameters. The parameters used in the experiments reported here are as follows:

1. Input variables are: error amplitude (ns) and error distribution (Uniform).

2. Accuracy-threshold (A_T).

The parameters used for the experiments reported below is as follows:

1. Input to variants: sinusoidal function $u(t) = 100.\sin(t)+100$ sampled at 0.1 s.

2. Voter-threshold value , $\varepsilon = 0.5$.

3. Accuracy-threshold value, $A_T = 0.5$.

4. All variants were perturbed using uniform error distribution with amplitude varies from 0 to 10.

226

The selected performance criteria are: normalised correct results (the ratio of correct outputs to the whole number of runs; n_c / n), normalised incorrect results (the ratio of incorrect results to the total number of tests,; n_{ic} / n) and normalised disagreed outputs (the ratio of disagreed results to the total number of runs; n_{na} / n). These measures may be expressed in percentage form.

The measure n_c / n represents the capability of a voter in producing correct results, higher values of this measure are desirable. The ratio n_{ic} / n is a measure of catastrophic outputs. From a safety point of view the smallest number of incorrect outputs which is consistent with small ratio of n_{ic} / n is desirable. This ratio can be a good performance criterion to evaluate a voter used in a safety critical system in which incorrect results are unacceptable. Finally the ratio of disagreed results of a voter to the total number of runs, n_{na} / n, is interpreted as benign outputs. These are errors successfully detected by the voter, which can be used to alert some higher level error management system. It must be noted that a 3-version programming system using inexact voter may produce *'no-result'* (a special code) output in the cases that i) the voter can not reach consensus upon the results of variants (such cases occur in majority and plurality voters) or ii) the built-in mechanism within a voter cannot produce an acceptable result in disagreement cases (for example, in smoothing voter, the smoothing mechanism can not generate a reasonable result). The *'no-result'* output can be used to initiate a safe shutdown of the system. It is obvious from safety viewpoint that generating a *'no-result'* output is much better than producing agreed but incorrect results. Hence, a tuple (n_c / n, n_{ic} / n, n_{na} / n) describes the performance of selected voters.

5. EXPERIMENTAL RESULTS

The results of voter comparison based on selected experimental scenarios defined in 4.3 in terms of performance criteria defined in 4.4 are presented. Since the behaviour of majority and plurality voters in a 3-version programming system is similar, only results for a majority voter are given. Each voter has been tested for 10000 runs with fixed accuracy threshold equal to 0.5. Figure 2 shows a comparative plot of n_c / n of voters versus error-amplitude. The figure indicates a non-linear (a negative exponential, approximately) relation between the increase in the amplitude of errors and decrease of n_c / n. The median voter has the largest ratio of n_c / n while majority voter has the smallest ratio and smoothing voter shows compromise behaviour between those two.

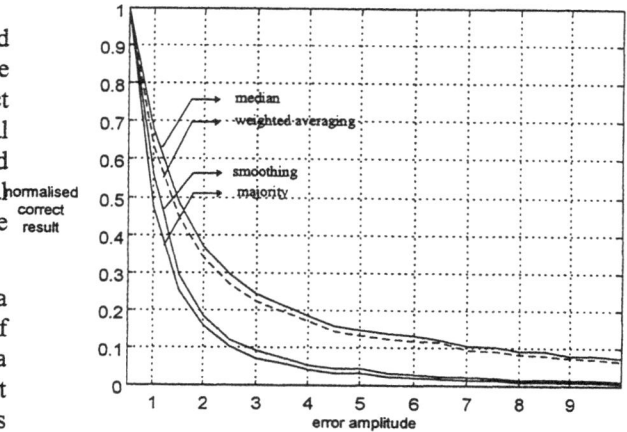

Figure 2. Normalised correct result vs. error amplitude

Figure 3 shows the plot of n_{ic} / n of voters versus error amplitude for 10000 runs. The large number of incorrect outputs for median and weighted averaging voters especially at the larger amplitude of error (error amplitude greater than about 1.5 as seen from figure 3) restricts their application in systems in which production of incorrect outputs could threaten the safety of the system

The majority voter produces the smallest number of incorrect outputs and smoothing voter has a behaviour between those two. All of the voters have, more and less, a close ratio of n_{ic} / n up to noise amplitude $\approx$ 1.5. The median and weighted averaging voters generate fewer incorrect results - comparable with those of the majority voter - only for small amplitude of error. This plot well illustrates two main group of voters :

1. Group including median and weighted averaging voters always produce output regardless of the validity of their inputs. This group generates larger number of incorrect results.

2. Group two includes voters that produce an output after some sort of processing of their input (checking validity of inputs, using of history, using some sort of estimation, etc.). This group generates smaller number of incorrect results compared with the first group.

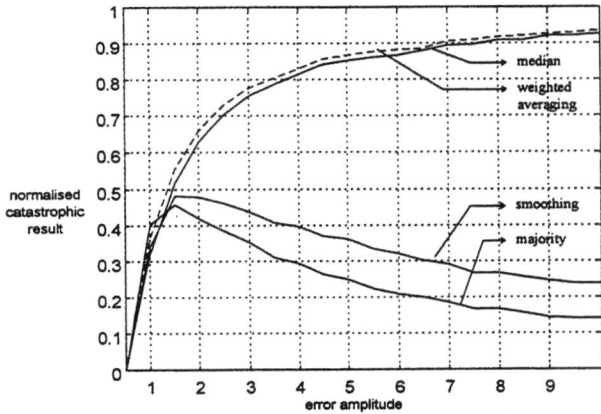

Figure 3. Normalised catastrophic result vs. error
amplitude

Figure 4 shows a plot of normalised disagreed (benign) results of voters; n_{na} / n, versus error amplitude. The median and averaging voters do not produce benign outputs, the majority voter produces the largest number of benign results and smoothing voter generates less benign results than the majority voter. The inability of median and weighted averaging voters to detect benign outputs makes them unsuitable for the safety critical systems.

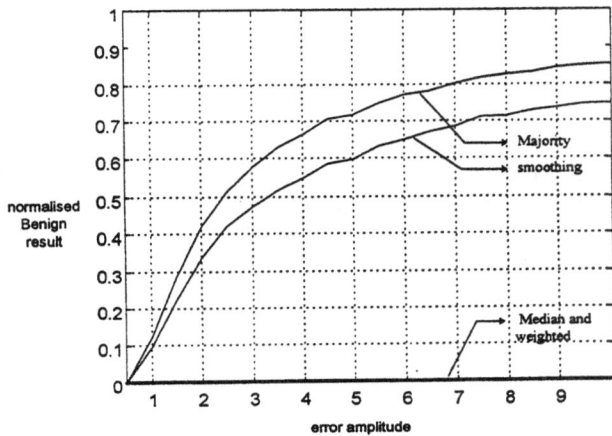

Figure 4. Normalised Benign result vs. error
amplitude

CONCLUSIONS

The error detection capabilities of the majority voter have been demonstrated. The majority voter can be said to be conservative in this sense. The median and weighted averaging voters, in contrast, can mask certain classes of double error, but at the expense of producing catastrophic errors in triple error scenarios. The smoothing voter is introduced with the objective of reaching a compromise between the result selection capabilities of the median voter and the conservative error detection properties of the majority voter. The results, presented in Section 5, show that the smoothing voter produces less benign errors and more correct results than the majority voter. In addition, the smoothing voter produces less catastrophic errors than the median voter. Selection of an appropriate voter for a specified system depends on the nature and requirements of that application. When, for example, a large number of correct outputs is required (here the number of catastrophic and benign errors are not considered), the median voter is the first candidate to employ. When an application requires the least possible incorrect outputs the majority voter is to be preferred. A trade off between these two cases can be achieved by means of the novel smoothing voter.

ACKNOWLEDGEMENTS

The authors gratefully acknowledge the support of UK EPSRC (under contract number GR/K64310). In addition the authors are grateful to the anonymous reviewers for their helpful comments.

REFERENCES

Bass, J. M. (1995). Voting in Real-Time Distributed Computer Control Systems. PhD thesis, Department of Automatic Control and System Engineering, The University of Sheffield, Sheffield, UK, Oct. 1995

Bass, J. M., G. Latif-Shabgahi and S. Bennett (1997). Experimental Comparison of Voting_Algorithms in Cases of Disagreement. *Proc. 23rd Euromicro Conference*, Budapest, Hungary, 1997. pp. 516-523

Blough, D. M., and F. G. Sullivan (1990). A Comparison of Voting Strategies for Fault-Tolerant Distributed Systems. *Proc. IEEE 9th Symp. On Reliable Distributed Systems*, 1990, pp. 136-145.

Gersting, J. L. (1991). A Comparison of Voting Algorithms for N-Version Programming. *Proc. 24th Anual Hawaii International Conference on System Sciences*, Vol. 2, pp. 253-262.

Lorczak, P. R., A. K. Caglayan and D. E. Eckhardt (1989). A Theoretical Investigation of Generalised Voters, *Proc. IEEE 19th Int. Symp. On Fault-Tolerant Computing Systems*, 1989, pp. 444-451.

Parhami, B. (1994). Voting Algorithms. *IEEE Trans. on Reliability*, Vol. **43**, No 4, pp. 617-629.

A HEURISTIC APPROACH TO FAULT TOLERANT
CONTROL OF UNKNOWN NONLINEAR SYSTEMS
USING NEURAL NETWORKS

J. R. Noriega and H. Wang

Department of Paper Science
UMIST
Manchester M60 1QD
U.K.

Abstract: A heuristic approach for the problem of fault tolerant control of unknown
nonlinear systems is discussed in this paper. The method uses a heuristically determined feedback function for the compensation of the system response and a neural
network model. It is assumed that changes in the system parameters can lead to
an increase in the magnitude of the residual signal. In this case, the residual is
formulated as the difference of the system output and the neural network model
output. A fault is normally associated with an unexpected increase in the residual
signal. Therefore, the residual is constantly monitored to detect the fault and to
start the compensation algorithm. In addition, the residual is fed back through a
compensation block. Thus fault tolerance is achieved by adjusting the control signal
of the failed system such that the residual signal approaches its original faultless
magnitude. The mathematical form of the compensation block is defined by a combination of experimentation and heuristic knowledge of the response of the system.
Copyright © 1998 IFAC

Keywords: Nonlinear systems, neural networks, fault detection.

1.INTRODUCTION

Stability and integrity of closed loop systems are
very important issues in industry because, besides
quality constraints, appropriate safety conditions
cannot be achieved without a stable, faultless
closed loop system. Some sources of instability
may eventually lead to irreparable damage to the
machinery or injury to the personnel in the plant.
In order to overcome these problems some methods have been developed that guarantee stability
and performance or at least a graceful degradation
of performance for some systems with known or
anticipated failure conditions. These methods are
known as fault tolerant control (FTC). To achieve
this aim researchers have adopted two main strategies known as reconfigurable and restructurable
control. By reconfigurable control it is implied
that a set of pre-designed controllers are available
to accommodate for anticipated failures. On the
other hand, restructurable control involves the on-
line computation of a new control that can accommodate for unanticipated failures.

The problem of redesign the controller such that

the aims of fault tolerant control are satisfied can take multiple forms, (Tsui, 1994; Jiang, 1994). Three levels of processing are proposed as a suitable framework for the automatic response to a fault, (Looze, 1985). These levels are: 1) detection and identification of failures; 2) determination of operating point conditions; and 3) trim, stabilise and regulate within the pre-determined operation conditions.

The methods use the matrices Q and R of the LQR method as redesign parameters. The matrix Q is kept constant while R is adjusted to maximise performance within bandwidth constraints. In (Howell, 1983) the characteristics needed to achieve a proper restructurable control under actuator failure are discussed. The technique of eigenstructure assignment has been used for reconfigurable and restructurable control by some researchers (Potts, 1981; Jiang, 1994). A reconfigurable control designed via an entire eigenstructure assignment is demonstrated in (Potts, 1981). The authors proceeded to transform the original system matrices with the Moor's algorithm. After observing the singular values for the transformed system, they proposed that a reduced system could be obtained by removing the least controllable state. The rejection of the less controllable state of the system reduces the average gain of the control matrix K. As a result, the possibility of actuator failure due to overloading is also reduced. In (Jiang, 1994) two cases are analysed, transient performance recovery and steady state performance recovery. In addition, two feedback mechanisms are studied, the state feedback and output feedback. The authors proved that for a full state feedback the stability of the reconfigured control is guaranteed and for the output feedback only a sufficient condition is derived. The eigenstructure of the controlled system is reassigned to resemble the original system eigenstructure. In the second case the difference of the outputs of the original and reconfigured systems are minimised by a proper choice of input weighting matrix in order to recover the original steady state performance.

Model following techniques provide a re-design reference of the desired performance of both faultless and faulty closed loop systems. The selection criteria for the elements of the weight matrices is problem dependent (Huang, 1990; Morse, 1990; Moerder, 1989).

In this paper it is assumed that the system is unknown and a neural network model of the faultless system is available. Two compensation functions are proposed and studied for different system parameter faults. The paper is organised as follows: section 2 briefly describes the neural network model of the system and the training method. Section 3 gives details of the fault toler-

ant control algorithm, the type of failures and the response of the compensated system.

2. NEURAL NETWORK MODEL

In this paper the unknown nonlinear dynamical system is described mathematically by the NARMAX framework as

$$\begin{aligned} y(t+1) \quad = \quad & f(y(t), y(t-1), \cdots, y(t-n), \\ & u(t), u(t-1), \cdots, u(t-m)) \quad (1) \end{aligned}$$

where y is the output of the system, u is the input, n and m are the known orders of the system and $f()$ is represents the overall nonlinear response of the system. The trained neural network model of the system is represented in general by

$$\begin{aligned} \hat{y}(t+1) \quad = \quad & \hat{f}(y(t), y(t-1), \cdots, y(t-n), \\ & u(t), u(t-1), \cdots, u(t-m)) \quad (2) \end{aligned}$$

where $\hat{f}()$ is the approximation of $f()$ as computed by the training of the weights of the neural network. The neural network model of the system is implemented by a two layer neural network. The structure of the neural network is illustrated in Figure 1.

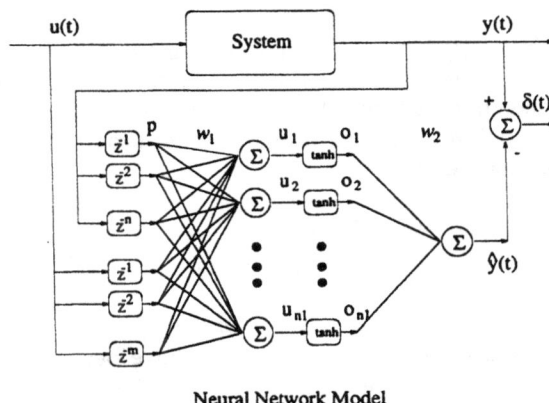

Neural Network Model

Fig. 1. Structure of the MLP for system modelling.

For our purpose, the training of the neural network must guarantee that the neural network model reproduces accurately the response of the faultless system. Therefore, the training of the neural network is performed using data from the available inputs and outputs of the faultless system.

3. FAULT TOLERANT CONTROL

The fault tolerant controller consists of an extra feedback loop whose function is to reduce the difference between the faulty nonlinear system and the neural network model. The proposed FTC is shown in Figure 2. In this case the nonlinear

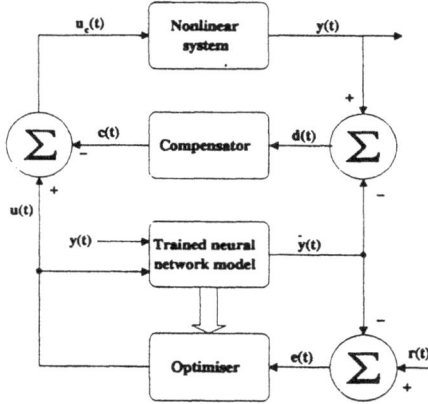

Fig. 2. Alternative fault tolerant controller.

system is controlled via the control algorithm described in (Noriega and Wang, 1998). The block called compensator forms an additional loop for the compensation of the faulty system. The compensator uses one of the following relations to recover the original system response.

$$c(t) = c(t-1) + kd(t) \qquad (3)$$

where $c(t)$ is the compensation signal, k is a constant found by experimentation and $d(t)$ is the modelling error which is expected to increase after the failure of the parameter. The second mathematical relation is

$$c(t) = c(t-1) + \\ ksign(d(t))(d(t) - d(t-1))^2 \qquad (4)$$

where $sign(d(t))$ gives -1 for a $d(t) < 0$ and $+1$ for a $d(t) > 0$. These relations have been tested and proved to work for particular faulty conditions using simulations and heuristic knowledge. There is, however, no analytical design procedure and stability analysis for this method at this time.

3.1 System and Failure Description

In general the systems modelled by the neural network are denoted in equation (2). This equation is rewritten to include the effect of the faults as

$$\hat{y}(t) = \hat{f}(f, y, u) + \omega \qquad (5)$$

where $f = (f_1, f_2, \cdots, f_\rho)$ is the vector of parameter faults, ρ is the number of the parameters of the system, $y = (y(t-1), y(t-2), \cdots, y(t-n))^T$ is the vector of measured outputs of the system, $u = (u(t-1), u(t-2), \cdots, u(t-m))^T$ is the vector of measured inputs of the system and ω is a noise

sequence. When no fault is present in the system, the dynamical response of the system is written as

$$\hat{y}(t) = \hat{f}(f_h, y, u) + \omega \qquad (6)$$

where $f_h = (1, 1, \cdots, 1)$. It is assumed that the fault in any of the parameters of the system can be represented by a multiplicative parameter f_i associated to the i-th parameter. Therefore, a faulty condition is represented by $f_i \neq 1$ which corresponds to a change in the value of the i-th parameter. This improved model allows us to describe changes in the parameters in general for simulation purposes.

3.2 Response to a Fault

The response of the fault tolerant control is illustrated by simulations. The faults are introduced as described before by changing one of the parameters of the system. The system chosen for these simulations is

$$\hat{y}(t) = \frac{a_0 y(t-1)}{a_1 + a_2 y^2(t-1)} + \\ (1 + abs(y(t-1)))u(t-1) \qquad (7)$$

where a_i are the constant parameters of the system. Any of these parameters can be subjected to a simulated fault simply by allowing that the desired parameter be changed as $a_{f_i} = f_i a_i$. The effect is to decrease the value of the affected parameter by a factor of $1 - f_i$. In this particular case the healthy value of the three parameters is assumed to be $a_i = 1$ for $i = 0, 1, 2$. Finally, y and u are the output and input to the system respectively.

Two experiments are proposed in order to illustrate the response of the fault tolerant control to a parameter failure. First, the effect of the size of the fault on the restructuration of the control is considered. Second, the ability to compensate for a particular fault is demonstrated. For the first experiment a fixed gain $k = 0.1$ is used and three sizes of the fault were chosen for the parameter a_0. These are $f_0 = 0.8$, $f_0 = 0.55$ and $f_0 = 0.2$. Figure 3 shows the response of the closed loop system without the compensation block. It can be seen that the neural network based controller described before fails to regulate the system output. This result is expected since the fault changes the response of the system. Consequently, the neural network model cannot follow the system output accurately and hence the closed loop system fails to compensate the system output. A more robust controller is implemented by using equation (3) in the compensation block. The response of the new system is shown in Figure 4 below. The response of the failing system is successfully compensated. The compensation block changes the input to the

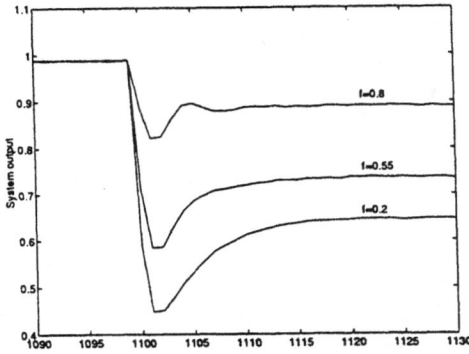

Fig. 3. Response to a failure without compensation.

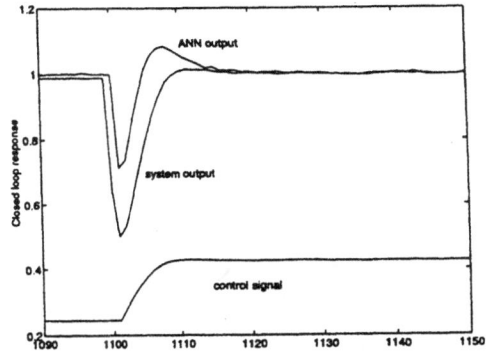

Fig. 5. Compensation of a fault in a_0.

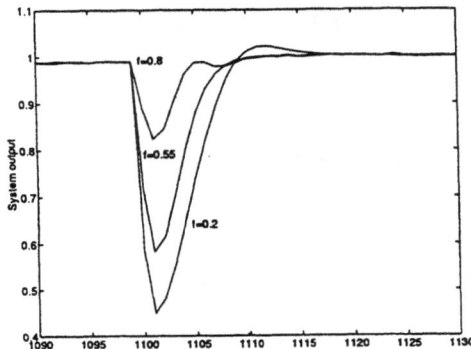

Fig. 4. Response to a failure with compensation.

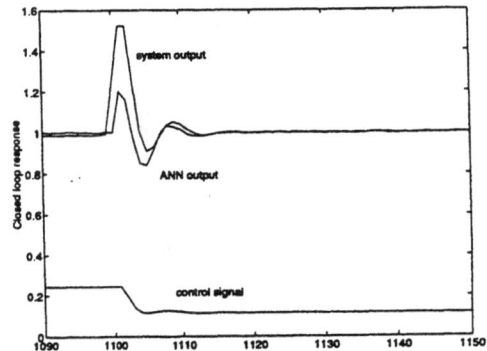

Fig. 6. Compensation of a fault in a_2

system only in order to reduce the difference between the neural network model and the system outputs. Such a change drives the system output closer to the faultless system response.

The second experiment requires changes in the three a parameters. In Figure 5 the fault size selected is kept constant at $f_i = 0.7$. It can be seen that the effect of the compensation is to increase the control demand in order to reduce the difference between the neural network output and the system output. The compensation of a fault in the parameter a_2 is illustrated in Figure 6. The compensation is again successful. In this case, however, the control demand is decreased in order to reduce the difference of the system output and the neural network output. In the case of the parameter a_1 the simulation results are presented in Figure 7. In this case, the automatic compensation method fails to compensate for the fault of parameter a_1. Therefore, it can be concluded that the selected compensation relations are limited for a subset of the possible faults in the system parameters.

In its present form, this method may be used to compensate for faults in a particular subset of parameters for anticipated faults.

4. CONCLUSIONS

In this paper a new fault tolerant controller for unknown nonlinear systems is proposed. The method is based on the use of a compensation function and a neural network healthy model. The neural network model is used to generate a residual signal for the detection and compensation of a faulty condition. In this case, the residual is defined as the difference between the system output and the neural network model output. On the other hand, the compensation function is determined partly by experimentation and partly from heuristic knowledge of the system. As a result, some faulty conditions can be satisfactorily compensated. However, an appropriate mathematical formulation of the stability of the compensated system needs to be developed.

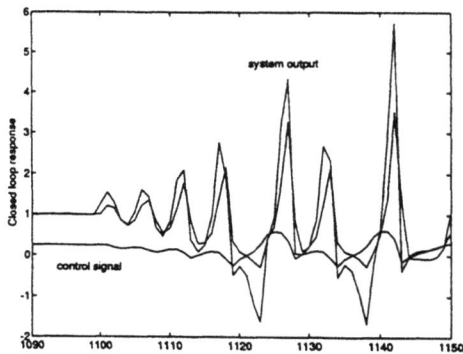

Fig. 7. Failed compensation of a fault in a_1.

ACKNOWLEDGEMENT

The first author gratefully acknowledges the financial support from CONACYT (México) which makes his work possible.

REFERENCES

Tsui, C-C, (1994) A General Failure Detection, Isolation and Accommodation System with model Uncertainty and Measurement Noise , *IEEE Transactions on Automatic Control*, Vol. 39, No. 11, pp. 2318-2321.

Ostroff, A.J., (1985) Techniques for Accommodating Control Effector Failures on a Mildly Statically Unstable Airplane , *Proceedings of the American Control Conference*, pp. 906-913.

Rattan, K.S., (1985) Evaluation of Control Mixer Concept for Reconfiguration of Flight Control System , *IEEE National Aerospace and Electronics Conference*, pp. 560-569.

Gao, Z. and P. J. Antsaklis, (1991) Stability of the Pseudo-Inverse Method for Reconfigurable Control Systems , *International Journal of Control*, Vol. 53, No. 3, pp. 717-729.

Looze, D.P. and J.L. Weiss and J.S. Eterno and N. M. Barrett, (1985) An Automatic Redesign Approach for Restructurable Control Systems , *IEEE Control Systems Magazine*, pp. 16-21.

Howell, W.E. and W.T. Bundick and R.M. Hueschen and A.J. Ostroff, (1983) Restructurable Controls for Aircrafts , *AIAA Guidance and Control Conference*, pp. 646-651.

Potts, D.W. and J. D'Azzo, (1981) Direct Digital design method For Reconfigurable Multivariable Control Laws for the A-7D Digitac 2 Aircraft , *IEEE National Aerospace and Electronics Conference*, pp. 1284-1291.

Jiang, J., (1994) Design of Reconfigurable Control Systems Using Eigenstructure Assignments , *International Journal of Control*, Vol. 59, No. 2, pp. 395-410.

Morse, W.D. and K.A. Ossman, (1990) Model Following Reconfigurable Flight Control System for the AFTI/F-16 , *AIAA Journal of Guidance*, Vol. 13, No. 6, pp. 969-976.

Huang, C.Y. and R.F. Stengel, (1990) Restructurable Control Using Proportional-Integral Implicit Model Following , *AIAA Journal of Guidance*, Vol. 13, No. 2, pp. 303-309.

Moerder, D. D. and N. Halyo, J. R. Broussard and A. K. Caglayan, (1989) Application of Precomputed Control Laws in a Reconfigurable Aircraft Flight Control System , *AIAA Journal of Guidance*, Vol. 12, No. 3, pp. 325-333.

Levin, A.U. and K. S. Narendra, (1993) Control of Nonlinear Dynamical Systems Using Neural Networks: Controllability and Stabilization , *IEEE Transactions on Neural Networks*, Vol. 4, No. 2, pp. 192-206.

Widrow, B. and M. A. Lehr, (1990) 30 Years of Adaptive Neural Networks: Perceptron, Madaline and Backpropagation , *Proceedings of the IEEE*, Vol. 78, No. 9, pp. 1415-1441.

Noriega, J.R. and H Wang, (1998) A Direct Adaptive Neural Network Control for Unknown Nonlinear Systems and Its Application , *IEEE Transactions on neural networks*, Vol. 9, pp. 27 - 34.

USING B-SPLINES TO REPRESENT OPTIMAL ACTUATOR DEMANDS FOR SYSTEMS FEATURING POSITION/RATE LIMITED ACTUATORS

Adrian G. Alford† and Chris J. Harris‡

*†Advanced Engineering Department, GKN Westland Helicopters Ltd,
Yeovil, Somerset BA20 2YB, UK. (email: whl.adv.eng@dial.pipex.com)*

*‡Image, Speech and Intelligent Systems (ISIS) Research Group,
Department of Electronics and Computer Science, University of Southampton,
Southampton SO17 1BJ, UK (email: cjh@ecs.soton.ac.uk)*

Abstract: The process of calculating optimal actuator demands for MIMO dynamic systems which feature position/rate limited actuators is investigated. Order 3 B-splines are identified as a good method for parsimoniously representing actuator demands. A least squares optimization algorithm is used to calculate the B-spline parameters necessary to match the plant and reference models' responses. The reference model is assumed to contain output and output rate constraints which are to be respected by the plant model. This technique is successfully applied to a nonlinear helicopter model and is able to respect the reference model's angular velocity (output) and angular acceleration (output rate) constraints at all times. *Copyright © 1998 IFAC*

Keywords: B-splines, actuating signals, limiting control actions, optimization, fuzzy logic, predictive control.

1. INTRODUCTION

The calculation of optimal actuator demands is a prerequisite process for model predictive control (MPC). MPC is an adaptive control approach which uses a representative plant model and an optimization routine to calculate a series of future actuator demands necessary to make the plant model respond as a reference model. The first actuator demand of this series is applied to the plant. Information from the real plant is used to update the plant model before the optimization process is repeated iteratively. This approach is known to possess desirable stability properties for nonlinear systems (Economou, 1986; Hunt, 1992; Hunt, 1992a; Keerthi, 1986; Mayne, 1990).

Pre-computed optimal actuator demands can also be used to train artificial neural networks (ANN) and neurofuzzy networks (NFN) to realise nonlinear optimal controllers.

This paper addresses the problem of finding an efficient method of calculating the sequence of actuator demands $u(t)$ necessary to minimise a plant model's response error $e(t)$. The plant model is assumed to be multi-input multi-output (MIMO) and to feature position/rate limited actuators. The response error $e(t)$ is defined as the plant model's response $y(t)$ subtracted from the reference model's response $y_{ref}(t)$ and the response's duration is assumed to be sufficiently long to permit the reference model to settle following a step input. The reference model is assumed to: encapsulate the desired dynamic behaviour of the plant; feature output and output rate constraints (which need to be respected by the plant model); have a similar dynamic order to the plant model; and be initialized to the same state as the plant model so that its output and output rate are initially identical. Peak response errors should be no larger than 1% of the demanded response. However, any response errors caused directly by *appropriate actuator saturation* (see Section 2.2 for definition) can be ignored. $u(t)$ needs to be smooth and continuous in order to facilitate the future mapping, using ANNs and NFNs, of sensor signals $y(t)$ to actuator demands $u(t)$. In contrast to *classical* optimal control, there is no explicit requirement to minimize the magnitude of actuator demands. However, the requirement for $u(t)$ to be smooth and continuous suggests that the high frequency content of $u(t)$ should be minimized.

Note, both electric and hydraulic actuators are commonly represented as first order lags possessing position and rate limits. It is assumed that the actuator's position limits are applied to the actuator's demand, thus preventing the actuator from being forcibly driven against its physical travel limits.

Whilst determining optimal $u(t)$, the optimization process evaluates the sensitivity (Jacobian matrix) of plant response error $e(t)$ to $u(t)$. This calculation returns a matrix containing rows/columns of zeros every time an actuator position/rate limit is encountered. Thus a severe problem can be created for least squares optimization algorithms which attempt to invert the Jacobian matrix in the process of calculating the optimal direction in which to change $u(t)$.

2. METHOD

2.1 Actuator Demand Representation

It is very important to minimize the number of unknowns within any optimization problem because:

* the time taken to find the global minimum is typically proportional to the square of the number of unknowns;
* the reliability of least squares optimization algorithms are dependent upon the conditioning of the Jacobian matrix which tends to improve with fewer unknowns;
* the statistical certainty of the optimized parameters increases with fewer unknowns.

The number of unknowns can be minimized if $u(t)$ is represented parsimoniously. Each unknown parameter of this representation needs to control a distinct and independent feature of $u(t)$ since this directly improves the optimization problem's conditioning. These independent features should preferrably be uncorrelated [assuming that functions $f(t)$ and $g(t)$ are not always zero then they are uncorrelated over the range $0 \leq t \leq t_{max}$ if $\int f(t)g(t) \, dt$ equals zero].

B-splines (Dierckx, 1995; Brown, 1994; Brown, 1995) are a very powerful method of representing both univariate and multivariate functions. In fact, they can be used as a common mathematical framework for representing both fuzzy logic and NFNs (Harris, 1992). B-splines approximate univariate functions by linearly combining a number of piece-wise polynomial basis functions of order k. Thus, an individual actuator demand can be represented by

$$u_i(t) = w_1\mu_1(t) + w_2\mu_2(t) + ... + w_n\mu_n(t)$$
(1)
$$for \ 0 \leq t \leq t_{max} .$$

Each basis function $\mu(t)$ is compact in the time domain and is continuous up to the $k-2$ derivative [a function $f(t)$ is compact over the range $0 \leq t \leq t_{max}$ if $f(t)=0$ for $t \leq t_1$ and/or $t \geq t_2$ where $t_1 > 0$ and $t_2 < t_{max}$]. The basis functions are rectangular, triangular and bell-like in shape for $k=1$, $k=2$ and $k \geq 3$ respectively (see dashed lines in figure 1 for the $k=3$ case). Naturally, any linear combination of these basis functions is also continuous up to the $k-2$ derivative. Hence, continuity can be guaranteed when $k \geq 2$ and smoothness can be guaranteed when $k \geq 3$. The parameter range of interest is subdivided by a number of knots $(\lambda_0, \lambda_1, \lambda_2, \cdots \lambda_{n-k+1})$ such that each basis function is k subdivisions wide and overlaps with its $k-1$ nearest neighbouring basis functions (when $k=1$ the basis functions do not overlap and are uncorrelated). Note that $0 = \lambda_0 < \lambda_1 < \lambda_2 < \cdots < \lambda_{n-k+1} = t_{max}$ and that each basis function is scaled such that $\mu_1(t) + \mu_2(t) + \cdots + \mu_n(t) = 1$ for $0 \leq t \leq t_{max}$. Each knot location can be loosely interpreted as a sample point between which the actuator demand is constructed via polynomial (of order k) interpolation. Clearly as the number of knots increase so does the ability to represent the high frequency content of $u(t)$.

To summarise, order 3 B-splines are used to represent $u(t)$ because:

* $u(t)$ is linearly dependent upon the unknown parameters $w_1, w_2, \cdots w_n$;
* the representation is guaranteed smooth and continuous when $k \geq 3$;
* by limiting the number of knots and their closeness to one another, the high frequency content of $u(t)$ can be regulated;
* the basis functions are optimally scaled;
* the basis functions are compact and are minimally correlated (in the case of $k=1$ they are completely uncorrelated).

When a nonlinear plant model (containing position/rate limited actuators) attempts to mimic a reference model which is being subjected to large step demands, the actuator limits are almost certainly to be encountered. These limits are most likely to be hit when t is small. During this initial period: the shape of the actuator demand $u(t)$ is not critical as long as it has sufficient magnitude to cause the actuator to saturate in the appropriate sense; and $e(t)$ is insensitive to all B-spline parameters. Hence, knot placing strategies which allow knot spacing to decrease as $t \to 0$ are not pursued. In addition, as $t \to t_{max}$, the reference model, plant model and actuator demand transients decay. Hence, knot placing strategies which allow knot spacing to decrease as $t \to t_{max}$ are not pursued. For the above reasons, a regular knot spacing strategy is adopted.

2.2 Cost Functions (C) and Optimization Algorithms

$$\min_{w} \quad C(w)$$

$$where \quad C(w) = \sum_{0}^{t_{max}} \left(\| W_y \; e(w,t) \|_2^2 \right) \qquad (2)$$

$$= \sum_{0}^{t_{max}} \left(\| W_y \left(y_{ref}(t) - y(w,t) \right) \|_2^2 \right)$$

The cost function is developed from a simple sum-of-squared-errors function, see equation (2). In (2), w is a vector of B-spline parameters which parsimoniously describes $u(t)$ and W_y is a diagonal matrix which normalizes all response errors so that they lie the range $[-1,+1]$. The motivation for expressing the problem in a least squares fashion is that fast and reliable optimization algorithms are available to solve them. The Levenberg-Marquardt least squares algorithm (Gill, 1981; Grace, 1996) is used since it offers a unique blend of robustness and efficiency.

It is assumed that the mapping from $u(t)$ to $y(t)$ is continuous (this is normally the case if the plant model represents a noiseless physical system) hence the cost function $[w \rightarrow C(w)]$ is also continuous. Note, cost function continuity is assumed by most optimization algorithms. However, the presence of actuator position/rate limits within the plant model creates regions of zero gradient and discontinuities in the cost function's first derivative.

An actuator activity term is added to the cost function to prevent the optimization process from getting stuck when actuator limits are encountered - leading to

$$C(w) = \sum_{0}^{t_{max}} \left(\| W_y \; e(w,t) \|_2^2 + \| \lambda W_a \; a(w,t) \|_2^2 \right). \qquad (3)$$

Where W_a is a diagonal matrix which normalizes all actuator activity so that it lies in the range $[-1,+1]$ and the diagonal matrix λ defines the relative weighting between response errors and actuator activity. The term $a(w,t)$ represents actuator activity and is defined as the difference between actuator demands $u(w,t)$ and actuator positions $x_a(t)$. The above definition of actuator activity only penalizes actuator rate and actuator demands which are outside the actuator's position limits. Absolute actuator position $x_a(t)$ is not by itself a good indicator of actuator activity since, for nonlinear systems, trim points can often involve actuators being significantly displaced from their central position.

λ is not fixed since this generates a conflict between optimizing w to minimize response errors $e(t)$ and actuator activity $a(t)$. Ultimately, $\|\lambda\| \ll 1$ is required so that the minimization of $e(t)$ is not significantly compromised. However, $\|\lambda\| \gg 0$ during the early stages of the optimization process (when $e(t)$ is high) prevents premature actuator saturation and avoids associated optimization problems. Hence, an approach is adopted whereby the diagonal elements of λ are scheduled with peak normalized response error (on a per-axis basis), thus $\|\lambda\| \rightarrow 0$ as $e(t) \rightarrow 0$. Scheduling λ with peak response error appropriately increases λ in response to brief periods of high response error caused by brief periods of inappropriate actuator saturation. However, this scheduling technique requires a modified cost function since actuator position/rate limits prevent $e(t) \rightarrow 0$. Note also that the efficiency of least squares optimization algorithms are generally higher for small residual problems in which $C(w_{optimum})$ is small. Hence, $e(t) \rightarrow 0$ and $\|\lambda\| \rightarrow 0$ are very desirable features.

By reinitializing the reference model (on a per-axis, per-time interval basis) to the plant model's state every time an actuator appropriately saturates, it becomes possible for $e(t) \rightarrow 0$. *Appropriate actuator saturation* is simply defined as: the position/rate limit which should reduce the response error's magnitude. Thus, the reference model is slaved to the plant model if the actuator's position/rate limits prevent the plant model from following the reference model. Assuming that the reference model's output and output rate constraints are imposed during reinitialization then $e(t) \rightarrow 0$ implies that the same constraints are being respected by the plant model.

Assuming that the B-spline parameters w_1, w_2, w_3 correspond in some way to the magnitude of an actuator demand at times t_1, t_2, t_3 (where $t_1 < t_2 < t_3$) then the differences $\Delta w_{21} (= w_2 - w_1)$ and $\Delta w_{32} (= w_3 - w_2)$ are proportional to the rate of change of actuator demand. The actuator's rate ultimately matches the actuator demand's rate which, if equal to a certain value (the rate limit), reduces the value of the cost function through reference model slaving. However, if $w_1 < w_2 < w_3$ [a similar argument applies when $w_1 > w_2 > w_3$] then this reduction in cost function can arise if w_2 increases (due to its relationship to w_1) or when w_2 decreases (due to its relationship to w_3) clearly a recipe for multiple (local) minima. This source of local minima can be eliminated by making the optimization process find $w_1, \Delta w_{21}, \Delta w_{32}, \cdots \Delta w_{n\,n-1}$ instead of directly finding $w_1, w_2, \cdots w_n$. Hence, the B-spline's parameter vector is differential encoded.

The cost function described thus far is likely to contain discontinuities, generated every time reference model slaving is enabled. However, by employing fuzzy logic to control the extent of reference model slaving, the cost function can become truly continuous. Fig. 1 defines the fuzzy variable *position limit* {positive position limithit, no position limit hit, negative position limit hit} using piecewise quadratic membership functions. Fuzzy variables *rate limit* {positive rate limit hit, no rate limit hit, negative rate limit hit} and *response error* {positive response error, zero response error, negative response

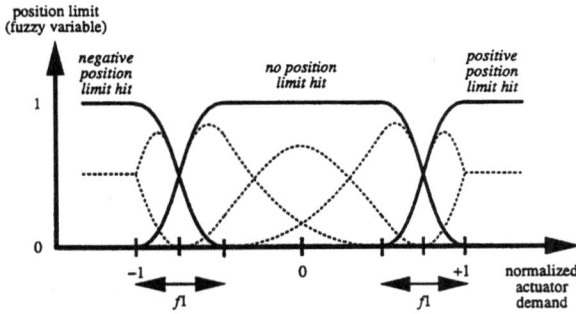

Fig. 1. Fuzzy variable definition

error} can be similarly defined. Note that these membership functions can be constructed by combining a number of adjacent B-spline basis functions (shown as dashed lines). The parameters f_1, f_2 and f lie in the range [0, 1] and control the level of smoothing applied to the definitions of *position limit*, *rate limit* and *response error* within reference model's slaving logic.

It is assumed that the *OR* operator is performed by summing fuzzy membership functions (followed by bounding to ensure that the result is within the range [0, 1]) and that the *AND* operator is performed by multiplying fuzzy membership functions. The degree of reference model slaving is assumed to vary from 0 (=no slaving) to 1 (=total slaving).

The use of fuzzy logic based on piece-wise linear membership functions ensures that the cost function becomes truly continuous. However the use of fuzzy logic based on piece-wise quadratic membership functions has the additional benefit of removing many of the discontinuities in the cost function's first derivative. Note that: f_1 is assumed to be equal to f_2; values of f_1 and f_2 greater than 0.1 are not considered because these significantly compromise the response (and stability) associated with the reference model; values of f_3 greater than 0.01 are not considered in order to permit 100% reference model slaving to occur before peak normalized response errors exceed 1%.

3. SIMULATION RESULTS

The methodology development in Section 2 is applied to a 17 state helicopter model. This helicopter model is unstable, cross-coupled, nonlinear and features position/rate limited actuators. Actuator demands for lateral cyclic, longitudinal cyclic and tail rotor are constructed from order 3 B-splines with regularly spaced knots. A total of 19 basis functions (6, 6 and 7 for the respective actuator demands) are required so that peak response errors are less than 1%. The collective actuator is assumed to be directly controlled by the pilot. Collective actuator demands generate significant roll, pitch, yaw rate cross-couplings and hence are used to inject disturbances into the other axes.

The associated reference model defines the roll rate, pitch rate and yaw rate response necessary for good handling qualities (ADS33D, 1994). The reference model consists of three single axis models each of which has second order dynamics. Thus, like the helicopter model, it is not possible to instantaneously affect output (angular velocity) or output rate (angular acceleration). Zero roll/pitch/yaw cross-couplings and zero response to disturbances (eg. collective inputs) are also specified. The reference model is constrained to produce roll, pitch and yaw rates within the range of ±1.57rad/s, ±0.63rad/s, ±1.05rad/s respectively and roll, pitch and yaw accelerations within the range of ±3.32rad/s², ±1.22rad/s², ±2.27rad/s² respectively.

Single axis step demands for roll rate, pitch rate, yaw rate and collective actuator demand are used as test cases. Various magnitudes of step input are used (±10%, ±25%, ±50% and ±100%) such that there are 32 test cases. The –100% roll and +100% rate yaw tests are notable because the former requires the lateral cyclic actuator to be near saturation for the response's duration whilst the latter requires the tail rotor actuator to be saturated for the response's duration.

Simulation runs last 1.276s (based on the longest time constant in the reference model multiplied by five) and use an integration step-size of 0.006125s (based on the shortest time constant, within either the helicopter or reference model, divided by five). A 20 iteration limit is also imposed per test manoeuvre. Note, if the B-spline's parsimonious representation of $u(t)$ is not used then the optimization process needs to deal with $624(=3t_{max}/\delta t)$ unknown parameters!

3.1 Cost Function Based on Equation (2)

When the cost function (2) is applied to the helicopter model: peak response errors are as high as 14.9% (–50% yaw rate test); and the reference model's angular acceleration limits are not respected, fig. 2 illustrates what happens during the –100% yaw rate test [solid line=helicopter model, dashed line=reference model, dotted line=limits, shaded regions=actuator saturation]. In addition, six out of the 32 test cases cause problems for the optimization process (iteration limit and/or poor-conditioning warnings). This is probably due to: the relatively high global minimum of this cost function,

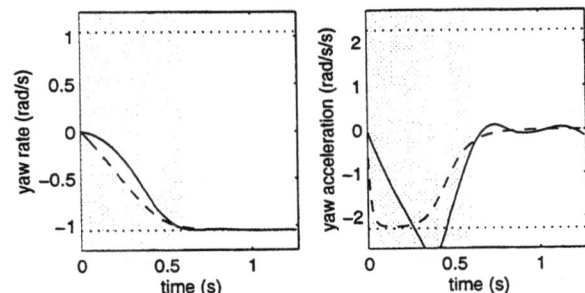

Fig. 2. -100% yaw rate response using equation (2)

leading to the optimization algorithm operating inefficiently; and the first derivative discontinuities and areas of zero gradient.

On the positive side, simulations reveal that this simple cost function:

* is continuous in value;
* possesses a single minimum;
* achieves 1% peak response errors for all manoeuvres which do not involve actuator saturation;
* does correctly saturate the tail rotor actuator for the entire duration of the +100% yaw rate test;
* does correctly drive the lateral cyclic actuator so that it was nearly saturated for the entire duration of the -100% roll rate test.

3.2 Cost Function with Reference Model Slaving

The next cost function to be evaluated is based on (3) and features: reference model slaving; actuator activity scheduling with peak response error; and a differentially encoded B-spline's parameter vector. To a large extent, this cost function solves the problem described in Section 1: peak response errors (including cross-couplings) are no higher than 1.2% (+100% roll rate test); and the reference model's angular acceleration limits are respected, fig. 3 illustrates the -100% yaw rate test.

Five out of the 32 test cases result in optimization process hitting the iteration limit. This is mainly due to the discontinuities in the cost function caused by the reference model's slaving logic. Fig. 4 illustrates the cross-section through the cost function, for the -100% roll rate test, associated with the Δw_{21} lateral cyclic actuator demand B-spline parameter.

3.3 Progressive Reference Model Slaving

The final cost function to be evaluated is the same as the last except that it uses a fuzzy interpretation of *appropriate actuator saturation* within the reference model's slaving logic. This results in reference model being progressively slaved to the helicopter model as the

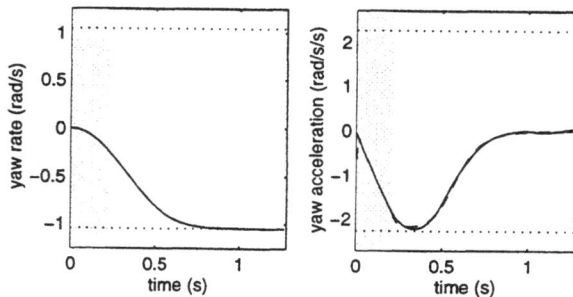

Fig. 3. -100% yaw rate response using the reference model slaving strategy

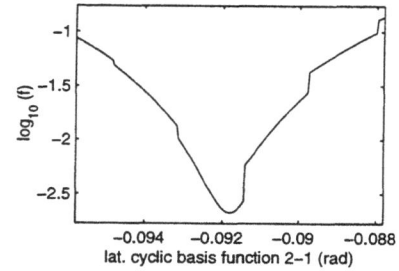

Fig. 4. Cost function cross-section using crisp logic to control reference model slaving

actuator's limits are approached. The following statement is an example of the fuzzy logic used to control reference model slaving.

* ·The degree of roll axis slaving is equal to (positive lateral cyclic position limit hit AND positive roll response error) OR (positive lateral cyclic rate limit hit AND positive roll response error) OR (negative lateral cyclic position limit hit AND negative roll response error) OR negative lateral cyclic rate limit hit AND negative roll response error).

Table 1 shows how the optimization process is affected by varying the fuzziness and overlap of the piece-wise quadratic membership functions used within reference model's slaving logic. These investigations indicate that setting $f_1 = f_2 = 0.02$ and $f_3 = 0.001$ is a good choice for the helicopter model. This level of fuzziness is sufficient to smooth the cost function (see Fig. 5) but has no perceptible effect on any time response. Additional investigations with the helicopter model indicate that the use of piece-wise quadratic membership functions within the reference model's slaving logic is numerically beneficial over piece-wise linear membership functions.

Table 1 Affect of fuzziness on the optimization process

$f_1 (=f_2)$	Number of times iteration limit encountered or warnings issued		
	f_3=0.0001	f_3=0.001	f_3=0.005
0.0001	6	4	7
0.001	5	4	5
0.002	4	4	5
0.005	4	2	4
0.01	5	3	2
0.02	0	1	2
0.05	1	1	3
0.1	1	1	1

This cost function is able to completely solve the problem stated in Section 1: peak response errors (including cross-couplings) are no higher than 1%; and the reference model's angular acceleration limits are always respected. Only one out of the 32 test cases

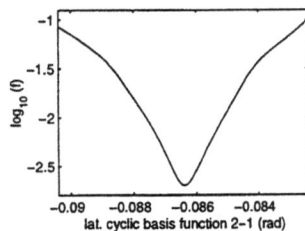

Fig. 5. Cost function cross-section using fuzzy logic to progressively slave reference model

(~50% yaw rate) results in the optimization process hitting the iteration limit. The reason for this is unknown since the cost function appears to be smooth and well behaved.

4. CONCLUSIONS

The process of calculating optimal actuator demands for MIMO dynamic systems which feature position/rate limited actuators is investigated. The optimization process attempts to make the plant model respond like a reference model. The reference model encapsulates the desired dynamic behaviour of the plant as well as output and output rate constraints.

Order 3 B-splines, with regularly spaced knots, are identified as a very good method for parsimoniously representing actuator demands. This technique constrains actuator demands to be smooth/continuous and optimally conditions the optimization problem.

A cost function without an actuator activity term features large areas of zero gradient in which the optimization algorithm is likely to get trapped. The introduction of an actuator activity term (defined as the difference between actuator position and actuator demand) solves this problem. This term should be scheduled with peak normalized response error (on a per-axis basis) so that actuator activity weighting is: large away from the global minimum (thus preventing the optimization process from getting stuck); and small in the vicinity of the global minimum (thus enabling the optimization process to focus on minimizing response error during the final stages of optimization).

By slaving the reference model to the plant model (on a per-axis, per-time interval basis) when appropriate actuator limits are hit: the cost function's global minimum reduces to near-zero; and the reference model's output and output rate limits are respected by the plant model. Unfortunately, this technique produces a discontinuous cost function with local minima (as a result of the B-spline parameter vector's relationship to actuator travel rates). However, these local minima can be eliminated if the optimization process determines a differentially encoded version of the B-spline's parameter vector (that is, differences between adjacent parameters are found rather than the parameters

themselves).

The cost function's discontinuities can be eliminated if a fuzzy interpretation of appropriate actuator saturation is used to progressively slave the reference model. Fuzzy logic which utilizes piece-wise linear membership functions makes the cost function continuous. However, fuzzy logic which utilizes piece-wise quadratic membership functions is superior since it also eliminates many of the cost function's first derivative discontinuities. A small degree of fuzziness is sufficient to smooth the cost function without having a noticeable effect on the plant model's time response.

REFERENCES

ADS33D: Aeronautical Design Standard, Handling Qualities Requirements For Military Rotorcraft. United States Army Aviation and Troop Command, July 1994.

Brown, M. and C. Harris (1994). *Neurofuzzy Adaptive Modelling and Control*. Prentice Hall, London.

Brown, M., Z. Abed and C.J. Harris (1995). Using the Neurofuzzy Approach to Building Dependable Neural Network Systems. In: *Neural Networks Producing Dependable Systems*. ERA Report 950973, UK.

Dierckx, P. (1995). *Curve and Surface Fitting with Splines*, Monographs on Numerical Analysis. Clarendon Press, Oxford, UK.

Economou, C.G., M. Morari and B.O. Palsson (1986). Internal Model Control, 5, Extension to Nonlinear Systems. In: *Industrial Engineering Chemical Process Design Developments* 25, pp.403-411.

Gill, P.E., W. Murray and M.H. Wright (1981). *Practical Optimization*. Academic Press Inc. (London) Ltd.

Grace, A. and M.A. Branch (1996). *MATLAB Optimization Toolbox User's Guide*. The MathWorks Inc., USA.

Harris, C.J., (1992). Comparative Aspects of Neural Networks and Fuzzy Logic for Real-Time Control. In: *Neural Networks for Control Systems*. Peter Peregrinus Ltd, ISBN 0863412793.

Hunt, K.J. and D. Sbarbaro (1992). Studies in Neural Network Based Control. In *Neural Networks for Control and Systems*. Peregrinus Ltd, ISBN 0863412793.

Hunt, K.J., D. Sbarbaro, R. Zbikowski and P.J. Gawthrop (1992a). Neural Networks for Control Systems: A Survey. *Automatica* 28, pp.1083-1112.

Keerthi, S.S. and E.G. Gilbert (1986). Moving Horizon Approximations for a General Class of Optimal Nonlinear Infinite-Horizon Discrete-Time Systems. In: *Proceedings of the 20th Annual Conference on Information Science and Systems*, pp.301-306. Princeton University, USA.

Mayne, D.Q. and H. Michalska (1990). Receding Horizon Control of Nonlinear Systems. *IEEE Transactions on Automatic Control* 35.

A STUDY OF PARALLELISM OF FUZZY CONTROL ALGORITHMS FOR NEUTRON POWER CONTROL

Daniel Vélez-Díaz [1], Jorge S. Benítez-Read [1], and Kishan K. Kumbla [2]

[1] *Instituto Nacional de Investigaciones Nucleares*
Gerencia de Ciencias Aplicadas
Apartado Postal 18-1027, Col. Escandón 11801, D.F., México
Tel. +(5)-329-7200 Ext. 2431 / 2432 Fax. +(5)-329-7301
e-mail: dvd@nuclear.inin.mx; jsbr@nuclear.inin.mx

[2] *University of New Mexico*
Department of Electrical and Computer Engineering
Albuquerque, New Mexico 87131

Abstract: This paper presents a study on the feasibility to process in parallel some of the stages of fuzzy rule based control algorithms (FRBC). Three FRBC's are considered for the control of neutron power in a research nuclear reactor: the first FRBC is a two-input one-output fuzzy controller; the second one is a three-input one-output FRBC; the third controller is a two-input one-output FRBC in which a boolean selection of the output membership functions (OMF) is made according to three different zones of the reactor power level. Computer simulations determine the time required by each of the FRBC's to process the input to give an adequate ouput (a controller cycle), under different degrees of parallelism. Comparisons of the main features of the parallel and non-parallel (sequential) FRBC's are presented. *Copyright © 1998 IFAC*

Keywords: Control algorithm, Fuzzy control, Nuclear reactor, Parallel computation, Parallelism, Sequential control algorithm.

1. INTRODUCTION

The development of technology and innovations in nuclear systems can be directed to improve the operational record of reactors such as stability, fuel burning, and robustness. Higher functional reliability and deeper degree of automation can be achieved in nuclear power plants using control systems and algorithms, first developed in research reactors.

The Nuclear Centre of Mexico has, among its key facilities, a research nuclear reactor. The instrumentation and the operating console have been practically unchanged since the reactor's installation in 1968 (DeGroot, 1968). A new reactor's operating console has been designed and built (González-Marroquín, et al., 1995), to replace the original one. The new console and its associated systems are expected to be licensed and installed in the reactor by the end of 1998. For this new console it is required the availability of reliable control algorithms for the automatic operation mode. To comply with this requirement, different algorithms to regulate and control the reactor's neutron power, based on advanced control theories and knowledge-based expert systems, have been proposed and studied.

A promising solution to the neutron power control problem is fuzzy logic control, which uses information in the same manner as human experts, and it does not require the complex mathematics associated with classical control theory (Jamshidi, et al., 1993). A fuzzy rule based control algorithm (FRBC) consisting of two inputs and one output has been designed and simulated. The results are promising in the sense that the neutron power setpoint is attained without the presence of overshoot, the safety conditions on the reactor period are maintained, and a stable power level can be sustained for long periods of time (Benítez-Read and Vélez-Díaz, 1996). A modification to this algorithm introduces the novel characteristic of a boolean selection of different sets of output membership functions (OMF), according to different zones of neutron power level (Benítez-Read and Vélez-Díaz, 1998). Likewise, another FRBC algorithm, using three inputs and one output, presents dynamic characteristics comparable to the two-input one-ouput FRBC with boolean selection of OMF, although the time required to attain full power level is a little bit longer (Benítez-Read and Vélez-Díaz, 1998).

Due to the fact that the computer operating system, which controls the complete operation of the reactor, would also implement a fuzzy control algorithm, the time required by these algorithms to process the input data has been estimated to be large compared to the dynamic response of the reactor. Therefore, an important issue is the search for techniques that could be used to reduce the time of the FRBC algorithms in a real-time implementation.

2. FUZZY CONTROL SYSTEM

The block diagram of the fuzzy control system is shown in Figure 1, where the reactor is the controlled system, $\rho_{ext}(t)$ is the system input variable (control variable), and $n(t)$ is the system output variable (controlled variable).

One of the crisp input variables to the fuzzy controller is the reactor period (T), which represents the time required by the power, at any instant of time, to be increased by a factor "e", considering that the neutron power rate of change is maintained constant. The other crisp input variable to the fuzzy controller is the normalized percent of the neutron power deviation from its final level (or setpoint), called %PD. Each pair of crisp input variables to the fuzzy controller produces a crisp output variable, m_ρ, that represents the rate of change of external reactivity. Once m_ρ is defined, it is used to obtain the

external reactivity, $\rho_{ext}(t)$ (control variable). Also, as can be observed, the system output variable $n(t)$ can be an special third crisp input.

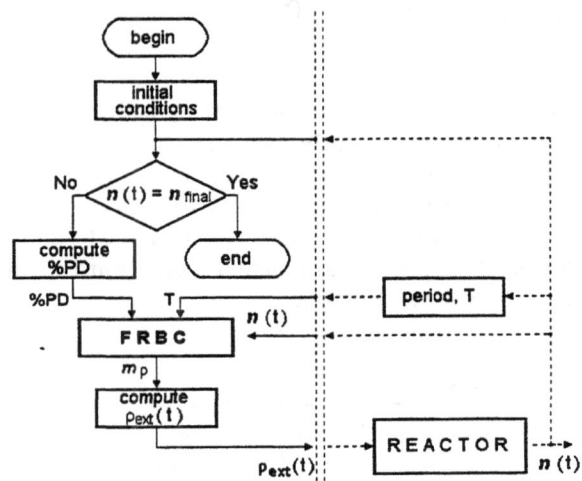

Figure 1. Block diagram of the fuzzy control system.

3. DIFFERENT TYPES OF FRBC

The fuzzy controllers designed and simulated for the reactor model can be defined as follows: a two-input/one-output fuzzy controller (FRBC1), a three-input/one-output fuzzy controller (FRBC2), and a two-input/one-output with a crisp selection of a set of output membership functions fuzzy controller (FRBC3). The crisp variable $n(t)$ is used as a third input on the FRBC2 or as boolean selection input on the FRBC3, as is illustrated in Figure 2.

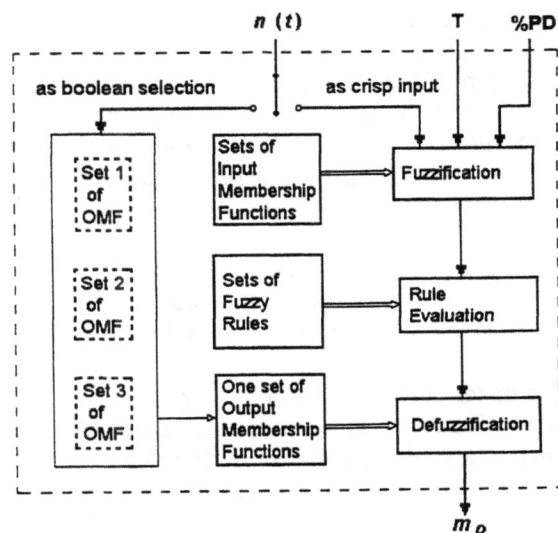

Figure 2. Main blocks of a Fuzzy Rule Based Controller (FRBC).

3.1 FRBC1: Two input / one output.

This typical fuzzy controller uses a set of membership functions for the crisp variable T, and a set of membership functions for the input %PD. This fuzzy controller produces a response, shown in the Figure 3 as $n_1(t)$.

At low neutron power levels, there occur relative large increments on the neutron concentration in the reactor core. Hence, in order to maintain the reactor period above the reactor's scram level (system shutdown), small positive rate of change values of the external reactivity, m_ρ, must result from the fuzzy controller.

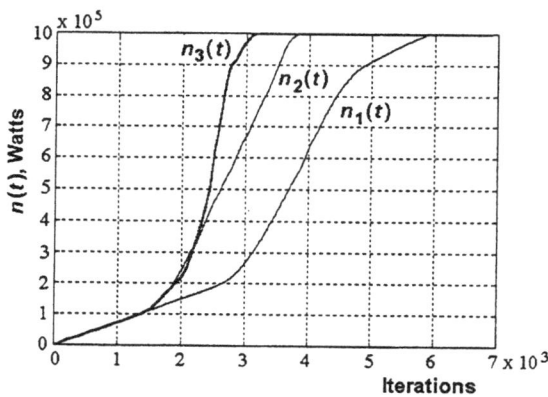

Figure 3. Reactor power responses: $n_1(t)$ using FRBC1, $n_2(t)$ for FRBC2 and $n_3(t)$ for FRBC3.

3.2 FRBC2: Three input / one output.

This fuzzy controller adds the neutron power level, $n(t)$, as a third input. This crisp input is fuzzified using three membership functions. The other two inputs, reactor period T and normalized power deviation %PD, use five membership functions each. This fuzzy controller, using seventy five fuzzy rules, produces a response $n_2(t)$, as shown in Figure 3. This controller attains the desired power level faster than the FRBC1.

3.3 FRBC3: Two input / one output with crisp selection of the output membership functions.

The instantaneous neutron power level, $n(t)$, used as an input to the fuzzy controller, increases considerably the number of fuzzy rules. Since the relative concentration of neutrons in the reactor core vary according to power regions of the reactor operation, the variable $n(t)$ can have other uses. In this study, $n(t)$ has been used in the FRBC3 controller to select one of the groups of output

membership functions (OMF). The response $n_3(t)$ in Figure 3 is determined by this fuzzy controller.

4. PARALLELISM IN THE FRBC's

Assuming that the time used by the control algorithm to process the input data, which come from the reactor model, is small compared to the variation of the state variables of the reactor, the simulation results of the closed-loop control system show a satisfactory behavior in the ascent and regulation of the reactor power, as can be observed in Figure 3. However, the FRBC implemented in sequential form in the reactor control computer, consumes a period of time of the order of seconds in carrying out an iteration. A possible solution to this problem is the use of parallel computing or the use of other kind of fuzzy control algorithms by the control system.

Table 1. Computational blocks for FRBC1.

Block	W_1/DOP_1	W_4/DOP_4	W_{12}/DOP_{12}
1	56 / 56	14 / 14	5 / 5
2	21 / 21	6 / 6	2 / 2
3	50 / 25	14 / 7	6 / 3
4	8 / 4	2 / 1	2 / 1
5	4 / 4	1 / 1	1 / 1
6	1 / 1	1 / 1	1 / 1
7	1 / 1	1 / 1	1 / 1

Table 2. Computational blocks for FRBC2.

Block	W_1/DOP_1	W_4/DOP_4	W_{12}/DOP_{12}
1	67 / 67	17 / 17	6 / 6
2	47 / 47	12 / 12	4 / 4
3	150 / 75	38 / 19	14 / 7
4	8 / 4	2 / 1	2 / 1
5	4 / 4	1 / 1	1 / 1
6	1 / 1	1 / 1	1 / 1
7	1 / 1	1 / 1	1 / 1

The fuzzy control algorithm can be seen as a sequence of blocks, in which all operations in a block should finish before the next block can begin its task. In some of these blocks, parallel computing can be implemented with different degrees of parallelism. Using this method for real-time implementation, a drastic reduction can be expected in the time required by the algorithm to give an outcome.

The parallelization of the algorithms requires to know: the computational load (number of floating-point operations) in each block for each processor, represented by W_k, and the degree of parallelism DOP_k, that represents the number of parallel routes that can be established within that block. The subscript k means the actual number of processors. Parallelism in the FRBC1 using 1, 4 and 12 processors is shown in Table 1. The FRBC2 is parallelized according to Table 2, and Table 3 shows the corresponding for FRBC3.

Table 3. Computational blocks for FRBC3.

Block	W_1/DOP_1	W_4/DOP_4	W_{12}/DOP_{12}
1	41 / 41	11 / 11	4 / 4
2	21 / 21	6 / 6	2 / 2
3	50 / 25	14 / 7	6 / 3
4	20 / 16	5 / 4	2 / 1
5	4 / 4	1 / 1	1 / 1
6	1 / 1	1 / 1	1 / 1
7	1 / 1	1 / 1	1 / 1

5. RESULTS AND COMPARATIVE ANALYSIS

The curve labeled $n_1(t)$ and shown in Figure 3, is obtained when the FRBC1 is used. Likewise, $n_2(t)$ and $n_3(t)$ are the neutron power responses when FRBC2 and FRBC3 are used, respectively.

Computing Time with one processor (CT_1). The computing time for the FRBC1 with one processor (FRBC1 CT_1) has the same median value as the FRBC3 CT_1. This is due to the great similarities between the two algorithms. The FRBC2 CT_1 has a longer median value; this result is expected because the number of fuzzy rules is increased from twenty five to seventy five. These results are shown in Figure 4. The performance of the three FRBC's with respect to the total time required to attain the desired power level (1 MW) is as follows (Figure 3): FRBC3 is the fastest of the three controller, followed by FRBC2, whereas FRBC1 is the slowest.

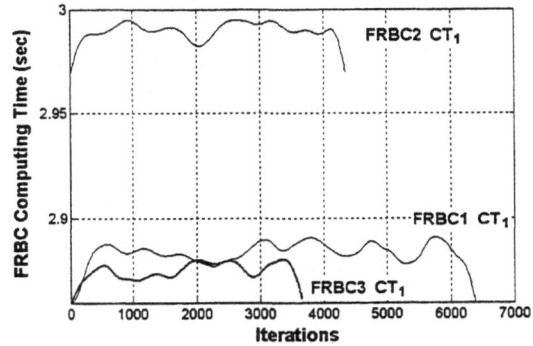

Figure 4. Computing Time for FRBC1, FRBC2, and FRBC3, using 1 processor.

Computing Time with four processors (CT_4). The results obtained from the three FRBC's using four processors are shown in Figure 5. The computing times are labeled as: FRBC1 CT_4, FRBC2 CT_4 and FRBC3 CT_4. Again, the median values of FRBC1 CT_4 and FRBC3 CT_4 are similar, while the corresponding for FRBC2 CT_4 is larger. The total computing time for FRBC3 is almost half of the FRBC1's.

The computing times with twelve processors are shown in the Figure 6. The results obtained are very close to those for four processors.

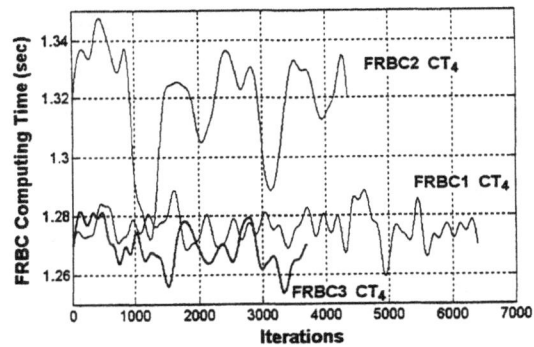

Figure 5. Computing Time for FRBC1, FRBC2, and FRBC3, using 4 processors.

Comparative analysis. The median values of the computing times for FRCB1 are nearly the same than the FRBC3's, using one, four and twelve processors; however, the difference between the total computing time is very important. The time is reduced in approximately 50%. Both, the median value and the total computing time for FRBC3 are smaller than the FRBC2's. This is due to the number of fuzzy, as well as to the adaptive characteristics of the FRBC3. Comparing now the computing times between FRBC1 and FRBC2, the results are: the median value of FRBC1 CT is smaller than the corresponding for FRBC2 CT

244

(due to the larger number of rules in FRBC2); the total computing time in FRBC1 CT is greater than FRBC2 CT (due to the longer time required by FRBC1 to attain the setpoint). The comparative results are shown in the Table 4.

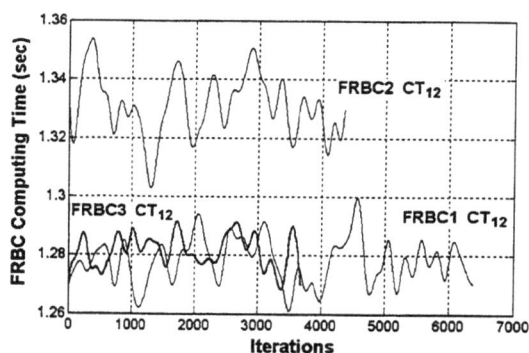

Figure 6. Computing Time for FRBC1, FRBC2, and FRBC3, using 12 processors.

Table 4. Comparative results of Computing Time.

FRBC Computing Time	Median Computing Time	Total Computing Time
FRBC1 CT_1	2.86 sec	28.16 min
FRBC2 CT_1	2.97 sec	19.04 min
FRBC3 CT_1	2.86 sec	15.38 min
FRBC1 CT_4	1.27 sec	24.92 min
FRBC2 CT_4	1.32 sec	16.79 min
FRBC3 CT_4	1.27 sec	13.58 min
FRBC1 CT_{12}	1.27 sec	24.96 min
FRBC2 CT_{12}	1.33 sec	16.91 min
FRBC3 CT_{12}	1.27 sec	13.71 min

REFERENCES

Benítez-Read, J.S. and Vélez-Díaz, D. (1996). *Neutron power control in a research reactor using a fuzzy rule based system.* Soft Computing with Industrial Applications, Vol. 5, pp. 53-58, TSI Press Series.

Benítez-Read, J.S. and Vélez-Díaz, D. (1998). *Comparative study of fuzzy control algorithms for a nuclear reactor.* 1998 World Automation Congress (accepted).

DeGroot, M. N., (1968). *Triga Mark III reactor: Instrumentation maintenance handbook.* Document No. GA-8585, Gulf General Atomic, Inc.

González-Marroquín, J. L., Rivero-Gutiérrez, T., and Sáinz-Mejía, E., (1995). New operating console of the Triga Mark III reactor: Design and implementation, (In Spanish). *Proc. of the VI Intl. Annual Congress of the Mexican Nuclear Society,* pp. 122-128, Huatulco, Oaxaca, México, Sept. 17-20.

Hwang, K. and Xu, Z.,(1996). Scalable Parallel Computers for Real-Time Signal Processing. *IEEE Signal Processing Magazine,* **Vol. 13**, No. 4, July.

Jamshidi, M., Vadiee, N., and Ross, T. J., (Editors), (1993). *Fuzzy Logic and Control: Software and Hardware Applications.* Prentice-Hall, Inc., Englewood Cliffs, New Jersey.

Vélez-Díaz, D. and Benítez-Read, J.S. (1997). *Fuzzy system to control the neutron power with different sets of output membership functions.* Proc. of the 7th IFSA (Intl. Fuzzy Systems Assoc.) World Congress, Prague, Czech Republic, June 25-29. Academia Prague, ISBN 80-200-0633-8, Vol. 4, pp. 132-136.

FUZZY-NEURAL MODEL FOR NONLINEAR SYSTEMS IDENTIFICATION[1]

Ieroham Baruch*,** and Elena Gortcheva**

*Institute of Information Technologies,
Bulgarian Academy of Sciences, Akad. G. Bonchev str.,
BAS bl. 29 A, Sofia 1113, BULGARIA
Fax:(+359-2)72-04-97; E-mail: ips@bgearn.acad.bg
**CINVESTAV--IPN, Ave. IPN No 2508,
A.P. 14-470 Mexico D.F., MEXICO
Fax: (+52-5)754-76-01; E-mail: gortche@ctrl.cinvestav.mx

Abstract: A state space representation of both continuous and discrete time mathematical models of Recurrent Neural Networks (RNN) are given in two layer Jordan canonical architecture, and a new improved Back-Propagation (BP) type learning method, is proposed. Some topology improvements, are suggested. The proposed RTNN model is linear in small and nonlinear in large, which permits to apply all well known state- and output linear systems design methods. The obtained RTNN model is incorporated in a rule-based fuzzy system, giving the possibility to approximate and to identify a complex nonlinear plants. Simulation results of nonlinear systems identification by the proposed fuzzy-neural system, using RNN BP learning, are given. Copyright © 1998 IFAC

Keywords: fuzzy modelling, neural network models, Jordan canonical form, nonlinear models, identification, system rule-based systems.

1. INTRODUCTION

Fuzzy systems (FS) are shown to be a general function approximators (Babushka and Verbruggen, 1997; Kosko, 1992) and as such they can be applied to nonlinear regression systems identification. The most popular Takagi-Sugeno, model, cited in (Babushka and Verbruggen, 1997), can incorporate in its consequent part a crisp mathematical function of the model input. This part could be a state-space model of the dynamic process as well. So, for different zones and regimes of the nonlinear process, the rule-based fuzzy systems approach could be used as a system coordinator of different state-space process models, corresponding to different rules. Now it is possible to use different identification techniques to obtain these state space models for different process zones and operation regimes. The Neural Network models are one of the possible models for systems identification.

The Neural Network (NN) modelling and application to system identification, prediction and control was discussed for many authors (Connor et al., 1994; Narendra and Parthasarathy, 1990; Pham and Yildirim, 1995; Sadharsanan and Sundareshan, 1991; Sastry et al., 1994; Yip and Pao, 1994). Mainly, two types of NN models are used: Feedforward (FFNN) and Recurrent (RNN). The main problem here is the use of different NN mathematical descriptions according to its application. For example, Narendra and Parthasarathy, 1990, applied NN for system identification and control, Sadharsanan and Sundareshan, 1991, used RNN for control systems

---

[1] This paper has been supported by CINVESTAV-IPN, Mexico and by the NSF- MSET, Bulgaria, Grant I-611/96.

synthesis by quadratic cost minimisation, Kosko, 1992, described different NN and FS for solution of control problems, Yip and Pao, 1994, solved control problems by RNN one-step ahead prediction, Connor et al., 1994 applied Nonlinear Autoregressive Moving Average (NARMA) RNN model for robust time series prediction, Pham and Yildirim, 1995, applied Jordan RNN for robot control, Sastry et al., 1994, and Tsoi and Back, 1994, applied also different RNN models to solve identification and control problems in their work. Baruch et al., 1996, in their previous paper applied the state-space approach to describe RNN in an universal way, defining a global linearized RNN model, named Recurrent Trainable Neural Network (RTNN) and study its stability by means of the first stability law of Liapunov. The paper proposed goes ahead in global RTNN model improvement dedicated to preserve RNN model stability during learning. The improved RTNN model is aimed to identify nonlinear multivariable dynamic processes. Three types of processes are suggested: processes, nonlinear on their output, state and input.

Simultaneously with the improving of RTNN topology and mathematical description, some advanced researches has been done on the methods of its learning (Sastry et al., 1994; Tsoi and Back, 1994). For this aim, three basic RNN models are considered. The first one was given by Hopfield and described by Kosko, 1992, Tsoy and Back, 1994.. This model have feedbacks from each NN output to each NN summator. There exists some difficulties of NN stability preserving during the learning. The second RNN model is the so called Flat RNN topology (Yip and Pao, 1994), which is similar to the generalised autoregressive model NARMA (Connor et al., 1994). Unfortunately, the research did not consider the multivariable case and the possibility of RNN learning. An interesting work in this field has been done by Sastry et al., 1994, which proposed a method for RNN learning, based on the Back-Propagation (BP) learning algorithm. The authors defined two types of neurones: memory neurones and network neurones, and applied two type of activation functions and two modifications of the Bpalgorithm, depending on the place of the neurone in the Network. The lack of universality here is evident. The third RNN model - the most promising one, was called Jordan RNN architecture (Pham and Yildirim, 1995). There are some more modifications of these architectures, described and criticised by Tsoi and Back, 1994. The Jordan architecture is more specific and contains three layers. In the hidden layer, there exists single unity feedback on each neurone and there do not exists cross connections. The drawback here is that the dynamic behaviour of that RNN could be only exponential, the feedback weights are not trainable, and due to the unity feedback, the RNN could be unstable. The aim of the paper is to describe a new method to learn the proposed improved RTNN

architecture. The method proposed represents an improved temporal version of the BP learning algorithm (Baruch et al., 1996).

The problem of RTNN application to nonlinear dynamic systems identification is also considered. The RTNN model proposed is easy to apply for control systems design, because it is linear in "small" and nonlinear in "great". As this model is described in a state-space form, it could be easily incorporated in one Fuzzy rule-based identification and adaptive control hierarchical system (FRBS).

2. DESCRIPTION OF THE IMPROVED RTNN MODEL

The continuous RNN linearized mathematical model, defined by Baruch et al., 1996, and written in state-space form is given by the equations:

$$\dot{x} = Ax + Bu \qquad (1)$$

$$y = Cx \qquad (2)$$

where x - is a n state vector; u - is a m- input vector; y - is a l- output vector; A is a full (mxm) constant state matrix, which diagonal elements represent the passive decay rate dynamic of each neurone and the other elements represent the lateral feedback between neurones; B is a (nxm) input matrix; C is a (lxn) output constant matrix. This mathematical model is sufficient for dynamic point of view, because if that linearized model is stable - as the sigmoid function S(x) is a single decreasing and bounded, than the nonlinear model will be stable due to the first stability law of Liapunov.

The discrete-time RNN linearized model could be described by means of the same state-space mathematical model, as it is:

$$X(k+1) = GV(k)+HU(k), \quad k=0,1....N \qquad (3)$$
$$Y(k) = CX(k) \qquad (4)$$

where X, Y, U are state, output and input n, l, m - vectors respectively; k is discrete - time integer variable; G, H, C are constant matrices of compatible dimensions. That matrices depends on the period of discretization To, as it is:

$$G(To) = ATo; \ H(To)= B(To) \qquad (5)$$

The RNN linearized state-space model (1), (2) could be transformed into Jordan Canonical Form (JCF), using similar canonical transformation:

....
$$v = Jv + Du, \quad x = Tv \qquad (6)$$

248

$$y = Fv \qquad (7)$$

where v, y, u are respectively n, l, m - vectors; T is a $(n \times n)$ transformation matrix; J is a block-diagonal constant state matrix; D, F are also constant matrices with compatible dimensions. The transformation could be done using eigenvalue/eigenvector method for diagonalization of the state matrix A. This similar canonical transformation not affects the dynamic input/output behaviour of the RNN model, but it affects only the state vector x, optimizing the RNN topology in some sense. The RNN model will be stable if system eigenvalues have negative real parts. Than it is easy to analyse the NN model controllability, observability and identifiability. Last concept, taken from systems theory, give us the possibility to check if the obtained global RNN model could be learned or not. From the block structure of D and F, corresponding to the block structure of J, we can conclude that: if the input matrix D has zero blocks - the RNN model is uncontrollable, and if the transpose of the output matrix F has zero blocks - the RNN model is unobservable (if one of both occurs the RNN model is unidentifiable). The analysis of the RNN model controllability / observability (identifiability / learnability) and stability give us the possibility to obtain full information of the dynamic behaviour and learning capability of the global RNN.

The same properties of the discrete-time state-space RNN model could be studied in a similar way, by means of canonical transformation into JCF (Baruch et al., 1996). The discrete-time RNN model will be stable if system eigenvalues are inside the unit circle. The mathematical model of the discrete-time RNN in JCF has the same description as (6), (7), but given for the discrete-time integer variable k. Using the same notation as above, we can write:

$$V(k+1) = JV(k)+DU(k), \quad X(k)=TV(k) \qquad (8)$$
$$Y(k) = FV(k) \qquad (9)$$

where the constant matrices J, D, F depends on T_o.

Based on the Jordan RNN canonical state-space representation, it is easy to describe the continuous and discrete-time RTNN model, as it was defined by Baruch et al., 1996.

Let us form two layers in the continuous and discrete-time versions of the Jordan RNN canonical representation (6), (7) and (8), (9), introducing two vector valued sigmoid functions and two new variables w, z, (W, Z) in the hidden and output layers.

Then the equations (6), (7) and _(8), (9), introducing the sigmoid function in the hidden and output layers, gives us the two layer architecture of the Jordan RTNN. The continuous and discrete-time JCFs of RTNN model are:

$$\dot{v} = Jv + Du, \ w = S(v), \ y = Fw, \ z = S(y) \qquad (10)$$

$$V(k+1) = JV(k)+DU(k), \ W(k)=S[V(k)],$$
$$Y(k) = FV(k), \quad Z(k)=S[Y(k)] \qquad (11)$$

where: w, z, (W, Z) are ne variables with respectively n, l- dimensions and $S(x)$ is a vector-valued sigmoid function, given by:

$$S'(x) = [s(x_1), s(x_2),...,s(x_j)] \qquad (12)$$

and the element of S is:

$$s(inp) = 1/[1+\exp(-inp)]; \ inp = \sum_i (d_i x_i + d_{io}) \qquad (13)$$

Here: inp is the input of the sigmoid function; d_i, d_{io} are trainable constant weights of the RTNN; $S'(x)$ signifies a vector transpose of $S(x)$.

The main advantages of the proposed two layer Jordan Canonical Form (JCF) RTNN architecture, defined by (1), (2) are:

1. It is described in state-space form (SISO or MIMO) and could serve as an one-step ahead state predictor/estimator;

2. The RTNN model is nonlinear in large and linear in small, so the matrices J, D, F obtained as a result of learning could be used for analytical design of linear state/output control laws. By means of a similar transformation the JCF could be transformed into Luenberger's Canonical Form which is easy to use for pole assignment design of control systems . The matrices J, D could be used for an optimal control systems design with quadratic performance index. The matrices J, D, F also could be used for an optimal P, PI, PID control systems design . Finally, the matrices J, D, F could be used in an adaptive iterative square-root algorithm for optimal control with quadratic cost criterion ;

3. The RTNN could solve optimal control problems itself by means of NN mapping (Sadharsanan and Sundareshan, 1991).

The RTNN two-layer architecture contains hidden and output layers. The output layer is a BP one and the hidden layer is a recurrent Jordan canonical. It was assumed that each Jordan block of it has only (1x1) or (2x2) dimension. To preserve the RTNN stability during the training, it is necessary to impose some restrictions on the model feedback, introducing a sigmoid vector function in it, which changes equations (10) and (11) in the form:

$$v = S(Jv) + Du, \quad w = S(v), \quad y = Fw, \quad z = S(y) \quad (14)$$

$$V(k+1) = S[JV(k)] + DU(k), \quad W(k)=S[V(k)],$$
$$Y(k) = FW(k), \quad Z(k)=S[Y(k)] \quad (15)$$

Another improvement of the RTNN architecture is to facilitate its realisation, approximating the sigmoid function s(inp) with a saturation:

$$sat(inp) = \begin{cases} +1, & inp \geq +1 \\ inp, & 0 \leq inp < +1 \\ 0, & inp < 0 \end{cases} \quad (16)$$

Both improvements of the RTNN architecture should be experimented by simulation examples during the learning.

3. RTNN LEARNING

The most common used **BP** updating rule, applied for the two layer **RTNN** canonical model, is the following:

$$Q_{ij}(k+1)=Q_{ij}(k) + \eta \Delta Q_{ij}(k) \quad (17)$$

where: Q_{ij} is the **ij**-th weight element of each weight matrix in the RTNN model to be updated; ΔQ_{ij} is the weight correction of Q_{ij}; η is the learning rate parameter.

Note that the **RTNN** model weight matrices here are denoted by **Q** for the sake of generality.

The weight corrections of the updated matrices in the discrete-time **RTNN** model, described by equation. (6), are given as follows:

- For the output layer:

$$\Delta F_{ij}(k) = [T_j(k)-Z_j(k)] \, Z_j(k) \, [1-Z_j(k)] \, W_i(k) \quad (18)$$

where: ΔF_{ij} is the weight correction of the **ij**-th elements of the (lxn) learned matrix F; T_j is a **j**-th element of the target vector; Z_j is a **j**-th element of the output vector; W_i is an **i**-th element of the input vector of the output layer, i.e. the hidden layer output.
- For the hidden layer:

$$\Delta D_{ij}(k) = R \, U_i(k) \quad (19)$$

$$\Delta J_{ij}(k) = R \, V_i(k-1) \quad (20)$$

$$R = F_i(k) \, [T(k)-Z(k)] \, W_j(k) \, [1-W_j(k)] \quad (21)$$

where: ΔD_{ij} is the weight correction of the **ij**-th elements of the (mxn) learned matrix D; F_i is a row vector of dimension (1xl), taken from the transposed matrix F'; [T-Z] is a (lx1) output error vector, through which the error is backpropagated to the hidden layer; U_i is an **i**-th element of the input vector; U; V_i is an **i**-th element of the vector V; ΔJ_{ij} is the weight correction of the **ij**-th elements of the l (nxn) block-diagonal matrix J under learning; R is an auxiliary matrix with compatible dimensions. Note that the matrix elements of **0** and **1** values will not be updated. The same equation for RTNN learning may be applied for the continuous-time case, given by equation. (10).

An improvement of the **BP** updating algorithm (8) is to introduce a momentum term, proportional to the past (**k-1**)-th weight correction, as it is:

$$Q_{ij}(k+1) = Q_{ij}(k) + \eta \, \Delta Q_{ij}(k) + \alpha \, \Delta Q_{ij}(k-1) \quad (22)$$

where: α is a momentum learning rate parameter.
This correction is appropriate to perform in the case when significant error-function oscillations occur. A lot of experiments of learning with different rates of learning η and α has been done. The experiments show that the optimal combination of these learning parameters is obtained when the following inequality condition yields:

$$r_{max} < sqrt(\eta^2+\alpha^2) < 1; \quad r_{max} = max|\lambda_i| \quad (23)$$

where λ_i is the maximum eigenvalue of the object under test.

The experimentally obtained low (14) shows that the circle with radius r, depending on the values of η and α ($r=sqrt(\eta^2+\alpha^2)$), must contain all eigenvalues of the state matrix J of the linearized discrete-time **RTNN** model (11).

Another improvement of the **RTNN** learning algorithm, successfully applied for the **BP** learning of discrete-time **RTNNs** consider unimportant units pruning and non-useful connections removing. Both methods remove the units or the weights, whose outputs or values tend to be zero. Simultaneously with the nodes pruning, a weight fixing could be applied.

Both learning algorithms, performing weights pruning and fixing lead to exclusion of weights or nodes from the process of learning. There are two possibilities: to fix some weights or to fix the whole node. The first is more efficient then the second because in this case we do not need to compute the node error. The improved learning algorithm for RTNN was tested with several linear and nonlinear dynamic objects. The topology improvements are also carefully studied.

4. FS RULE-BASED COORDINATION MODEL

Let us assume that the unknown system $y = f(x)$ generates the data $y(k)$ and $x(k)$ measured at k, $k-1$,.., then the aim is to use this data to construct a deterministic function $y = F(x)$ that can serve as a reasonable approximation of $y = f(x)$ in which the function f is unknown. The variables $x = [x1,...,xp]' \in \aleph \subset \Re^P$ and $y \in Y \subset \Re$ are called regressor and regressand, respectively. In FS modelling, the function F is represented as a collection of if-than fuzzy rules (Babuchka and Verbruggen, 1997), represented by the statement:

IF antecedent proposition then consequent proposition

The Takagi-Sugeno model, cited by Babuchka and Verbruggen, 1997, is a mixture between a linguistic and mathematical regression model. The rule antecedents describe the fuzzy regions in the input space. The rule consequent is crisp mathematical function of the inputs. The FS model have the most general form:

$$R_i: \quad \text{If } x \text{ is } A_i \text{ then } y_i = f_i(x), i=1,2,.., \qquad (24)$$

Having in mind the linearized version of the RTNN model (15), this fuzzy model could be rewritten in the form:

$$R_i: \quad \text{If } x(k) \text{ is } J_i \text{ and } u(k) \text{ is } D_i$$
$$\text{then } \begin{cases} x_i(k+1) = J_i x(k) + D_i u(k) \\ y_i(k) = F_i x(k) \end{cases} \qquad (25)$$

So, it is clear that this fuzzy model combines a fuzzy antecedent proposition with a deterministic mathematical equation describing the consequent part of the rule (Babuchka and Verbruggen, 1997). Now, it is possible to incorporate the RTNN model (23) into the FS model (25), using the linear part of the RTNN model in the consequent proposition part of (25) and the restrictions, expressed by the saturation of (23) to construct the antecedent proposition part of (25). The biases, obtained in the process of BP learning of the RTNN model could be used to form the membership functions, as they are natural centers of gravity, [10], for each variable. The number of rules could be optimized using the mean-square error (MSE) of RTNNs learning, which is a natural measure of the precision of the approximation of the nonlinear object model (MSE is up to 5%). The structure of the entire identification system contains a fuzzyfier a FRBS, and a set of RTNN models. The system does not need a defuzzyfier, because RTNN models are crisp limited linear state-space models. A possible adaptive control system, could contain also a set of controllers incorporated in a FRBS, designed on the base of the obtained set of RTNN models.

5. SIMULATION RESULTS

Let us consider the following nonlinear discrete-time system described by the difference equations:

$$h(k) = u(k) \exp(-3|x1(k)|) \qquad (26)$$
$$y(k+1) = -0.025\,y(k) -0.3\,y(k-1) + h(k) \qquad (27)$$

The nonlinear function $h(k)$ is taken from the paper of Babushka and Verbruggen, 1997, where a graphic of the exponential term is given and it is approximated by seven membership functions. The state-space structure of the RTNN model gives us the possibility to approximate the entire dynamic nonlinear model by only two RTNN models, each of fifth order, operating in the slightly recovered intervals $(-1, 0.2)$ and $(-0.2, 1)$, for which only two membership functions have to be designed. Some simulation results for different RTNN models of the same nonlinear process are shown on figures. The Fig.1 shows an unsuccessful attempt to simulate the object by one RTNN model. The error obtained for both training and generalisation phase is more than 20% (the NN generalisation signal is a combination of harmonic signals, described in Baruch et al., 1996). Fig. 2 shows a successful result, obtained for the first RTNN model learning of an output object process signal cut in the first output signal level interval $(-1, 0.2)$.

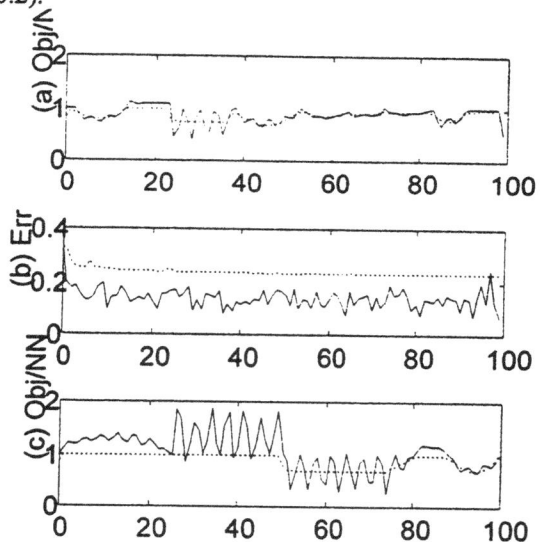

Fig. 1. Full RTNN model (20% LER, 23% GER).

Each figure has the following common items:
(a,c) Outputs of the RTNN (dashed line) and of the Object (solid line) during the last epoch of training;
(b) RTNN error per epoch during all the period of learning (solid line) or generalisation (dashed line);
LER signifies learning error, GER signifies generalisation error. All experiments has been done with the following parameters - $\eta=0.2$, $\alpha=0.9$, epoch size N=100 cycles and number of epochs n=100. The value of the output offset is a=1.

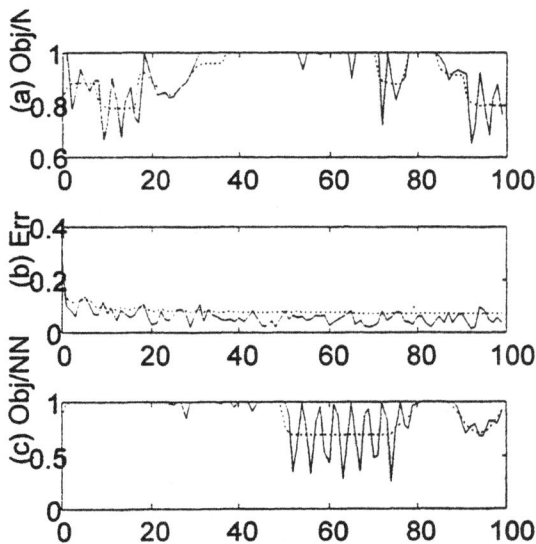

Fig. 2. First RTNN model (5% LER, 7% TER).

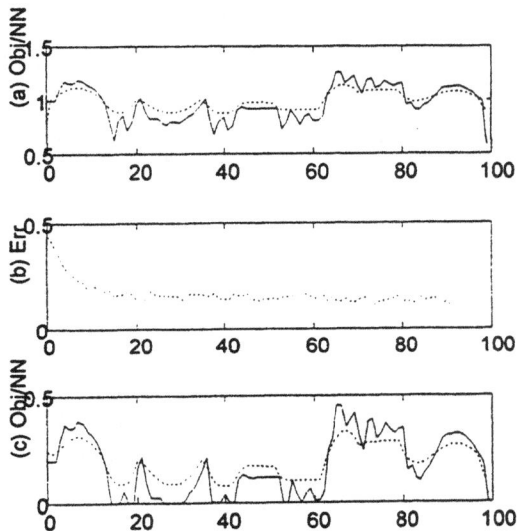

Fig. 3. Combined model. a). Object output and the combined signal of the two RTNN outputs; b) total error LER=8%; c). second RTNN model - cut output signal of the object and RTNN output.

Fig. 3. Illustrates the work of both **RTNN** models and gives its common learning error. Fig. 3.c. illustrates the learning of the second **RTNN** model. The object output is cut in the second level interval (-0.2,1).

CONCLUSIONS

An unified state space representation of both continuous and discrete-time mathematical models of **RTNN**, is given in two layer Jordan RNN. The stability, observability, controllability, learnability conditions, are given. The paper suggests to improve the **RTNN** topology introducing a sigmoid vector function in the **RTNN** feedback and saturation instead of a sigmoid function, to preserve the NN stability and to enhance its realisation. A new improved **BP RTNN** learning algorithm, is proposed

and experimented by appropriate examples. The RTNN model is incorporated in a **FRBS** model of the identified complex nonlinear process. The structure of the fuzzy-neural system is described and a simulation results of a two **RTNN** nonlinear model identification, are also given.

REFERENCES

Babushka, R. and H.B.Verbruggen (1997). Fuzzy Modelling: Principles, Methods and Applications. In: *Proc. of the Int. Workshop on Intelligent Control INCON'97, Oct. 13-15, 1997, Sofia, Bulgaria*, ISBN 954-920-7-1, (C by UAI- Bulgarian Union for Automation and Informatics, Ed.), pp. 1-23.

Baruch, I., I. Stoyanov and E. Gortcheva (1996). Neural Network Models of Dynamic Processes: Stability and Learning. In: *Proc. of the 3-rd Int. Symp. MMAR'96, Sept. 10-13, 1996, Miedzyzdroje, Poland*, (S. Banka, S. Domek, Z. Emirsajlov, Ed.), **Vol. 3**, pp. 1169-1174, TU of Szczecin, Poland.

Connor, J.T., R. Martin and L. Atlas (1994). Recurrent NN and Robust Time Series Prediction. *IEEE Transactions on NNs*, **Vol. 5**, No. 2, pp. 240-253.

Kosko B. (1992). Neural Networks and Fuzzy Systems (A Dynamical Systems Approach to Machine Intelligence). *Prentice-Hall Int. INC, Englewood Cliffs, N.J.*

Narendra, K.S. and K. Parthasarathy (1990). Identification and Control of Dynamic Systems using Neural Networks. *IEEE Transactions on NNs*, **Vol. 1**, No. 1, pp. 4-27.

Pham, D.T. and S. Yildirim (1995). Robot Control using Jordan NNs. In: *Proc. of the Internat. Conf. on Recent Advances in Mechatronics, Istanbul, Turkey, Aug. 14-16, 1995*, (O. Kaynak, M. Ozkan, N. Bekiroglu, I. Tunay, Ed.), **Vol. II**, pp. 888-893, Bogaziçi University Printhouse, Istanbul, Turkey.

Sadharsanan S. and M.Sundareshan (1991). Exponential Stability and a Systematics Synthesis of a Neural Networks for Quadratic Minimization. *IEEE Transactions on NNs*, **Vol. 4**, No. 5, pp.599-613.

Sastry, P.S., G. Santharam and K.P. Unnikrishnan (1994). Memory Networks for Identification and Control of Dynamical Systems. *IEEE Transactions on NNs*, **Vol. 5**, No. 2, pp. 306-320.

Tsoi, A.C. and A.D. Back (1994). Locally Recurrent Globally Feedforward Networks: A Critical Review of Architectures. *IEEE Transactions on NNs*, **Vol. 5**, No. 2, pp. 229-239.

Yip, P.P.C. and Y.H. Pao (1994). A Recurrent Neural Net Approach to One-Step Ahead Control Problems. *IEEE Transactions on SMC*, **Vol. 24**, No. 4, pp. 678-683.

AUTHOR INDEX

www.ingramcontent.com/pod-product-compliance
Lightning Source LLC
Chambersburg PA
CBHW082305210326
41598CB00028B/4446